INTERVENTIONS: NEW STUDIES IN MEDIEVAL CULTURE
Ethan Knapp, Series Editor

The Politics of Ecology

LAND, LIFE, AND LAW IN MEDIEVAL BRITAIN

EDITED BY

Randy P. Schiff

and Joseph Taylor

THE OHIO STATE UNIVERSITY PRESS

COLUMBUS

Library of Congress Cataloging-in-Publication Data

The politics of ecology : land, life, and law in medieval Britain / edited by Randy P. Schiff and Joseph Taylor.

    pages cm. — (Interventions: new studies in medieval culture)

 Includes bibliographical references and index.

 ISBN 978-0-8142-1295-0 (cloth : alk. paper) — ISBN 0-8142-1295-6 (cloth : alk. paper) —

 1. Literature, Medieval—History and criticism. 2. Political ecology—England. 3. Biopolitics—England. 4. Law, Medieval. I. Schiff, Randy P., 1972– editor. II. Taylor, Joseph, 1977– editor. III. Series: Interventions (Columbus, Ohio)

 PN671.P65 2015

 809'.89410902—dc23

2015030905

Cover design by Christian Fuenfhausen

Text design by Juliet Williams

Type set in ITC Galliard

Printed by Thomson-Shore, Inc.

9 8 7 6 5 4 3 2 1

Contents

_৵৩৵৩৵৩_

Part III • Politics, Affect, and Life

Illustrations

⟨❧❦❧⟩

Acknowledgments

THE EDITORS would like to thank The Ohio State University Press and, specifically, editor Eugene O'Connor for his vigilance and forthrightness in helping us bring this volume to press. We are grateful to series editor Ethan Knapp and to former director Malcolm Litchfield for their labor and encouragement in this endeavor. The anonymous reader for Ohio State and George Edmondson made numerous helpful suggestions to improve the individual essays within the collection as well as the project's larger arguments. We, of course, must thank each of our contributors, whose critical work we have had the pleasure of editing. We would also like to acknowledge Elizabeth Scala for helpful suggestions in the volume's early stages and Jeffrey Jerome Cohen, both for useful early comments and for the opportunity to offer two sessions related to the volume's arguments in his "Ecologies" thread at the 2012 New Chaucer Society Congress in Portland, Oregon. We are grateful for funds provided by the University of Alabama in Huntsville's Humanities Center that contributed to the production of this book and to friends and colleagues who have proffered compelling discussions that further enhanced our thinking about its topics. We would like to offer thanks to Eileen Joy and Punctum Books for permission to publish Kathleen Biddick's essay in our volume. Her work originally appeared in her book, *Make and Let Die: Untimely Sovereignties* (Brooklyn: Punctum Books, 2015). Finally, we would like to thank our families for their love and support in pursuit of this collection's completion, especially Maki Becker, Duncan and Desmond Schiff, and Laura, Ellie and Annabel Taylor.

Introduction

The Politics of Ecology

LAND, LIFE, AND LAW IN
MEDIEVAL BRITAIN

Randy P. Schiff
and Joseph Taylor

IF MEDIEVAL literary studies is, like so many fields, currently conditioned by an ecological turn that dislodges the human from its central place in materialist analysis, then why now focus on the law? Is not the law the most human, if not indeed *the* human, institution? In proposing that all life in medieval Britain, whether animal or vegetable, was subject to the same legal machine that enabled all claims on land, are we not ignoring the ecocritical demand that we counteract human exceptionalism and reframe the past with inhuman eyes? *The Politics of Ecology* answers these questions by infusing biopolitical material and theory into ecocentric studies of medieval life. Understanding the biological and the legal as mutually constituted, a biopolitical ecocriticism avoids an uncritical anthropocentrism by insisting that all biological entities precede, pass through, and are conditioned by law. Responding to recent expansions of ecocriticism, as well as to work in the related fields of animal studies and object-oriented ontology, this collection enters into current ecotheoretical conversations by exploring the literary and political entanglements produced by the triple helix of land, life, and law.

Our focus on law, despite appearances, does not privilege the human end of the Cartesian divide; rather, it exposes the fallacy of such binaries. Providing an especially capacious model of the ecological, Timothy

Morton explores what he calls the "mesh," which figures the boundless, borderless interconnection of all material entities that a new ecological thought must begin to envision and analyze.[1] Yet, life in this mesh—whether animal, vegetable, or indeed inorganic—is not completely unmoored from the realities of the Symbolic—or at least not yet! This is not to say that law always bounds and restricts life in the medieval mesh but that legality is part of its very fabric. In medieval Britain, law pervades numerous life-networks in vast and complex ways for which ecocriticism cannot solely account. For this reason, we suggest a hermeneutic that explores the enmeshed relations of all biological life. If Thomas Lemke rightly reads biopolitical inquiry as insistent on dispelling the illusion that humans are essentially "free beings" whose characteristics are due to being socialized rather than being "in large part biologically conditioned,"[2] then a biopolitical lens helps expose the co-constitution of the legal and the biological in an acutely law-saturated medieval West.

The fundamental interpenetration of land, life, and law can be seen with special clarity in notions of the medieval forest. For no other field of literary study does the forest have as fraught and—for biopolitical analysis—as illuminating a cultural significance than for the Western Middle Ages. The medieval forest is simultaneously a space of absolute law and its seeming opposite—utter wildness. The *Middle English Dictionary* highlights these two primary, yet paradoxical meanings in its first two definitions of "forest." Definition 1a in the *MED*, understanding the forest as "a large tract of uninhabited, or sparsely inhabited, woodland; a wilderness," brings us, in terms of literary life, within the wild spaces into which errant knights or exiled maidens slip off as they leave the civilized court spaces of romance. Definition 2a in the *MED*, according to which the forest is "a wooded tract belonging to a ruler, set apart for hunting; a royal forest" or "a wood enclosed by walls; a park," places us fully within an artificially constructed, legal space.[3] The interpenetration of these two seemingly opposite views of sylvan space—simultaneously wild woods and highly regulated forest—plays a central role in medieval Western ecohistory. What is more, the duality of the term "forest" in medi-

---

1. For Timothy Morton's influential image of the mesh as a means of imagining the interconnected nature of all things, see *The Ecological Thought* (Cambridge, MA: Harvard University Press, 2010), 28–38.

2. Thomas Lemke, *Biopolitics: An Advanced Introduction* (New York: New York University Press, 2011), 17.

3. *Middle English Dictionary*, s.v. "forest," http://quod.lib.umich.edu/m/med/.

eval Britain illustrates biopolitics' central concern, the entanglement of what the classical Greeks kept separate—*zoē*, or the simple fact of living, embodied in all flora and fauna outside of the *polis*, and *bios*, or political life, as a living caught up in law.[4] Much as the medieval forest shows us the interrelation of land, life, and law through its essence as a legal space that is not a static property but a vibrant ecosystem whose vegetation and animals were managed both for economic and recreational exploitation, so will a variety of medieval literary landscapes, viewed through a biopolitical lens, reveal dynamic zones thoroughly defined by the law.

To see law as exterior to the environment it encounters would merely repeat the problematic moves of early ecocritics, who viewed environment as an outside to be drawn into relation with literature. In a seminal passage from the landmark collection, *The Ecocriticism Reader,* Cheryl Glotfelty defines ecocriticism as

> the study of the relationship between literature and the physical environment. Just as feminist criticism examines language and literature from a gender-conscious perspective, and Marxist criticism brings awareness of modes of production and economic class to its reading of texts, ecocriticism takes an earth-centered approach to literary studies.[5]

Glotfelty's pioneering definition of the field is specifically political—understandably so, given ecocriticism's foundations in environmentalist activity. But, as Leerom Medevoi has recently pointed out, each of these critical modes juxtaposed with ecocriticism posits literature as a subject "looking outward" to or registering an "outside" of itself, including the environment, a realization that proves more difficult for medievalists who desire ecocritical reading to disclose the material life of literature as literally part of the landscape.[6] Early ecocriticism's assumption of nature's exteriority is best evidenced in so-called deep ecology, a mode that sought to declare the environment oppressed and abused by modernity. As Ursula Heise explains, "Deep ecology foregrounds the

---

4. On classical Greek distinction of the "natural" life of *zoē* and the political life linked exclusively with *bios*, especially as it is formulated in Aristotle, see Giorgio Agamben, *Homo Sacer: Sovereign Power and Bare Life*, trans. Daniel Heller-Roazen (Stanford: Stanford University Press, 1998), 1–3.

5. Cheryl Glotfelty, "Introduction: Literary Studies in an Age of Environmental Crisis," in *The Ecocriticism Reader: Landmarks in Literary Ecology*, ed. Cheryl Glotfelty and Harold Fromm (Athens: University of Georgia Press, 1996), xix.

6. Leerom Medovoi, "The Biopolitical Unconscious: Toward an Eco-Marxist Literary Theory," *Mediations* 24, no. 2 (2010): 123–38.

value of nature in and of itself, the equal rights of other species, and the importance of small communities."[7] The discourse of deep ecology that powerfully shapes first-wave ecocriticism has deep roots in an American environmentalist movement haunted by, and yet complicit in, exceptionalism. George Perkins Marsh, whose 1864 work *Man and Nature* is often cited as the first ecologically grounded text of the environmentalist movement, established the new science's human-exceptionalist paradigm: all human "action" is seen as unnatural, since it "derange[s]" the "harmonies" of nature."[8] However much Marsh's work may seem caught up in an ancient exceptionalism active since at least *Genesis,* referring to the natural world as subject to disorder due to the arbitrariness of man, whose "Creator" will call him forth "to enter into its possession,"[9] the bulk of the study is devoted to scientifically informed assessments of conservation problems in various geographical zones. By systematically linking human destructivity with humanity's special status as "living in physical nature," but "not of her,"[10] Marsh answers in advance what he calls "the great question" in the closing sentence of his treatise— "whether man is of nature or above her."[11] While deep ecology certainly rejects Marsh's theological confidence in humanity's preordained dominion over the natural world, it shares Marsh's insistence that true nature exists utterly outside of the sphere of human action.

The deep-ecological view that a critical response to a state of environmental emergency entails holding fast to a conceptualization of nature's exteriority to an exceptionalist humanity has profoundly shaped much ecocritical work.[12] In his monumental study of Henry David Thoreau, Lawrence Buell makes explicit the implications of this crisis for critical reading as part of a larger ethics: "If . . . western metaphysics and ethics need revision before we can address today's environmental crisis, then environmental crisis involves a crisis of the imagination, the amelioration of which depends on finding better ways of imaging nature and

---

7. Ursula K. Heise, "The Hitchhiker's Guide to Ecocriticism," *PMLA* 121, no. 2 (2006): 507.

8. George Perkins Marsh, *Man and Nature: Or, Physical Geography as Modified by Human Action,* 1864, ed. David Lowenthal (Cambridge, MA: Belknap Press, 1965), 18.

9. Ibid., 36.

10. Ibid., 36–37.

11. Ibid., 465.

12. See, for example, Richard Kerridge and Neil Sammells, who introduce ecocriticism as a discipline born of emergency conditions, with its judgments of texts' "ideas" and "coherence" reduced to their status as "responses" to "environmental crisis" in *Writing the Environment: Ecocriticism and Literature* (London: Zed Books, 1998), 5.

humanity's relation to it."[13] Buell's call for "better ways" illustrates the degree to which humankind, for early ecocritics, must relate or respond to nature as an external phenomenon. Thus, humans look upon nature from outside—from a particular, privileged position. The explicitly environmentalist political thrust of deep ecology has produced some of ecocriticism's most notable work, but its insistence on detaching the human from nature, along with its persistent desire for pure wilderness and other untouched, idealized natural spaces has severely limited the potential range of ecocritical study. Deep ecology's desire for an apolitical landscape has been seminally called into question by William Cronon, who highlights the constructedness of the "wilderness" concept as a "landscape of authenticity."[14] As Greg Garrard explains in describing the constructivist critique that has shadowed deep-ecological strains of ecocritical inquiry, "the choice between monolithic, ecocidal Modernism and reverential awe is a false dichotomy."[15] Subsequent ecocritics have responded by opening up their discourse and hermeneutic to a wider set of apparatuses. Developed in reaction to deep ecology, one mode of critique associated with the environmental justice movement involves a rejection of environmentalist efforts to portray humanity as utterly detached from nature, in favor of ecooriented analyses of social and political inequalities.[16] John Gamber illuminates the epistemological starting point of such criticism as the "simple assumption that people are natural" and insists that all aspects of human society, ranging from material productions such as cities and toxic waste to cultural constructions such as ethnic identity and government policy, are part of the biological "matrix" within which we have "evolved" and that is subject to ecocritical analysis.[17] One particularly valuable strain of environmental justice criticism brings race to the full attention of an ecocritical studies that ignores urban spaces in

13. Lawrence Buell, *The Environmental Imagination: Thoreau, Nature Writing, and the Formation of American Culture* (Cambridge, MA: Belknap Press, 1995), 2.

14. William Cronon, "The Trouble with Wilderness, or, Getting Back to the Wrong Nature," in *Uncommon Ground: Rethinking the Human Place in Nature,* ed. William Cronon (New York: W. W. Norton & Co., 1995), 80.

15. Greg Garrard, *Ecocriticism* (London: Routledge, 2004), 71.

16. On environmental justice criticism and activism, see Joni Adamson, Mei Mei Evans, and Rachel Stein, eds., *The Environmental Justice Reader* (Tucson: University of Arizona Press, 2002); Stacy Alaimo, *Bodily Natures: Science, Environment, and the Material Self* (Indianapolis: Indiana University Press, 2010); and Luke Cole and Sheila Fisher, *From the Ground Up: Environmental Racism and the Rise of the Environmental Justice Movement* (New York: New York University Press, 2000).

17. John Gamber, *Positive Pollutions and Cultural Toxins: Waste and Contamination in Contemporary U.S. Ethnic Literatures* (Lincoln: University of Nebraska Press, 2012), 1.

favor of green, often sparsely inhabited (or empty) spaces.[18] Other modes of ecocritical theory that have moved sharply away from deep ecology include ecofeminism, which highlights the manifold ways in which environmental discourse and activism have been limited by masculinist perspectives that are structurally bound up with the very acquisitiveness and destructivity fueling environmental crisis,[19] and postcolonial ecocritical studies, which explore imperialist exploitation of nature and alternative, indigenous modes of engaging with the environment.[20] Finally, another crucial trend in current ecocriticism can be seen in deconstructive critiques of the concept of nature at the heart of first-wave environmentalism, with critics like Morton seeing nature as a transcendent notion that obstructs our engagement with a fully material environment.[21]

The Politics of Ecology positions itself fully within the wake of such critiques of deep ecology, and the critics collected within the volume are informed by these various modes of rejecting efforts to detach human productions (whether physical or ideational) from the natural world. In highlighting the political, we aim to incorporate the key insights of environmental justice, which holds that a commitment to understanding and engaging with environmental issues can go hand in hand with keen interests in race, gender, and economics. Even as this collection highlights the ways in which all aspects of political life are bound up with material systems, we also seek to avoid having a sense of ecology that is so inflated as simply to signify anything—to, in Medovoi's words, theorize

---

18. See, for examples, Michael Bennett and D. W. Teague, eds. *The Nature of Cities: Ecocriticism and Urban Environments* (Tucson: University of Arizona Press, 1999); and Michael Bennett, "Anti-Pastoralism, Frederick Douglass, and the Nature of Slavery," in *Beyond Nature Writing: Expanding the Boundaries of Ecocriticism*, ed. Karla Armbruster and Kathleen R. Wallace (Charlottesville: University of Virginia Press, 2001), 195.

19. For important ecofeminist studies, see Stacy Alaimo, *Undomesticated Ground: Recasting Nature as Feminist Space* (Ithaca, NY: Cornell University Press, 2000); Carolyn Merchant, *The Death of Nature: Women, Ecology, and the Scientific Revolution* (San Francisco: Harper and Row, 1990); Val Plumwood, *Feminism and the Mastery of Nature* (London: Routledge, 1993); and Noel Sturgeon, *Ecofeminist Natures: Race, Gender, Feminist Theory and Political Action* (New York: Routledge, 1997).

20. For insightful postcolonial ecocriticism, see Elizabeth DeLoughrey and George B. Handley, eds., *Postcolonial Ecologies: Literatures of the Environment* (Oxford: Oxford University Press, 2011); Graham Huggan and Helen Tiffin, *Postcolonial Ecocriticism: Literature, Animals, Environment* (New York: Routledge, 2010); and Laura Wright, *Wilderness into Civilized Shapes: Reading the Postcolonial Environment* (Athens: University of Georgia Press, 2010).

21. See, for example, Timothy Morton's embrace of ecocritique in *Ecology without Nature: Rethinking Environmental Aesthetics* (Cambridge, MA: Harvard University Press, 2009), 13.

the environment so broadly as to make it "appear to include anything and everything."[22] By foregrounding the law, insofar as it is at the basis of scientific understandings of a rule-governed world and yet at the heart of various cultures' understanding of themselves as political jurisdictions, we will be able to harness the power both of human-oriented environmental justice methodologies and of the variously posthumanist eco-materialisms that are currently vigorously informing ecocriticism—and medieval literary texts prove to be an especially powerful area to conduct such inquiry.

Medieval literary studies has been one of the most active sites for the adoption, refinement, and reformulation of ecocritical theory. Insofar as it has always been enmeshed in, shaped with, determined through, and literally written on the environment, medieval British literature offers a compelling framework through which to pursue a simultaneously biopolitical and ecological inquiry. One clear means through which the living environment is not in merely external relation to medieval literature, but rather constitutes its very material existence, involves a key medium through which much work reaches us—the animal skins used to inscribe parchment. Demonstrating the complex entanglement of literary work and living nature through analysis of the parchment industry and its large-scale conversion of animal skins into writing surfaces, Bruce Holsinger pointedly declares, "Medieval literature is, in the most rigorously literal sense, nothing but millions of stains on animal parts."[23] To such a patently materialist explanation that recommends medieval literature to biopolitical analysis, we can also add a highly ideational one—the centrality of allegory in medieval literary imagination, which suffuses so much work with the life-saturated worlds of forests and seascapes. In his reading of the "unmappable" settings of early ecocentric literature disseminating from around the Irish Sea, Alfred Siewers, for example, highlights the ways in which medieval literary texts instinctively "integrat[e] aspects of spiritual, imaginative, and natural realms of human life and the physical environment, including wilderness and animals, and that permits shape-shifting as well as transport through time and space."[24] Siewers, here, renders the medieval literary landscape into the threshold between the "fact" of nature and its cultural construction, between, as he says, "the

---

22. Medovoi, "Biopolitical Unconscious," 129.

23. Bruce Holsinger, "Of Pigs and Parchment: Medieval Studies and the Coming of the Animal," *PMLA* 124, no. 2 (2009): 619.

24. Alfred K. Siewers, *Strange Beauty: Ecocritical Approaches to Early Medieval Landscape* (New York: Palgrave Macmillan, 2010), 5.

biological and the imaginary, the body and the environment, the subjective and the objective."[25] In this sense, landscape, rather than acting as a
boundary, constitutes the linkage that melds the literary into life and vice
versa.

Within the last decade, medieval studies has witnessed a striking proliferation of ecocritical work and a continuing diversification of methodologies. Marking a significant intervention in the writing of pre- and
early modern environmental history, Lisa Kiser and Barbara Hanawalt's
interdisciplinary essay collection *Engaging with Nature: Essays on the
Natural World in Medieval and Early Modern Europe* offers both literary and historical models for exploring environments and animals in a
more strictly material sense.[26] More recently, Siewers places a section
on "Medieval Natures" at the very heart of his collection *Re-Imagining
Nature: Environmental Humanities and Eco-Semiotics.*[27] In his transdisci-
plinary collection *Prismatic Ecology: Ecotheory Beyond Green,* medievalist
literary critic Jeffrey Jerome Cohen creates a lively space for reimagining
ecological criticism, in which a multiplicity of colors represents both the
desire and the need to move beyond the green modes of reading which,
rooted in deep ecology and other early ecocritical methods, tend—
sometimes unwittingly—to reaffirm a nature-culture divide. Looking at
a late-fourteenth-century manuscript illumination of a painting, Cohen
meditates on the fact that the artist's very materials—his paints—come
from "precise combinations of pulverized minerals, juice pressed from
harvested berries, oak gall boiled in water and mixed with powdered
egg shells, common ash, rare pollen, acidic urine."[28] For Cohen, "color
is not some intangible quality that arrives belatedly to the composition
but a material impress, an agency and partner, a thing made of other
things through which worlds arrive."[29] Much as an economy of animal
life and death was woven into the thriving production of parchment, so
were manifold vegetable and mineral elements bound up in a biologically
grounded textual world.

---

25. Ibid.

26. Lisa Kiser and Barbara Hanawalt, eds. *Engaging with Nature: Essays on the Natural World in Medieval and Early Modern Europe* (Notre Dame: University of Notre Dame
Press, 2008).

27. Alfred K. Siewers, ed., *Re-Imagining Nature: Environmental Humanities and
Eco-Semiotics* (Lanham, MD: Bucknell University Press, 2013), 109–60.

28. Jeffrey Jerome Cohen, "Ecology's Rainbow," in *Prismatic Ecology: Ecotheory
beyond Green,* ed. Jeffrey Jerome Cohen (Minneapolis: University of Minnesota Press,
2013), xvi.

29. Ibid, xvi.

Medieval literary studies also recommends itself to biopolitically inflected ecocriticism due to the vital links with animal studies and objected-oriented ontology being currently maintained by medievalists. Animal studies has a long history in medievalist scholarship, producing rich critical intersections with ecocriticism in the work of such scholars as Lisa Kiser and Susan Crane. Challenging the critical commonplace of Chaucer's seemingly anthropomorphic birds in a 2006 lecture to the New Chaucer Society, Crane observes that the "communicative, resourceful animals of romance express, in highly imaginative terms, a widespread conviction that humans and animals share contiguities beyond their mere physicality,"[30] contiguities, Crane argues, that deserve attention along- side the typical scholarly concerns of romance. Karl Steel's *How to Make a Human: Animals and Violence in the Middle Ages* has served to fur- ther complicate this discourse in productive ways. Addressing the "vio- lence against animals through which humans attempt to claim a unique, oppositional identity for themselves," Steel illustrates the violent man- ner by which "the human" is produced as an effect rather than a cause of the domination of animals; his book works, as he says, to "inter- rupt, historicize, and re-open the supposed givenness of the 'natural' categories of human and animal."[31] In either a remarkable coincidence or canny understanding of medievalists' posthuman proclivities, a 2009 issue of *PMLA* included both a series of essays on theories and methodol- ogies of "Animal Studies" and of "Medieval Studies in the Twenty-First Century."[32] A 2009 issue of the journal *Postmedieval,* edited by Steel and Peggy McCracken, also documented an "animal turn" in medieval stud- ies, more specifically, its engagement in a posthumanism growing more pervasive in the humanities broadly.[33] Lesley Kordecki's *Ecofeminist Subjectivities: Chaucer's Talking Birds* and Crane's more recent *Animal Encounters: Contacts and Concepts in Medieval Britain* have continued to diversify medieval ecocriticism, infusing the field with gender- and animal studies-inflected currents, respectively.[34]

---

30. Susan Crane, "For the Birds," *Studies in the Age of Chaucer* 29 (2007): 23–41.

31. Karl Steel, *How to Make a Human: Animals and Violence in the Middle Ages* (Co- lumbus: The Ohio State University Press, 2011), 15–16.

32. See *PMLA* 124, no. 2 (2009): 361–69, 472–563, 564–75.

33. *Postmedieval* 2, no. 1 (2011).

34. Leslie Kordecki, *Ecofeminist Subjectivities: Chaucer's Talking Birds* (New York: Palgrave, 2011); Susan Crane, *Animal Encounters: Contacts and Concepts in Medieval Britain* (Philadelphia: University of Pennsylvania Press, 2012); see also Joyce Salisbury, *The Beast Within: Animals in the Middle Ages* (London: Routledge, 2010); Carolynn Van Dyke, ed. *Rethinking Chaucerian Beasts* (New York: Palgrave, 2012); the colloquium on

Medieval literary critics have also contributed significantly to object-oriented ontology and its investigations of *things* beyond the human and animal. This work stems, in part, from Bruno Latour's theorization of actor-networks and assemblages and has been furthered by political scientist Jane Bennett's *Vibrant Matter: A Political Ecology of Things*.[35] As Bennett claims, such an attention to the material hopes, in part, to "dissipate the onto-theological binaries of life/matter, human/animal, will/determination, and organic/inorganic" and "to sketch a style of political analysis that can better account for the contributions of nonhuman actants."[36] Medieval literary critics have considerably advanced and complicated such object-oriented analysis. In a 2013 issue of *Postmedieval*, for example, authors were tasked with answering Bennett's challenge to "explore the impress and agency of the nonhuman"[37] in compelling analyses of the four elements: air, water, fire, and earth. The *Postmedieval* volume emphatically defines an ecomaterialism that posits "an attempt to re-describe human experience so as to uncover more of the activity and power of a variety of nonhuman players amidst and within us."[38]

In marshaling the resources of vibrant ecomaterialist work going on in medievalist criticism, the *Politics of Ecology* insists on orienting such materialist analysis toward the law. While much of the ecomaterialist work surveyed above moves criticism toward the important task of bringing human and nonhuman actors into a single vista, it too rarely sees law as a force transcending human control. At the same time, much of the excellent work on medieval law and literature eschews ecomaterialist questions.[39] This volume brings together these disjointed fields of study.

---

animalia in *Studies in the Age of Chaucer* 34 (2012); and the symposium on "Animal Theories" in *New Medieval Literatures* 12 (2010).

35. See Bruno Latour's *Reassembling the Social: An Introduction to Actor-Network-Theory* (Oxford: Oxford University Press, 2005) and his *We Have Never Been Modern*, trans. Catherine Porter (Cambridge, MA: Harvard University Press, 1993); Jane Bennett, *Vibrant Matter: A Political Ecology of Things* (Durham: Duke University Press, 2010).

36. Bennett, *Vibrant Matter*, x. Here, Bennett appropriates the term "actant" from Latour, who views it as "a source of action that can be either human or nonhuman" (*Vibrant Matter*, viii); see Bruno Latour, *Politics of Nature: How to Bring the Sciences into Democracy*, trans. Catherine Porter (Cambridge, MA: Harvard University Press, 2004), 237.

37. Jeffrey Jerome Cohen and Lowell Duckert, "Howl," *Postmedieval* 4, no. 1 (2013): 3.

38. Jane Bennett, "The Elements," *Postmedieval* 4, no. 1 (2013): 109.

39. For excellent work on medieval law and British literature, in which ecomaterialist questions only rarely surface, see Emily Steiner and Candace Barrington, eds., *The Letter of the Law: Legal Practice and Literary Production in Medieval England* (Ithaca, NY:

While the emphasis on biopolitics that inspired this collection may seem to fly in the face of the recent medievalist trend to erode anthropocentrism, we maintain that the conviction that legal force saturates all relations among animals, plants, and their sustaining environments provides a unique vista onto the politics of ecology in medieval Britain.

A Politics of Ecology?

Ecological study and biopolitics have long partnered to explore humanity and its environment. Swedish political scientist Rudolph Kjellen arguably coined the term "biopolitics" in his 1916 study, *The State as a Form of Life,* where he set about theorizing the state as a living entity with instinct and natural drives. In his later work, *Outline for a Political System,* Kjellen explained his organicist biopolitics as analogous to biology, conflating natural, physical, and cultural life into the Greek term *bios.*[40] In the 1960s, biopoliticians began to approach political behavior through the methods and concepts of the biological sciences. This so-called naturalist biopolitics sought to describe and explain behavior in working toward "a politics consistent with biological exigencies."[41] But such biopolitical theorists failed to meld successfully the discourse and methodologies of biology with the social sciences' focus on culture. Worse, biopolitical inquiry made its own work more difficult by frequently positing nature as a distinct system that shapes politics by acting on a population or being acted upon by the population or its government. In so doing, scholars working in and through a naturalist biopolitics, as Lemke observes, "prolong the very dualism of nature and society whose continuing existence they also bemoan," even as they epistemologically depend—not unlike early ecocritics focused on an alien nature—on an "external observer who objectively describes certain forms of behavior and institutional processes."[42] This second wave of biopolitical inquiry more rigorously theorized life, both by complicating stories of its origins via focus on genetic and reproductive technologies and by illustrating the manner by which it becomes

Cornell University Press, 2002); Richard Firth Green, *A Crisis of Truth: Literature and Law in Ricardian England* (Philadelphia: University of Pennsylvania Press, 2002); and Wendy Scase, *Literature and Complaint in England, 1272–1553* (Oxford: Oxford University Press, 2007).

40. Roberto Esposito, *Bíos: Biopolitics and Philosophy,* trans. Timothy C. Campbell (Minneapolis: University of Minnesota Press, 2008), 17.

41. Lemke, *Biopolitics,* 17.

42. Ibid., 20, 17.

the focus of political action. Such naturalist biopolitical theorists, however, remained trapped in a discourse for which life is always objectively on the outside.

Michel Foucault breaks from these early biopolitical orientations. In the final session of his 1975–76 lectures at the Collège de France, Foucault defines biopolitics' last domain as "control over relations between the human race, or human beings insofar as they are species, insofar as they are living beings, and their environment, the milieu in which they live."[43] Foucault's biopolitical work marks a philosophical turn away from individualized existence and toward a more generalized life. Foucault rejects previous models for biopolitics—both the organicist/naturalist approaches that saw biology informing the political and the politicist model assuming politics' extension to include life. Instead, Foucault sees a political schism in which the biological and the political meld into "a constellation in which modern human and natural sciences and the normative concepts that emerge from them structure political action and determine its goals."[44] Anticipating the fallacy of wilderness in ecocriticism—of the pure, untouched landscape to which man might retreat—Foucault clarifies that his "environment," though it includes the "geographical" and the "climatic," is never natural: rather, it "has been created by the population and therefore has effects on that population."[45] Reorienting the focus of his following year's lectures (compiled in *Security, Territory, and Population*) from the disciplinary measures of the state to its conservationist and regulatory practices, Foucault nevertheless does not claim that security replaces law and discipline: rather, security helps them function more efficiently while contributing new apparatuses to increase the productive possibilities of society. These *dispositifs*, as Foucault terms them, must account for the natural environments within which populations operate.

Foucault's biopolitical emphasis on a generalized "environment" rather than a sharply bordered "territory" is crucial to our study. As Medovoi explains, Foucault eschewed limiting analysis to merely legally circumscribed territories and instead conceived of the "environment" as a more fundamental context, thereby moving biopolitical inquiry from merely static "juridical" spaces in which laws are applied only accord-

---

43. Michel Foucault, *Society Must Be Defended: Lectures at the Collège de France 1975–76*, ed. Mauro Bertani and Alessandro Fontana, trans. David Macey (New York: Picador, 2003), 245.

44. Lemke, *Biopolitics*, 33.

45. Foucault, *Society Must Be Defended*, 245.

ing to the extent of "jurisdiction" to a more capacious "environment" structured by "the regularities of life and its biological requirements."[46] Foucault explains this shift himself:

> Discipline works in an empty, artificial space that is to be completely constructed. Security will rely on a number of material givens. It will . . . work on site with the flows of water, islands, air, and so forth. . . . This given will not be reconstructed to arrive at a point of perfection, as in a disciplinary town. It is simply a matter of maximizing the positive elements, for which one provides the best possible circulation, and of minimizing what is risky and inconvenient, like theft and disease, while knowing that they will never be completely suppressed.[47]

While discipline manifests its power in the sterile space of the prison, for example, security attempts to manage, as much as it can, a population and its environment, the landscapes, waterways, and mere oxygen that might be momentarily controlled but that are ultimately elusive to power. Foucault, thus, demonstrates how land, life, and law conflate into a mesh within which neither one or another occupies a center that looks out to a relationship with the other.

Foucault generally places the threshold of biopolitics in the eighteenth century when a liberal art of governing began consciously to shift away from the binaries inherent in sovereign power—normative laws that connote right and wrong, and disciplinary and other structures that affirm the capable and incapable—and toward a governmentality for which normativity is merely a point of reference rather than a firm boundary. Hence, individual and community interests are not so much defined by prohibitive laws as they are regulated by perpetually calculating risks: as Foucault asserts, "Security [*is* the] principle of calculation."[48] The Italian philosopher Giorgio Agamben disagrees with Foucault's historical timeline for when natural life becomes caught up in state calculations. Addressing Foucault's analyses of state techniques for managing subjects, and of the technologies of the self by which individuals nevertheless maintain privacy, Agamben asks, "Where in the body of power, is the zone of indistinction (or, at least, the point of intersection) at which

---

46. Medovoi, "Biopolitical Unconscious," 129.

47. Michel Foucault, *Security, Territory, Population: Lectures at the Collège de France 1977–1978*, ed. Michel Senellart, trans. Graham Burchell (New York: Picador, 2007), 19.

48. Michel Foucault, *The Birth of Biopolitics: Lectures at the Collège de France, 1978–1979*, ed. Michel Senellart, trans. Graham Burchell (New York: Picador, 2010), 65.

techniques of individualization and totalizing procedures converge?"[49] If Foucault fails to locate the point at which natural life and political life converge, Agamben finds *zoē* (natural life) infused in *bios* (political life) in Aristotle's very definition of man from his *Politics* as a "living animal with the additional capacity for political existence."[50] "Correcting," then, as he might say, Foucault's biopolitical inquiries, Agamben contends that "Western politics is a biopolitics from the very beginning."[51] In this temporal shift—and in his readings of premodern phenomena from burlesque romance to Western monasticism—Agamben's work invites vigorous participation in biopolitical discussion by scholars of premodern literature.

Crucial to our study is Agamben's understanding of nature's relationship to sovereignty. With Thomas Hobbes's political theory in mind, he illustrates how "sovereignty . . . presents itself as an incorporation of the state of nature in society or, if one prefers, as a state of indistinction between nature and culture, between violence and law, and this very indistinction constitutes specifically sovereign violence."[52] This is played out in sovereignty's management of bodies, which is the primary task of any sovereign power, whether modern or premodern. Agamben's thesis depends on the concept of the state of exception, the space where law has been absented by the sovereign in an act that is paradoxically ascribed by law itself. Here, natural life is excluded while, at the same time, it affirms the law's power; it is, thus, exclusively included or inclusively excluded. Beyond the law, Agamben finds both the natural life of the banned subject, the sacred man who is no longer protected by law, and the sovereign, who decides on the exception.

From the standpoint of the juridical—and this was certainly true for medieval jurists—natural law was always *a priori* to positive law. In his monumental study of medieval political theology, Ernst Kantorowicz explains that the sovereign "was bound to Natural Law not merely in its transcendence and metalegal abstraction, but also in its concrete temporal manifestation, which included the rights of clergy, magnates, and people—a very important point in an England which relied predominantly

---

49. Agamben, *Homo Sacer*, 6.

50. The quote is Foucault's oft-repeated synopsis of Aristotle from the *Politics*, see Michel Foucault, *The History of Sexuality: Volume I, An Introduction*, trans. Robert Hurley (New York: Vintage Books, 1980), 143.

51. Agamben, *Homo Sacer*, 181.

52. Ibid., 35.

on unwritten laws and customs."[53] Emphasizing the state of exception's grounding in natural law, Michael Wilks argues that, for medieval thinkers, there will

> always remain cases which the existing law does not cover, or cases of emergency or special circumstances in which it would be detrimental to the common good, to the *status republicae,* to enforce the law as it stands. In these cases equity (which may be equated with *justitia* or natural law) demands that the law should be ignored: it has temporarily ceased to conform to the standards of ultimate rightness which give it validity and force. A "case of necessity" thus becomes seen as an occasion when natural-divine law, which transcends positive law, is directly involved. Consequently it is sacred duty for the ruler as animate divine natural law to override the provisions of the common law of the community.[54]

The very state of exception, then, that Carl Schmitt famously emphasizes as the defining attribute of sovereignty emerges as derived from the natural—for Aristotle and for medieval legists, from a divine nature, while for modern thinkers of natural law, from a self-evident "principle of life defined as rationally pursued."[55] Nature, in this sense, is not external to state power (weighing in from the outside to right a wrong); rather, in the very case of medieval states of exception, natural law—like bare life—is at the very center, literally the foundation on which an exceptional politics is built.

Further evidence of the profound relevance of medieval literature—which so often allegorizes and naturalizes power relations—to biopolitical inquiry can be found in the fact that Agamben turns to a medieval literary depiction of the werewolf to illustrate the material entanglement of sovereignty, nature, and the state of exception. Discussing Marie de France's twelfth-century *Bisclavret,* Agamben notes, "The transformation into a werewolf corresponds perfectly to the state of exception, during which (necessarily limited) time the city is dissolved and men enter into

---

53. Ernst H. Kantorowicz, *The King's Two Bodies: A Study in Medieval Political Theology,* rev. ed. (Princeton, NJ: Princeton University Press, 1997), 148–49.

54. Michael Wilks, *The Problem of Sovereignty in the Late Middle Ages: The Papal Monarchy with Augustinus Triumphus and the Publicists* (Cambridge, UK: Cambridge University Press, 1963), 217.

55. Amy Swiffin, *Law, Ethics, and the Biopolitical* (London: Routledge, 2011), 31. For Carl Schmitt's seminal statement that "sovereign is he who decides on the exception," see *Political Theology: Four Chapters on the Concept of Sovereignty,* rev. ed., 1934, trans. George Schwab (Chicago: University of Chicago Press, 2005), 5.

a zone in which they are no longer distinct from beasts."[56] While this zone is a state of being outside the law—the life of the sacred man, *homo sacer*—its embodiment is simultaneously the forest and its seeming opposite. Marie's Bisclavret is hunted as an animal until being saved by the uncannily human licking of a king's boots, while his return to human form in the king's bedroom discloses the intimate connection of woodlands and court. For Agamben, the life of Bisclavret is "not a piece of animal nature without any relation to law and the city . . . [but] rather, a threshold of indistinction and of passage between animal and man, exclusion and inclusion."[57] Bisclavret embodies the intertwining of man and animal, forest and *polis,* nature and law, all of which are crucial to a biopolitical understanding of sovereignty's power over life. The werewolf's banning and return thus shatters the illusion of an absolute nature-culture divide, reminding us that the Hobbesian state of nature both "constitutes and dwells within" the law.[58] In developing Foucault's biopolitical inquiry by juxtaposing *zoē* and *bios,* Agamben inspires ecocritical analyses that envisage a living web traversing the animal and vegetable worlds, with all life equally enmeshed with the law.

Tensions abound in the study of law and nature in the British Middle Ages, in part, because this period witnessed the advent of environmental devastation in Britain—species decimation due to overhunting, deforestation, and laws of enclosure—long before the Industrial Revolution of the late-eighteenth and nineteenth centuries. These complexities are evident even in the period's most canonical author, Geoffrey Chaucer. In her 2004 essay, "Ecochaucer: Green Ethics and Medieval Nature," Sarah Stanbury claims that Chaucer's nature "orchestrates human actions without demanding much in return by way of guardianship," a revelation that seems to resonate with ecocritical desires to "expose the historical and cultural practices by which we have prioritized culture over nature."[59] Yet, as Stanbury suggests, this "enfiefing of nature in the Middle Ages . . . may have had dire historical consequences for land use by exonerating her human subjects from custodial responsibilities."[60] The point is provocative for the very double bind that it illustrates: nature is both

---

56. Agamben, *Homo Sacer,* 107.

57. Ibid., 105.

58. Ibid., 106.

59. Sarah Stanbury, "Ecochaucer: Green Ethics and Medieval Nature," *Chaucer Review* 39, no. 1 (2004): 13.

60. Ibid.

empowered and imperiled by Chaucer in ways that anticipate modern environmental exploitation and our contemporary thrust to recenter nature within politics, culture, and religion. This duality is noted in a recent survey of medieval ecocriticism, where Vin Nardizzi ponders a pervasive and potentially generative idea that the Middle Ages might be "the era where our ongoing ecological crisis first began." What is more, as Nardizzi provocatively claims, "the rehearsal of this ecological story constitutes a foundational form of ecotheory."[61]

The Politics of Ecology offers its own intervention in medieval ecological readings by focusing on the acute significance of law and sovereignty in the ecomaterialists's mesh that encompasses all entities: far from a vague model simply celebrating a unity of an impossible diversity of things, the mesh becomes, through a biopolitical lens, a structure with particular, unequally powerful kinds of connectivity. Our volume's ecoreadings with a biopolitical consciousness do not seek to reinvigorate a nature/culture divide or anthropocentric renderings of the political objectification of ecological subjects by a sovereign power. Rather, the essays herein seek to interpret both the manner by which law aims to conflate these spaces into a single object of domination and how human and nonhuman actants resist power's jurisdictional claims—how, ultimately, power comes to realize its own vulnerability in a grander network of life, agency, presence, and embodiment in the medieval world.

What, however, does sustained literary critical analysis of power's vulnerability ultimately mean for those wrestling under the weight of sovereignty and biopolitical rule, whether medieval or modern? We would argue that attention to medieval intersections of land, life, and law opens up a space for clarifying biopolitics' most pressing concern: whether and how to bring together *bios* and *zoē*, authentic life and bare life, in a manner that opposes the thanatopolitical thrust of sovereignties across time. For such an affirmative biopolitics, Italian philosopher Roberto Esposito explains, "Norm and life cannot mutually presuppose one another because they are part of a single dimension in continuous being."[62] In his own work on affirmative biopolitics, Agamben takes medieval monasticism as a case study. Defining what he calls "form-of-life," Agamben explains:

> We must return ever anew to contend with it as its undeferrable task: how
> to think a form-of-life, a human life entirely removed from the grasp of

---

61. Vin Nardizzi, "Medieval Ecocriticism," *Postmedieval* 4, no. 1 (2013): 113.
62. Esposito, *Bíos,* 185.

the law and a use of bodies and of the world that would never be substantiated into an appropriation. That is to say again, to think life as that
which is never given as property but only as a common use.[63]

The cenobitic life, thus, conflates norm and life, universal and particular,
into a "form-of-life" within which use is freed from appropriation by an
outside law or sovereign, and within which the violent separation of political subjects into categories of oppression is neutralized. This is not open
rebellion against the sovereign and the state, or their laws of property;
rather, "form-of-life performs," in Agamben's words, a "special form of
negligence" that ultimately renders sovereign power "inoperative."[64] Further engagement with premodern texts will yield more examples of a so-
called affirmative biopolitics. Even when the chapters in this volume do
not specifically seek such examples, they, nonetheless, aim to clarify the
ways that modern biopolitical thought has been shaped by engagements
between power and life in the medieval period, with none more important than the law-saturated world of the medieval forest that provides our
volume with its first critical frame.

    63. Giorgio Agamben, *The Highest Poverty: Monastic Rules and Form-of-Life,* trans.
Adam Kotsko (Stanford, CA: Stanford University Press, 2013), xiii. Both Esposito and
Agamben, among others, take their cues on affirmative biopolitics from Gilles Deleuze's
idea of "pure immanence," though they certainly also look back to Deleuze's early study
of Baruch Spinoza's philosophy. Pure immanence preoccupied Deleuze throughout his
career. In his final essay, "Immanence: A Life," Deleuze professes a collective singularity—rather than a population of individuals tied to subjectivity, a fabricated independence
that biopolitics repeatedly reveals to be a false notion. He finds such singularity, for example, in infants, who tend to all look alike, and yet they perform singular gestures—a
smile, a coo, a grimace—that are, in Deleuze's words, "pure event freed from the ac-
cidents of internal and external life." The affirmative potential of such singularity, or in
Esposito's development of this idea as the "impersonal," is captured in Deleuze's use
of the indefinite article (a life) that "is in itself: it is not in something, to something; it
does not depend on an object or belong to a subject." On Deleuze and immanence, see
Pure Immanence: Essays on A Life, trans. Anne Boyman (New York: Zone Books, 2001),
26–28; Deleuze and Félix Guattari, *A Thousand Plateaus: Capitalism and Schizophrenia,* trans. Brian Massumi (Minneapolis: University of Minnesota Press, 1987), especially
253–63; and Deleuze and Guattari, *What is Philosophy?* trans. Graham Burchell and Hugh
Tomlinson (New York: Verso, 1994), especially 35–60. See also Roberto Esposito, *Third
Person: Politics of Life and Philosophy of the Impersonal,* trans. Zakiya Hanafi (New York:
Polity Press, 2012).
    64. See Giorgio Agamben, *Profanations,* trans. Jeff Fort (New York: Zone Books,
2007), 75, 86.

Biopolitics and Forest Law

Medieval literary studies proves to be an especially fertile disciplinary site for infusing ecocriticism with the insights of biopolitical analysis. In bringing together incisive medievalist work on the critical zone where land, life, and law thoroughly interanimate one another, the *Politics of Ecology* identifies three areas of acute biopolitical interest, with the first being a very familiar one to scholars following the ecocritical turn. Even as current ecocriticism breaks decisively from deep ecology's tendencies both to devalue human alterations to the environment and to privilege the verdant spaces of green politics, the forest nevertheless looms large in ecocritique—and it is precisely in the medieval woodlands that our volume begins. The medieval forest provides ecocriticism with a unique body of evidence for repositioning the study of green spaces within a bio-politically framed world. In contrast to the problematic wildernesses of deep ecology, medieval woodlands powerfully illustrate the complexities by which wild and civil spaces mutually encode one another, revealing the saturation of land and life by the force of law.

Much recent work has built upon the vital duality of the medieval forest as both wild and regulated, though the preponderance of such analysis has concentrated on the post-Norman world. Usually beginning with the Norman project of afforesting massive swathes of British land after the conquest of England, and pursuing such analysis into the late medieval subdivision of much of this land into parks for aristocrats and gentry, scholars have convincingly demonstrated why the medieval forests of Britain so compellingly instantiate both natural and legal histories.[65]

Recognizing that the medieval forest shows particularly clearly the material interrelation of land, life, and law in medieval Britain, the first section of our collection pursues insights generated by the juxtaposition of biopolitics and forest law. In "Biopolitics in the Forest," Karl Steel explicates the complex webs of forest law and exception. While a poacher could murder deer—and be punished for it—forest law in fact demanded the king's cervicidal tendencies. Yet, this exceptional space could push back against the sovereign's seemingly limitless power. Steel probes not what the sovereign can kill but what he must "manage" to

---

65. For thoroughgoing surveys of the relationship between woodland and British forest and park regulations, see John M. Gilbert, *Hunting and Hunting Reserves in Medieval Scotland* (Edinburgh: J. Donald, 1979); S. A. Mileson, *Parks in Medieval England* (Oxford: Oxford University Press, 2009); and Charles R. Young, *The Royal Forests of Medieval England* (Philadelphia: University of Pennsylvania Press, 1979).

realize his "fantasies of control in the forest."[66] Thinking of the forest in biopolitical, rather than strictly sovereign, terms enlivens the relationship between life and law. In doing so, Steel questions biopolitics' own anthropocentrism and the paradoxical concerns of the medieval forest simultaneously demanding violence against, and protection of, fauna. These competing ethics challenge attempts to distinguish between medieval and modern governmentalities and indeed between sovereignty and biopolitics altogether, for the appearance of the sovereign is already the birth of biopolitics. Forest law best illustrates biopolitics' limits because the human and nonhuman bodies it hopes to manage not only resist but reconfigure the law to different ends. This is evidenced for Steel by the deer carcass that could both incriminate unwitting-yet-proximate wanderers and be blamed itself for dying improperly (without known cause). The sputtering biopolitical machine born of inquiries into the carcass's death, the distribution of its parts—its skins and antlers, its sometime rancid meat—and its very sacredness (neither saleable nor able to be discarded) accentuate the forest as not merely a zone of law and violence but also "a network of interdependence and struggle."[67]

Jeanne Provost in chapter 2 reveals the sovereign to be vulnerable to the very hunt that signifies his authority. Reading numerous forest court cases alongside *The Wedding of Sir Gawain and Dame Ragnelle*, Provost argues that the sovereign asserts his power over animals through the hunt while, at the same time, he presents his own body as meat to be devoured by the office of kingship. The conflation of the human and the cervid is pronounced in the poem when Arthur's killing of a deer is juxtaposed with the king's vulnerability to the watchful hunter, Sir Gromer Somer Joure. Yet, as Provost finds, the poet of the *Wedding* proposes an alternative attention to life through a loathly lady, Ragnelle, who embodies the prolific forest communities and ecosystems challenging the narrow legal encompassing of the King and his deer. Ragnelle's bodily *copia* provokes a sympathy for life that undoes the value-laden management that forest law performs and forces the sovereign to recognize his own embodied connections with vast networks of the living.

In "The Physician and the Forester: Virginia, Venison, and the Biopolitics of Vital Property," Randy P. Schiff analyzes a biopolitical agenda driving the discourse of Geoffrey Chaucer's Physician. After using prosopopoeia to appropriate the authority of Nature, the Physician proceeds to

---

66. See chapter 1 of this volume, 37.
67. Ibid., 45.

expose a politics of death that dooms Virginia. The Physician's heroine is victimized as much by her father, who calls forth the Roman law of *vitae necisque potestas,* as by the state, whose corrupt judge Apius redefines Virginia as a slave—and thus, as essentially vulnerable and killable. By embedding a discussion of deer whose life and death are subject to the law within a "network of asides devoted to the regulation of children,"[68] the Physician portrays Virginia's fate as a biopolitical matter for the state. The Physician carves out a space for a medically informed alternative to the thanatopolitics of the family and the state, presenting himself as the only one who might one day answer Virginia's poignant question of whether anyone can offer a "remedye" (VI.236). In deploying "remedye," a word that can "refer both to a legal or medical mechanism," the Physician indicts an "older world guided only by the laws of fathers," even as he offers, Schiff suggests, the alternative vision of a "modern world guided by medical practitioners who can transform the law of the state into a biopolitics focused on managing and multiplying life rather than merely policing it."[69]

Networks, Land, and Objects

This volume's emphasis on law's engagement with life and land implies that our chapters concern only human stories, yet our broader narrative is one in which the sovereign and his law struggle to effect power over other human *and* nonhuman actors. As our chapters on biopolitics and the forest show, medieval law and the sovereigns who sanctioned it might have hoped to subject all life and land to their will, but they could, at best, only manage the complex networks of flora and fauna, of humans and nonhumans, in their midst. Foucault's explanation of a biopolitical "milieu" touches on the expanse of subjects and environments that confront power's controlling efforts in the Middle Ages. Replacing his earlier concept of a "people" with "population" in his later lectures, Foucault defines the latter as "a multiplicity of individuals who are and fundamentally and essentially only exist biologically bound to the materiality within which they live."[70] This materiality or "milieu" includes "rivers, marshes, [and] hills" that are givens—that is, they cannot always be controlled

---

68. See chapter 3 of this volume, 86.
69. Ibid., 87.
70. Foucault, *Security, Territory, Population,* 21.

or coerced, routed or removed—and security must work with this elusive material, as well as collectives of individuals and constructed spaces (houses, for instance). These elements, together, constitute a new sort of territory different from that which disciplinary mechanisms could simply condition or control. The natural elements of the milieu always promise to evade, exceed, or surprise the sovereign's capabilities. Medieval networks of circulation included both natural and artificial landscapes, objects, life, and power.

If our first cluster of essays probed the medieval forest and its dense legal frames, then this second array of chapters attends to networks that move across and through varying landscapes of land, life, and law. In particular, our authors concern themselves with objects that circulate within these entangled spaces. Medievalists have long attended to "things," but recent studies of medieval objects have clarified further the ways that these objects created circuits of commerce and state control over the larger populations within which they were distributed. Claire Sponsler, for example, explains the cultural and political anxieties born of new sartorial fashions taking hold across England:

> In response to this fear, discourses of sartorial control sought to restrict access to these commodities. In so doing, they treated dress as a proxy for the socially constructed bodies beneath, seeking to regulate subjects by prescribing what might be worn. Through what could be described as an early form of commodity fetishism, dress effectively displaced individual subjects as the target of authoritative surveillance, a move that quite deliberately mistook commodities for living bodies.[71]

The retracted line between the living and the lifeless, the fluid and inert, bodies of humans and objects that we find in sumptuary laws vanishes as well for medieval Europe's most prominent objects: relics. As Patrick J. Geary contends, relics are hybrid commodities, both persons and things that were "bought and sold, stolen or divided, much as any other commodity was."[72] The markets for such goods comprise what Kellie Robertson views as "networks among donors, patrons, merchants, worshippers, and (occasionally) thieves."[73] The chapters within this second cluster

---

71. Claire Sponsler, *Drama and Resistance: Bodies, Goods, and Theatricality in Late Medieval England* (Minneapolis: University of Minnesota Press, 1997), 5–6.

72. Patrick J. Geary, *Living with the Dead in the Middle Ages* (Ithaca, NY: Cornell University Press, 1994), 169.

73. Kellie Robertson, "Medieval Things: Materiality, Historicism, and the Premodern Object," *Literature Compass* 5/6 (2008): 1071.

engage these boundaries at the place of their disappearance between the sartorial, the relic, and the severed tree.

In "On the Line of the Law: The London Skinners and the Biopolitics of Fur," Michelle R. Warren tackles biopolitical networks through her analysis of the sumptuary discourse that centered, specifically, upon fur in late medieval London. Reading the fur of the doorkeeper's coat in Kafka's parable "Before the Law" alongside sumptuary laws regulating fur in the city, Warren finds a complex system of legal exceptions that ultimately profane the sovereign from the very legal space he labors to create. While attention is often paid to the social implications for class or rank to which these laws testify, Warren takes up sumptuary and craft regulations from the point of view of the furriers and skinners themselves. She argues, "Through the lens of biopolitics, sumptuary legislation turns a forest-based clothing material into a tool of urban hierarchies."[74] These guildsmen occupy an important place for a medieval biopolitical fabric within which they affected the lives of both animals—preying on their living bodies and relying on the commodities produced by their dead ones—and city-dwellers, who depended on the furriers' and skinners' forest-based clothing for participation in urban social systems. Although such sartorial artifice might be seen as excessive and unnatural, the political systems of the fourteenth and fifteenth centuries found these objects and their makers necessary collaborators. More than this, however, the Skinners Guild that is Warren's focus moved across hierarchical divisions, making both "ordinary and extraordinary furs" and crafting a commodities circuit that "operates without the sovereign at the center."[75] Through these networks, the figure of the mayor, elected by the great guilds of London, becomes a doorkeeper to the law, legislating nature into culture.

Medieval biopolitical networks, such as the ones that included the city of London, the forests from which the Skinner's Guild found their living skins, the materials they fashioned from these animals, and the laws that regulated their distribution, resonate in Actor-Network Theory and in the work of thing theorists. Jon Murdoch explains the actor-networks theorized by Bruno Latour, John Law, and Michael Callon, among others: "In many ways the theory makes the most of the Foucaultian [*sic*] insight that it is not power *per se* that is important but the various materials, practices, discourses in which power is both embedded

---

74. See chapter 4 of this volume, 117.
75. Ibid., 121.

and transported."[76] As Latour himself asserts, "Purposeful action may not be the properties of objects, but they are not properties of humans either. They are properties of institutions, of apparatuses, and of what Foucault calls *dispositifs*."[77] Governmentality's suffusion over individual bodies and populations is evident in actor-networks' enmeshment of individuals (actors) and collectives (populations of humans, nonhumans, and objects), of vibrant networks and static emplacement.[78]

In "Saintly Ecologies: Tracing Collectivities in the Life of King Oswald of Northumbria," Mary Kate Hurley examines actor-networks in Bede's *Historia* and Ælfric's *Lives of Saints* in order to understand what sorts of agency medieval texts grant humans and nonhumans. Hurley finds collectivities of humans and nonhumans emerging around the land associated with the Northumbrian king and saint, Oswald. The cross Oswald erects at Heavenfield, where he defeated Cadwalla in 634 CE, becomes a site of worship and, with the moss that grows upon it, an object of healing. The cross, then, constitutes not merely a fashioned object but a mediator between the saint, his followers, and the environment itself. As Hurley explains, history—specifically Oswald's life, works, and death—coalesce to sanctify the soil at Maserfield where he falls in battle in 642 CE. This soil heals the animal and the human, provokes its own transportation by human actors to other spaces, and is believed to save whole structures from fire, all of which point to an agential network highlighting the interconnection of human and nonhuman actors in the literary narratives of Oswald's life and death. The cross, the moss, and the soil do not comprise individual objects of healing distributed separately in autonomous acts of human narrative; rather, these actants—including the written histories themselves—are intertwined within a network, an ecology of holiness, that illustrates the interdependency of power, land, and life.

As the chapters in this volume illustrate, a melding of biopolitical inquiry with ecocriticism captures the ways that ruling bodies confront their limitations in sprawling networks of life and material that they cannot fully contain, coerce, or overwhelm. It is a narrative through

---

76. Jon Murdoch, *Post-Structuralist Geography: A Guide to Relational Space* (London: Sage, 2006), 58.

77. Bruno Latour, *Pandora's Hope: Essays on the Reality of Science Studies,* trans. Catherine Porter (Cambridge, MA: Harvard University Press, 1999), 192.

78. Michael Callon and John Law, "After the Individual in Society: Lessons on Collectivity from Science, Technology and Society," *Canadian Journal of Sociology* 22, no. 2 (1997): 5.

which we are made increasingly aware that our orientations of sovereignty—where it waxes and wanes in a historical *telos*—have been misstated, that, rather than having never been *modern,* we have never been *postmedieval.* Latour contends that modernity itself disseminates from a fictive nature/culture (and language/reality) binary, and he finds, with a medievalist's cynicism, that the point of transition occurred in the seventeenth century, when modernity *finally* straightens out premodernity's "horrible mishmash of things and humans, of objects and signs."[79] Inherent in Latour's timeline, however, is the assumption that premodern collectivities offer a sociocultural moment through which modernity might relearn to think in multiplicities rather than binaries.

In a compelling reorientation of this volume's attention to medieval forest law as a progenitor of modern biopolitics, Kathleen Biddick's chapter, "Undeadness and the Tree of Life: The Ecological Thought of Sovereignty," proposes an exclusive inclusion of the thing itself, the tree as a *"res sacra"* that "can be cut, but may not be sacrificed."[80] Trees of life—including the Bronze-age inverted stump known as Seahenge and the carved images of the Ruthwell and Cloisters Crosses—help to shape the very heart of biopolitics, even in their sacredness, much as the premodern era itself remains exclusively included in modern biopolitical theory. These felled trees comprise what Biddick terms the "sovereign cut,"[81] a splicing of old sovereignty and biopolitics that simultaneously affirms the power of the king while threatening to profane him. In one of her examples, Biddick places the Cloisters Cross in the midst of the rivalry between Henry II and Thomas Becket. The cross's carving of a particular exchange between the high priest Caiphas and Pontius Pilate gestures to a state of exception inherent in Henry's Constitutions of Clarendon (1164)—and, thus, to the King's ability to "decide"—as well as to what Biddick analyzes as the making-Jew of Henry by Becket's circle—his being aligned with Caiphas. Biddick presses an affirmative biopolitical valence to such trees of life, tracking their "undeadness," or their spectral materialism, as what Jane Bennett (channeling Deleuze's pure immanence) calls "a life [that] names a restless activeness, a destructive-creative-force presence that does not coincide fully with any specific body."[82] As with the many carvings observed by Biddick on their trunks,

---

79. Latour, *We Have Never Been Modern,* 39.
80. See chapter 6 of this volume, 154.
81. Ibid., 160.
82. Bennett, *Vibrant Matter,* 54. See also 18n63.

these trees of life blur boundaries between their branches and the forms of matter they touch. Individual and subjected bodies beholden to the law fade into disengaged, recombinant strains of biopolitical DNA.

Our first cluster of chapters captures the ways that medieval forests become what Latour might term "obligatory passage points" through which legal networks circulate."[83] Affirmative biopolitics' aims are not so far removed from the collectivities that Actor-Network Theory proposes, also eschewing the subject-object dichotomy in favor of human/nonhuman assemblages. As Hurley (citing Latour) asserts in her chapter, if we "[break] down the fictive distinction between subjects and objects, we are better able to see that multiple agents might simultaneously build and shape a given environment, even one nominally wrought by humans."[84] The absenting of individual political subjects, including the sovereign himself, in favor of a collective of singularities, looks out onto a coming community of life-affirming humans and nonhumans.

Politics, Affect, and Life

In thinking of politics, affect, and life in a medieval biopolitical frame, the volume's final cluster builds upon the earlier attention to forest law and to object networks that illustrate the pervasiveness of enmeshed legality and biology. Emotions are never absent from sovereignty's machinations, even in its modern valence. Michael Hardt and Antonio Negri's extensive attention to labor, for example, illustrates how modes of production—in this case immaterial labor—function to create affective responses in the societies that consume their products.[85] Social constructivist theories of emotion have long asserted the cultural and political conditioning of subjects' emotional responses to phenomena. In his *Navigating Feeling*, William Reddy defines a "collectively constructed common sense," contending that "[a] normative style of emotional management is a fundamental element of every political regime, of every cultural hegemony."[86]

---

83. See Bruno Latour, *Science in Action: How to Follow Science and Engineers through Society* (Cambridge, MA: Harvard University Press, 1987), 103–44.

84. See chapter 5 of this volume, 133.

85. See Michael Hardt and Antonio Negri, *Multitude: War and Democracy in the Age of Empire* (London: Penguin, 2004).

86. William Reddy, *The Navigation of Feeling: A Framework for the History of Emotions* (Cambridge: Cambridge University Press, 2001), 114, 121.

Such "emotional regimes,"[87] as Reddy terms them are not, however, only a modern mode of coercion.

The Middle Ages knew such emotional conditioning as well, as demonstrated in the work of Barbara Rosenwein and Sarah McNamer, to name but two. Thinking medieval "feeling," McNamer, for example, enables us to find in *Sir Gawain and the Green Knight* a "forced gaiety that Arthur manufactures in his court" and to see that the poem demonstrates how "chivalric action rests on an affective foundation . . . [that is] not instinctive, like fear, but a complex, cultural script."[88] While the essays herein do not task themselves specifically with the history of emotions or affect theory, they demonstrate biopolitics' coercive conditioning of the affective in medieval Britain and beyond. This final cluster of essays culminates a larger argument of the volume—that the study of medieval networks of land, life, and law yields new maps of hope for grasping the origins of biopolitics, and, more important, for turning thanatopolitical governmentality into a politics of life. In so doing, these essays provide impetus for further discussion of just how medieval biopolitics informs the present, particularly how attention to medieval biopolitical thought and practice points us toward the fusion of *bios* and *zoē* in an affirmative politics of life.

In "Sovereign Ecologies: Managing the King's Bodies in Anglo-Norman Historiography," Joseph Taylor examines the work of twelfth-century historiographers, who were tasked with a complex cultural and political project—melding the Norman Conquest into what Robert Stein has called "the master narrative . . . of the rise of the Norman state in England" wherein "an English state is . . . both the precursor and preordained outcome of the story."[89] Such a genealogical narrative sought to condition an affective response that conceived Englishness as a cultural given, yet the turbulence of the early twelfth century wherein these his-

---

87. Ibid., xiii.

88. Sarah McNamer, "Feeling," *Middle English: Oxford Twenty-First Century Approaches to Literature*, ed. Paul Strohm (Oxford: Oxford University Press, 2007), 252–53. See also McNamer's *Affective Meditation and the Invention of Medieval Compassion* (Philadelphia: University of Pennsylvania Press, 2010), and Barbara Rosenwein's works, including her edited collection, *Anger's Past: The Social Uses of an Emotion in the Middle Ages* (Ithaca, NY: Cornell University Press, 1999), her monograph, *Emotional Communities in the Early Middle Ages* (Ithaca, NY: Cornell University Press, 2006), and her seminal essay, "Worrying about Emotions in History," *American Historical Review* 107 (2002): 821–45.

89. Robert Stein, *Reality Fictions: Romance, History, and Governmental Authority, 1025–1180* (Notre Dame, IN: University of Notre Dame Press, 2006), 89–90.

torians fashioned their texts overwhelms the very stories they tell. Specifically, the biological facticity of kings' dead and dying bodies in these accounts bespeaks an anxiety born from Henry I's succession crisis—a crisis as much of the king's flesh and blood as of his *corpus mysticum*—and the period of anarchy that followed. The result of this narrative programmatic is not Englishness but a questioning of sovereignty's power in the face of nature's grander network of life. Eric Santner has recently posed that the task of modern biopolitics is to manage the semio-somatic surplus of what Kantorowicz famously called the "King's Two Bodies." In Santner's reading, then, as sovereignty fades into the background of modern political power along a Foucauldian timeline, an excess of the king's symbolic body remains to be taken up by the people, the new purveyors of sovereign power. Biopolitics' task is, thus, to manage the people's sovereign flesh. Yet, twelfth-century historiographers such as William of Malmesbury illustrate how, in the very period wherein Kantorowicz sees theories of the two bodies arise, attention was driven exclusively to the dead and decomposing bodies of kings given over to the overarching power of nature. Historiography's management of the king's body (or bodies), then, does not so much generate affective connections to the newly re-presented English identity as provoke the haunting revelation that its very figurehead is nothing but fragile, organic material in a larger theater of life, and that the symbolic body of the king—and, perhaps more frightening, Englishness itself—was never really there in the first place.

In "Radical Conservation and the Eco-logy of Late Medieval Political Complaint," Stephanie Batkie takes Alan of Lille's *Omnis mundi creatura* as a point of departure for her study of political poems. While poets of the late-fourteenth and fifteenth centuries take nature, per Alan, as an idealized measure of their own living—man compared to an exterior truth—Batkie illustrates how these later works make immanent the natural world within the political struggles of the time. The combination of these two operations—nature as representation and nature as political resource—perfectly capture a medieval biopolitics that is "less a question of population control, as we see in Foucault's formulation, and more a question of harmonizing the natural and political orders through the intervention of the state."[90] In this sense, readers of medieval political allegory become "biopolitical agents" performing the affective labor of interpretation—regulating reader response—in order to restore simili-

---

90. See chapter 8 of this volume, 218.

tude between nature and the political order. The immanence of nature within these political moments can be employed to justify both loyalists' sentiments and dissidents' actions against a particular power and, further, to dull the boundaries between nature, politics, and literature. Batkie focuses on the Ricardian complaint poem "There is a Busch that is Forgrowe," which allegorizes several of Richard's most notorious ministers as either overgrown foliage—through *paronomasia*, Sir John Bushy is a bush, while Henry Green is green grass—or living sigils of their respective houses—Richard Fitzalan, the Earl of Arundel is a steed. For the poet, this flora has become "unnatural and unruly," and it is up to the reader or "anyone capable of cultivating virtue to return the land to its proper, productive state."[91] The "eco-logy" of "There is a Busch," then, uses similitude and analogy, in Batkie's words, to "produce a vision of governmentality" in late medieval England.

In the volume's final chapter, "Lost Geographies, Remembrance, and *The Awntyrs off Arthure*," Kathleen Coyne Kelly argues that "a 'good' environmentalism going into the future actually requires" not only exploring the "past of the natural world" but also "writing the stories that we tell ourselves about the natural world" and the "structuring grief" often found therein.[92] For her part, Kelly seeks the "lost geography" of the mythical Arthurian lake, Tarn Watheling, and her self-consciously ecomaterialist approach concerns not simply the past narratives of the Tarn but its very "thingness." Mourning the Tarn through its textual history, from Gervase of Tilbury's early-thirteenth century *Otia Imperialia* to late-fourteenth and fifteenth century northern Gawain romances to twentieth-century geological surveys, Kelly documents the waxing and waning of the lost lake's political and cultural identity—the ebb and flow of when it does and does not matter. If we comprehend how mourning and melancholia—as appropriate, even strategic, affective responses to ecological change—shape our environmental narratives, then we will aid our confrontation with, and regulation of, grief, as well as improve our future readings. Kelly poses here a powerful question: "Does, or could, mourning advance an environmentalist activist agenda, one in which 'biopower' is not the imposition of the state on all beings, their bodies, and their relations" but "instead is the willing and ethical embracing of *all* beings, bodies, and relations?"[93] Her transhistorical look at Tarn Watheling clarifies why we need not only "scientific insights into

---

91. Ibid., 222.
92. See chapter 9 of this volume, 236.
93. Ibid.

anthropogenic effects on ecosystems" but also synchronic and diachronic histories of "how humans have re-presented the natural world back to themselves in the arts."[94]

Kelly's reflections on the significance of the aesthetic world in ecomaterialist analysis provide a pertinent cue to recapitulate the overarching concern of the collection. In assembling literary and historical studies of legally saturated networks of life and land, the *Politics of Ecology* illuminates the biopolitics pulsing throughout medieval Britain. Uniquely situated to contribute to biopolitical inquiry due both to forest law's profound influence across multiple discourses and to the centrality of naturalized allegories in its narrative structures, medieval British literature provides an especially fertile corpus for investigating the literary history of ecopolitics. The *Politics of Ecology* seeks to provide a powerful space for the convergence of medievalist critics currently shaping fields such as ecocriticism, animal studies, and object-oriented ontology into a common concern with biopolitics that enliven and enrich the following investigations into literary intersections of land, life, and law in medieval Britain.

---

94. Ibid., 264.

PART I

Biopolitics and Forest Law

1

Biopolitics
in the Forest

❦

Karl Steel

Sovereignty

The word "forest" comes from the Latin *foris,* meaning "outside." The medieval English forest is not a sylvan but a legal space, defined both by the forest law and, as Richard fitz Nigel claimed in his twelfth-century *Dialogue of the Exchequer,* by being outside other laws. According to fitz Nigel, forest judgments depend only "on the decision of the King or of some officer specially appointed by him." There the king leaves behind "the jurisdiction of other courts of the realm" to enter into his "penetralia" [secret places] to enjoy his "maxime delicie" [greatest pleasure]; there he sloughs off "the anxious turmoil native to court," "naturalis libertatis gratiam paulisper respirant" [to take a little breath in the free air of nature].[1]

Fitz Nigel's characterization of the forest as a zone of absolute royal freedom may have been inspired by his own monarch, Henry II, who in 1175 reversed his promise not to execute a group of knights who had murdered a forester and made a controversial decision to subject even earls, barons, and clergy to the forest law. Fitz Nigel may even have recalled Eleanor's decision, after her husband's death, to bypass typical amnesty procedures when she spared some forest prisoners from pun-

---

1. Richard Fitzneale, *Dialogus de Scaccario,* ed. and trans. Charles Johnson, rev. ed. F. E. L. Carter and Diana E. Greenway (Oxford: Clarendon Press, 1983), 60.

ishment, for, as Carl Schmitt's *Political Theology* observes, the sovereign
shows her freedom, too, when she acts "as the graceful and merciful lord
who proves by pardons and amnesties [her] supremacy over [her] own
laws."[2] With models like these, it is no wonder that for fitz Nigel the
freedom of the forest is a freedom from law that is itself the law at its
most arbitrary, founded on nothing but the sovereign's own whim.[3] The
king in his *penetralia* is most himself, an obscene hedonist at the heart of
prohibition, and a master of life and death.

Early versions of the forest law threatened poachers with blinding,
castration, or even flaying and execution,[4] inspiring a crowd of eleventh-
and twelfth-century monks and courtiers to protest that the king loved
the forest animals more than his own human subjects.[5] The late eleventh-
century *Rime of King William,* for example, accuses the king of loving
stags "as if he were their father."[6] The king's love paradoxically encloses
his forest animals within his political order while also exposing them to
"an unconditional capacity to be killed."[7] For the king in the forest, all

2. Charles R. Young, "English Royal Forests under the Angevin Kings," *Journal of
British Studies* 12 (1972): 5–6. I quote the Schmitt from Kathleen Davis, *Periodization
and Sovereignty: How Ideas of Feudalism and Secularization Govern the Politics of Time*
(Philadelphia: University of Pennsylvania Press, 2008), 81.

3. For an extended commentary on the self-referentiality of the law, see Jacques Derrida, "Force of Law: The 'Mystical Foundation of Authority,'" in *Acts of Religion,* ed. Gil
Anidjar, trans. Mary Quaintance (New York: Routledge, 2002), 230–98.

4. For shifting trends in punishment, see Emma Griffin, *Blood Sport: Hunting in
Britain since 1066* (New Haven, CT: Yale University Press, 2007), 36–37; William Perry
Marvin, *Hunting Law and Ritual in Medieval English Literature* (Woodbridge: Boydell
& Brewer, 2006), 52–53; and Donald G. Stagg, *A Calendar of New Forest Documents,
1244–1344* (Winchester, UK: Hampshire County Council, 1979), 7. For flaying (likely
only a fantasized punishment, never implemented), see Pseudo-Cnut *Constitutiones de
Foresta,* ed. in Felix Liebermann, *Über Pseudo-Cnuts Constitutiones de foresta* (Halle: Max
Niemeyer, 1894), para. 22, 53. For blinding and castration, see the First Forest Assize
(1184) and, from later in the same year, the Assize of Woodstock, which explains that
punishment should be inflicted as was done in "tempore regis Henrici aui sui" [in the
time of King Henry, his ancestor], that is, Henry I; the 1198 assize repeats and clarifies this point by adding "ut amittant oculos et testiculos": Roger of Howden, *Chronica,* ed. William Stubbs (London: Longman, Green, Reader, & Dyer, 1868), IV.63. For
the first two assizes, see Nicholas Vincent's online editions at the *Early English Laws*
website: "First Forest Assize," *Early English Laws,* http://www.earlyenglishlaws.ac.uk/
laws/texts/ass-for/view/#edition,1/vu-image (accessed November 4, 2012); "Assize
of Woodstock," *Early English Laws,* http://www.earlyenglishlaws.ac.uk/laws/texts/
ass-wood/view/#edition,1/hv-image (accessed November 4, 2012).

5. Marvin, *Hunting Law,* 49–52, 63–67.

6. From the Peterborough Manuscript, *The Anglo-Saxon Chronicle,* ed. and trans.
Michael James Swanton (New York: Routledge, 1998), 221.

7. Giorgio Agamben, *Homo Sacer: Sovereign Power and Bare Life,* trans. Daniel Heller-
Roazen (Stanford: Stanford University Press, 1998), 85.

life was available for the king to spare or, especially, to slay; since, per Foucault, the sovereign demonstrates his supreme position by killing, "the right of life and death is always exercised in an unbalanced way: the balance is always tipped in favor of death."[8] Such killing is not simply killing but rather the sacrificial centerpiece of a ritual through which royalty establishes its claims to life, death, and territory. Thus, when William of Newburgh complained that Henry I and Henry II made little distinction between poachers and murderers, and expressed his outrage by coining "cervicidas" and "fericidarum"—killers of deer or of wild beasts, respectively[9]—out of the Latin *homicida,* he comes close to missing the point: while a poacher could be a *cervicida,* a king never could, because the sacrificial character of the king's killing of the deer within the forest had nothing to do with murder. When a king did it, this killing was not one the law prohibited but rather one that law explicitly demanded; the king, in his fantasy at least, has escaped from the law, its prohibitions, and its demands precisely by setting aside space for such killing.

Understood this way, the forest perfectly exemplifies what Giorgio Agamben calls a sovereign state of exception.[10] Agamben develops this concept from Schmitt's *Political Theology,* which identifies the state's autonomy in its "miraculous" capacity to suspend the law it itself embodies and establishes.[11] For Agamben, the medieval paradigm of this nonfoundational foundation of the sovereign appears in Marie de France's *Bisclavret,* where, in the forest, for a "time the city is dissolved and men enter into a zone in which they" either "are no longer distinct from beasts" or where, more accurately (because not every man could hunt) and less dramatically, there is just a "special proximity of sovereign and werewolf.[12] To illustrate how the spirit of the forest "dwells permanently in the city,"[13] or, in this case, the royal court, Agamben points to the king and Bisclavret's mutual affection, to their hunting together, and to Bisclavret's return to his human shape "on the very bed of the sovereign." Agamben also might have looked to Bisclavret's successive assaults, first

---

8. Michel Foucault, *Society Must Be Defended: Lectures at the Collège de France, 1975–76,* ed. Mauro Bertani and Alessandro Fontana, trans. David Macey (New York: Picador, 2003), 240.

9. William of Newburgh, *Historia Rerum Anglicarum,* ed. Richard Howlett (London: Longman, 1884), I.3, 30 and III.26, 280.

10. See Agamben, *Homo Sacer,* and *State of Exception,* trans. Kevin Attell (Chicago: University of Chicago Press, 2005).

11. For an efficient and clear discussion of Schmitt, see Davis, *Periodization and Sovereignty,* 78–83.

12. Agamben, *Homo Sacer,* 107.

13. Ibid.

against his estranged wife's lover at an assembly of the king's barons, and then against his wife, who has her nose bitten off when she pays the king a visit, notably after one of his hunts. In both cases, the royal court classifies Bisclavret's attacks as legitimate revenge, because, as they reason, he had never attacked anyone else. In other words, so long as Bisclavret spares the king and his men, his acts will have the force of law. As if to ensure that he, too, has his chance to harm her, the king, this tyrannical wolf-man, has the wife tortured until she is made to confess her guilt. After this confession, Marie neither kills off this character nor separates her from her new husband nor her sovereign. She is made to live on as a subject, a fearsome example to the others of what it means to try to get away from the werewolf of the law. Her female descendants attest to this continuing, unequal, inescapable, and arbitrary violence of sovereignty: some are born noseless, some not.[14]

But to take the forest as a sovereign space with the king at top or center is to flatter the king, or to fear him too much, or to trust too much in the possibility of making things better if only the tyrant could be eliminated. Fitz Nigel is also one of these flatterers. For that matter, so is John of Salisbury, when his *Policraticus* concentrates its hunting complaints on the tyranny of the hunter-king.[15] To frustrate the flatterers, we might simply observe that even post-Conquest medieval English "kings lacked the means, the instruments of government, to make absolutism a practical reality."[16] Moreover, fitz Nigel mischaracterizes the forest law: it was not actually outside the law but rather overlapped those already in force in the places it governed,[17] while the revision of the 1184 forest law in the same year of its issue suggests at least a pretense of legal limits to royal prerogative. This counterargument to sovereignty would be just a simple matter of naively and insufficiently contrasting fact to fantasy. We might also observe that the medieval forest was also, at least in chivalric and other narratives, a place of wonder and strangeness, the wilderness

---

14. For another medievalist development of Agamben's insights, see Emma Campbell, "Political Animals: Human/Animal Life in *Bisclavret* and *Yonec*," *Exemplaria* 25 (2013): 95–109. Readings like these could be developed further by reference to Count Robert II of Artois's pet wolf, which he allowed to ravage the countryside around his hunting preserve at Hesdin; see Sharon Farmer, "Aristocratic Power and the 'Natural' Landscape: The Garden Park at Hesdin, ca. 1291–1302," *Speculum* 88, no. 3 (2013): 655.

15. For discussion of *Policraticus* on hunting, see Marvin, *Hunting Law*, 64–67.

16. Wilfred Lewis Warren, *King John* (Berkeley: University of California Press, 1978), 243.

17. Marvin, *Hunting Law*, 55; Stagg, *New Forest Documents, 1244–1344*, 3–4; Young, "English Royal Forests," 6–7.

outside the law: but going out into this forest would mean leaving the law behind too quickly. The forest was neither absolutely constrained nor was it absolutely free. We should, therefore, complicate the forest by tracking what the sovereign must manage to realize his fantasies of control in the forest. To satisfy Foucault's often-repeated complaint that "in political thought and analysis, we still have not cut off the head of the king,"[18] the remainder of this chapter will thicken its account of the powers and efforts of the medieval English forest: first, by thinking of it in biopolitical terms to follow what trouble the sovereign gets into by trying to make the forest his; and then, by liberating biopolitics from its anthropocentrism by exploring how all fantasies of sovereignty must be negotiated with a multitude of other sovereigns, some human, but mostly not. This chapter will not remove royal heads but rather multiply them, discovering a host of sovereigns available for any who wish to pay them mind.

Biopolitics and Biopolitical Failure

In contrast to sovereignty, biopolitics aims to control not only death but also life itself. It acts not on individuals but on populations. It collects data, profiles criminals. It attempts to eliminate disease or at least to confine it to its proper limits. It encourages some groups to marry and breed while simultaneously sterilizing or liquidating others. Foucault arrives at a discussion of biopolitics in the last of his 1975–76 Collège de France lectures, but only after he traces the development of an anti-regal counterhistory of racial war and oppression, in which the aristocracy imagined that royal oppression began with their ancestors being defeated by foreign invaders whose descendants set themselves up as tyrants. According to Foucault, this counterhistory eventually solidifies into a fantasy in which the state (now encompassing royalty, aristocracy, and the whole population) imagines itself threatened by racial degeneracy, to which it responds by developing techniques of controlling, measuring, and channeling life to preserve the best of the state's particular, biological character. He argues that it is in this language of war, resistance, atavism, and breeding that biopolitics develops most fully.

Foucault's narrative splits the medieval from the modern, taking sovereignty as characteristic of the former and biopolitics of the latter. In

---

18. Michel Foucault, *The History of Sexuality: Volume 1, An Introduction*, trans. Robert Hurley (New York: Vintage Books, 1980), 88–89.

a volume largely directed at medievalists, it may be otiose to lay out, once again, the corrections Foucault's historical narrative requires. For example, when he neatly observes that "throughout the Middle Ages, judicial practice was a multiplier of royal power," whereas in the "new governmental rationality" of the modern era, "legal theory and judicial institutions no longer serve as the multiplier, but rather as the subtractor of royal power,"[19] Foucault sweeps aside the complications of Ranulf de Glanville's late twelfth-century *Treatise on the Laws and Customs of Realm of England,* which sometimes characterizes the king as a Roman despot, at once constituting and above the law, and at other times as himself guided by the laws and his lords.[20] Cataloging more failures of Foucault's medieval models would be tedious, ungenerous, and small-minded, especially as Foucault himself explains that despite the neatness of his narrative, "in reality you have a series of complex edifices in which, of course, the techniques themselves change and are perfected, or anyway become more complicated, but in which what above all changes is the dominant characteristic."[21] This, at least, even the most pedantic medievalist could admit: the disappearance of the king and the rise of the state, with its pretense to impersonal and immaterial existence, does change the dominant ways governmentality operates and presents itself to itself and its subjects. For this chapter, the mistake that matters is one to which Foucault never confessed, and which the criticism has only started to track: namely, his anthropocentrism. By leaving most life outside his attention, Foucault missed how biopolitics was the dominant characteristic of, at least, those legal spaces where the post-Conquest medieval English king thought he exercised his sovereignty most freely. The paired concerns of the medieval English forest, namely its attempt to monopolize legitimate violence and to oversee cervid wellbeing, frustrates attempts to seal off medieval and modern forms of governmentality and, indeed, sovereignty and biopolitics from each other, because in England, at least, the birth of the sovereign is also simultaneously the birth of biopolitics.

The biopolitical characteristics of medieval English forest are not hard to come by. First, because, as Barbara Hanawalt neatly points out, "royal foresters were England's first contact with a policing force that

---

19. Michel Foucault, *The Birth of Biopolitics: Lectures at the Collège de France, 1978–1979,* ed. Michel Senellart, trans. Graham Burchell (New York: Palgrave Macmillan, 2008), 8.

20. Ralph Turner, *Judges, Administrators, and the Common Law in Angevin England* (London: Hambledon Press, 1994), 77.

21. Michel Foucault, *Security, Territory, Population: Lectures at the Collège de France, 1977–1978,* ed. Michel Senellart, trans. Graham Burchell (New York: Palgrave Macmillan, 2007), 22.

had a regular beat and made routine rounds," descriptions of the police surveillance so important to biopolitical studies would do well to start in the medieval British forest, centuries prior to the development of a regular police force anywhere else in the realm.[22] Nowhere else in England were the king's violence and surveillance so professionally executed or were the tasks so carefully divided. Forests were the responsibility of wardens or keepers, to whom the crown would direct commands for wood or venison. The wardens could in turn charge foresters with the responsibility of patrolling the forest and apprehending lawbreakers. *Verderers*—men of rank elected to this position, whether they wanted it or not—answered directly to the king. They would keep records of violations of the forest law and could, in minor cases, serve as judges. Above these were the chief justices of the forest and finally the *justices in eyre,* summoned to wherever they were needed to ensure the forest law's execution. All collectively, in addition to their other duties, prevented forest inhabitants from making new clearings or erecting new buildings, ensured the protection of trees and undergrowth, and above all protected the exclusivity of hunting: they required that dogs who might have been used for illicit hunting be "expeditated," that is, that they have their paws mutilated; they caught and punished poachers with fines and imprisonment, or worse; and they investigated the deaths of boar and, in particular, cervids with the same concern, and techniques, used to investigate the deaths of humans.[23]

Furthermore, a biopolitical attention to knowing the hunting stock helps clarify the concern for classification in the Anglo-Norman and Middle English versions of William Twiti's *Art of Hunting,* both of which devote large sections of their short treatises to painstakingly listing or generating names for antlers of various sizes. The categorization might be recognized not just as natural science, nor just as the promulgation of insider knowledge,[24] nor even, more practically, just a way for hunters to "keep score" or distinguish between deer during the hunt. Certainly, the

---

22. Barbara A. Hanawalt, *Of Good and Ill Repute: Gender and Social Control in Medieval England* (New York: Oxford University Press, 1998), 192.

23. For this synthesis of the forest law, see J. Charles Cox, *The Royal Forests of England* (London: Methuen & Co., 1905), 17–24, Marvin, *Hunting Law,* 61–63, and Charles R. Young, *The Royal Forests of Medieval England* (Philadelphia: University of Pennsylvania Press, 1979), 74–92. For the inconvenience of serving as a verderer, see Young, "English Royal Forests," 13. For the legal "commonplace" of the similarities between techniques of human and animal death investigation in the twelfth and thirteenth centuries, see Stefan Jurasinski, "The Rime of King William and its Analogues," *Neophilologus* 88 (2004): 139.

24. See, for example, Anne Rooney, *Hunting in Middle English Literature* (Suffolk, UK: Boydell & Brewer, 1993), 18–19, and Marvin, *Hunting Law,* 102–6.

lists were likely used for all these purposes, but they were also in fact the work of a crank, a "long and unnecessarily repetitive account"[25] as one hunting scholar complains, in the sense that biopolitics, with its interest in managing everything, leads inevitably to excesses of bookkeeping. Following the imperative of power itself to always know more, such excesses divide normal from abnormal life, making life, on the whole, available for increasingly precise manipulation, and, as I will argue, inevitably multiplying the opportunities for biopolitical failure.

As biopolitical bookkeepers, forest professionals kept count of cervid populations and sought to keep them alive so that no one or nothing could kill them but the sovereign and his agents. While these counts lacked the statistical analysis of fully fledged biopolitics, they still treat subjects not as individuals but as populations, to be managed as a group, and capable of a collectively normal or abnormal life. Thus, for example, when Philip Okeover complained in 1448 that John Cokeyne killed 120 deer in his park and "lafte in seid parke but 5 dere alyve," his agents must either have known that the forest had initially 125 deer, or known that they could believably claim such knowledge in court.[26] Early fourteenth-century records of the Forest of Pickering required that keepers count both living deer and those that had died of disease, and one early sixteenth-century document distinguishes among the various kinds of diseases, including *wyppys, garget,* and the *rotte.*[27] To keep them from disease and unwanted death, forest professionals provided nurse-cows for orphaned fawns and protected all the deer during "fence moth" and the rut, while the deer were fawning.[28] During the winter, shelter and additional forage would be provided, and noncervid livestock expelled from the forest.[29] These latter acts are particularly biopolitical because they are not *charitable* distribution, the paradigmatic act of a merciful sovereign who could just as arbitrarily take resources away. They are, instead, acts

---

25. William Twiti, *The Middle English Text of The Art of Hunting,* ed. David Scott-Macnab (Heidelberg, DE: Winter, 2009), lxvii.

26. George Wrottesley, ed., "An Account of the Family of Okeover of Okeover, County Stafford, with Transcripts of the Ancient Deeds of Oceover," in *Collections for a History of Staffordshire,* vol. 7, n. s. (London: Harrison and Sons, 1904), 53. For remarks on accounting of deer herds in English parks, see Jean Birrell, "Procuring, Preparing, and Serving Venison," in *Food in Medieval England,* ed. C. M. Woolgar, D. Serjeantson, and T. Waldron (New York: Oxford University Press, 2006), 186–87.

27. Jean Birrell, "Deer and Deer Farming in Medieval England," *Agricultural History Review* 40, no. 2 (1992): 121–22.

28. Ibid., 116.

29. Ibid., 117.

inspired by an apparently selfless concern with a group's wellbeing that aims to maintain or increase the size of a healthy population.

One of Roberto Esposito's key insights is his description of an "immunitary paradigm," in which the biopolitical promotion of birth simultaneously operates with a thanatopolitics against the unworthy life that pollutes the social body. Poachers would seem to provide the medieval analogue of this unworthy life, except that punishments for poaching tended not be as severe as the law allowed, and fines tended to be tailored to the offender's ability to pay:[30] in this, at least, the law treated its subjects as individuals, not as a mass of vermin. The actions of the *luparii,* the forest's professional wolf hunters,[31] provide a better analogue. While humans have had a long antipathy toward wolves, the institution of professional wolf-hunters began in England only after the first Norman kings of Britain began to encompass huge swaths of their new land in the forest law.[32] At this point, wolves become, in the classic formation of sovereign power, a kind of life included within the law precisely by being systematically excluded from it. Notably, *luparii* and other medieval hunters of vermin employed traps and nets to accomplish this exclusion. Through the use of techniques more practical and direct than those used by the ritual of the elite hunt,[33] techniques themselves disdained by elite hunters,[34] they practiced a kind of ceremony of hunting without ceremony, a self-conscious *indifference* to the kind of killing that would mark out the victims as individuals or as meriting any kind of ritual or finesse. This was a killing that understood itself as only extirpation and thus was an exercise in a thanatopolitics against the kinds of forest life framed as unworthy of life or narratives of either mourning or celebratory death.

---

30. Young, *Royal Forests,* 103–7.

31. Aleksander Pluskowski, "The Wolf," in *Extinctions and Invasions: A Social History of British Fauna,* ed. T. P O'Connor and Naomi Sykes (Oxford: Windgather Press, 2010), 71–73; Robert Bartlett, *England under the Norman and Angevin Kings, 1075–1225* (Oxford: Clarendon Press, 2000), 671–72.

32. For antipathy against wolves, see Aleksander Pluskowski, *Wolves and the Wilderness in the Middle Ages* (Woodbridge, UK: Boydell & Brewer, 2006). For wolf extermination in earlier hunting preserves, arguably the first that associated kingship specifically with hunting, see Eric J. Goldberg, "Louis the Pious and the Hunt," *Speculum* 88 (2013): 618.

33. Edward of Norwich, *The Master of Game,* ed. William A. and F. N. Baillie-Grohman (Philadelphia: University of Pennsylvania Press, 2005), 61; Marvin, *Hunting Law,* 105.

34. Crane, *Animal Encounters,* 107; John Cummins, "Veneurs s'sen vont en paradis: Medieval Hunting and the 'Natural' Landscape," in *Inventing Medieval Landscapes: Senses of Place in Western Europe,* ed. John Howe and Michael Wolfe (Gainesville: University Press of Florida, 2002), 41.

In their aggregate, these techniques can all be classified as husbandry. This is exactly my point, namely, that husbandry is the scandalous foundation of a biopolitical analysis that has tended to be committed, more or less explicitly, to defending human particularity by trying to keep humans from being treated *like animals*. Foucault observes that "unlike discipline, which is addressed to bodies, the new nondisciplinary power [of biopolitics] is applied not to man-as-body but to the living man, to man-as-living-being; ultimately, if you like, to man-as-species"[35] and that in the sixteenth and seventeenth centuries we witness the "development of a medicine whose main function will not be public hygiene, with institutions to coordinate medical care, centralize power, and normalize knowledge."[36] Esposito writes that in modern biopolitics "life enters into power relations not only on the side of its critical thresholds or its pathological exceptions, but in all its extension, articulation, and duration" and calls this a "new rationality centered on the question of life."[37] The obvious modernist and humanist biases of these observations ought to be contested. When Foucault states that "man is to population what the subject of right was to the sovereign," or Esposito explains that biopolitics aims not only at "obedience but also at the welfare of the governed,"[38] their analysis might have gone even further had they said that biopolitics treats humans like livestock, or, more particularly, like the sovereign's livestock, which is to say, like venison.

I confess to have sprinted through my examples, in the hopes of sketching out areas for future, and I trust far more precise, research in this area, which could account, for example, for the distinctions between the management of forests, warrens, and chases. The remainder of this chapter will look not at the successes of biopower in managing life or producing certain kinds of life as knowable or proper but rather at biopolitics' inevitable frustrations. On its face, biopolitics looks more powerful than sovereignty. It reaches further and does more. To use a well-worn and dubious distinction from Agamben: biopolitics undercuts its subjects' political life, their *bios*, by directly managing their bare, biological life, their *zoē*.[39] Sovereignty is concerned with subjugation and land, with

35. Michel Foucault, *Society Must Be Defended*, 242.

36. Ibid., 244.

37. Roberto Esposito, *Bíos: Biopolitics and Philosophy*, trans. Timothy C. Campbell (Minneapolis: University of Minnesota Press, 2008).

38. Foucault, *Security, Territory, Population*, 79; Esposito, *Bíos*, 28.

39. For a philological critique of the role this supposedly foundational division plays in Agamben's political theory, his critique of Foucault, and their distinct historical narra-

the law and enforcing it, or, more precisely, with justifying the sovereign's position and maintaining it beyond all justification. Biopolitics, on the other hand, is concerned with bodies and what they do, which is to say, it is concerned with *everything*. This is its problem: the sheer size of biopolitics' remit frustrates it. Faced with the liveliness of all things, biopolitics can only try to catch up.

In stressing biopolitical failures and in promoting a disanthropocentric conception of biopolitics, I follow Cary Wolfe's *Before the Law: Humans and Animals in a Biopolitical Frame,*[40] which wrests Foucault from his anthropocentrism by arguing that the resistant bodies frustrating biopower need not be human subjects. Life of whatever sort will "burst through power's systematic operation[s] in ways that are more and more difficult to anticipate," a point, as Wolfe observes, indebted as much to Nietzsche as to Foucault.[41] For Wolfe, resistance does not have to be a deliberate choice of a rational agent; the resistant—an adjective that might be heard as a noun—may be resisting in ways it cannot ever possibly understand. Self-awareness and rationality, a true response versus a merely instinctual reaction, all shibboleths of a presumptively human subjectivity, need no longer be our primary concern. Biopolitics deals with bodies, and bodies resist; they have their own energies, sometimes capable of turning biopolitical interventions to other ends. In other words, the victim of the sovereign can only either die, be allowed to live, or escape, whereas the subject of biopolitics, even when it obeys and generates itself as a good subject, will still get out of hand by living in ways as yet unanticipated or unmanageable. Killing turns out to be the easiest thing a sovereign can do.

Wolfe discusses the resistance of factory-farmed animals, the way that their maltreated bodies strike back or, viewed more positively, the way

tives of the relation of sovereignty to biopolitics, see Jacques Derrida, *The Beast and the Sovereign,* ed. Michel Lisse, Marie-Louise Mallet, and Ginette Michaud, trans. Geoffrey Bennington, vol. 1 (Chicago: University of Chicago Press, 2009), 315–17. To disprove the supposedly fundamental distinction between *zoē* and *bios,* Derrida points to Agamben's own citation of Aristotle's *Metaphysics* 1072b, which speaks of a "*zoē aristē kai aidios,*" "a noble and eternal life," evidence that *zoē* too could mean a political or group life. The entry in Henry George Liddell and Robert Scott, *A Greek-English Lexicon,* revised by Sir Henry Stuart Jones, with the assistance of Roderick McKenzie (Oxford: Clarendon Press, 1940), s.v. "ζωή," bears out this critique; from Herodotus's account of the customs of the Amazons, definition A(3) is a "way of life."

40. Cary Wolfe, *Before the Law: Humans and Other Animals in a Biopolitical Frame* (Chicago: University of Chicago Press, 2013).

41. Wolfe, *Before the Law,* 32–33. For example, see Esposito's discussion of Nietzsche and life in *Bíos,* 78–109.

they get creative: keeping these animals alive long enough to be slaughtered profitably requires enormous amounts of antibiotics, which in turn leads to antibiotic-resistant pathogens, like MRSA, which spread with deadly efficiency from livestock to humans.[42] Here, as Wolfe observes, is the vicious circle of the "'immunitary' paradigm," in which techniques of defending life against death—by, for example, ensuring outsized national stores of livestock—create still more risk, which requires redoubled immunitary efforts, and so on.

My model of liveliness comes, as well, from Steve Shaviro, who advocates for a "deflationary" description of intention via George Molnar's concept of "physical intentionality," by which "physical powers, such as solubility or electrical charge, also have that direction toward something outside themselves that is typical of psychological attributes," even if this intentionality has no "semantic or representational content."[43] Likewise, I draw on Jane Bennett's "theory of distributive agency." From her, I take the recognition that "human intentions [are] always in competition and confederation with many other strivings,"[44] a "heterogeneous series of actants with partial, overlapping, and conflicting degrees of power and efficiency."[45] One of her chief examples is the American blackout of 2003, for which no single element can be wholly responsible: storms and climate change, capitalism and deregulation, and flows of various kinds of electricity interact to produce results that can't quite be predicted or reduced to a single reason. Through "a touch of anthropomorphism," Bennett keeps open the possibility of various nonhuman agencies in order to "catalyze a sensibility that finds a world filled not with ontologically distinct categories of beings (subjects and objects) but with variously composed materialities that form confederations."[46] As much as we might want to reserve all our blame for a sovereign—lobbyists, Enron, Dick Cheney, whoever—this would be a mistake. To "cut off the head of the king," we need to aim at more than just the king. It may seem to be mystification to recognize nonhuman agency in a supposedly human system, but it is just as much a humanist mystification to believe we're the only significant actors in town. Medievalists seeking historical jus-

42. Jim Wickens, "Sick as a Pig," *The Ecologist,* March 26, 2009, http://www.theecologist.org/investigations/food_and_farming/268866/sick_as_a_pig.html.

43. Steve Shaviro, "Panpsychism And/Or Eliminativism," *The Pinocchio Theory,* October 24, 2011, http://www.shaviro.com/Blog/?p=1012, and George Molnar, *Powers: A Study in Metaphysics,* ed. Stephen Mumford (Oxford: Clarendon, 2006), 63.

44. Bennett, *Vibrant Matter,* 32.

45. Ibid., 33.

46. Ibid., 99.

tification for this picture of nonhuman agency may be comforted by Caroline Walker Bynum's recent discussions of thirteenth-century and later medieval scholars who built on Aristotle's "encompassing [of] what we would call the living and the inanimate in one physics" to argue for "a sort of autonomy of actuality or desire in matter."[47] Other thinkers from various eras and regions could be listed: Spinoza, Deleuze and Guattari, Haraway, and so on. Taken as a whole, these thinkers evidence that the belief in the human monopoly on agency is just a culturally and historically specific option, dominant but not inevitable. This belief can do good analytical work (as in Marxist denaturalization and historicism), as can belief in the absence of *any* significant autonomy (as in some strains of psychoanalysis). But to break anthropocentric hermeneutic orbits, whether humanist or posthumanist, and to dismantle spirit/matter binaries wherever they might be found, such concentration of agency in humans alone or the dispersal of agency into an unending and unbreachable network of drives could be supplemented by other analytic models that recognize, even if just strategically, that the world comprises more than only human subjects and instinctual or inert objects, more or (generally) less akin to what we imagine ourselves to be.

This expanded attention allows us to better attend to how, in the medieval forest, deer, their carcasses, scavengers, parasites, and even the climate resist or frustrate the operations of biopower and indeed how they themselves operate, not only resisting but rather compelling humans to resist *them*. Recognizing these actors *as* actors rather than just as objects dislodges the notion of the forest as a zone of inextricably tangled violence and law centered on human sovereignty. With such work in mind, the forest can also be conceived of as a network of interdependence and struggle, in which the outsized gravity of the king must negotiate with other powers, each with its own pretense to sovereignty. In stressing sovereignty here, I follow a central argument of object-oriented ontology, namely, that objects are not fully absorbed in their networks or ecosystems, because their being always withdraws from full contact with any particular object or group of objects.[48] Things cannot know or use each other up entirely. Each is in some unfathomable way its own master. Graham Harman writes:

---

47. Caroline Walker Bynum, *Christian Materiality: An Essay on Religion in Late Medieval Europe* (Cambridge, MA: Zone Books, 2011), 236; see also 254.

48. For introductions to the field, see Graham Harman, *Towards Speculative Realism: Essays and Lectures* (Winchester, UK: Zero Books, 2010), and Levi R. Bryant, *The Democracy of Objects* (Ann Arbor, MI: Open Humanities Press, 2011).

It is not some special feature of the human psyche or human deeds that turns a thing into a caricature. This reduction belongs to any relation between any two objects in the universe, no matter what they may be. My perception of fire and cotton fails to use up the total realities of these beings, since they are describable at infinite length in a way I can never approach. . . . But more generally, the fire and cotton also fail to make full contact with each other when they touch, despite their uniting in a bond of destruction that takes no heed of the colors and scents that humans or animals may detect emanating from both of them. In other words, objects withdraw from each other and not just from humans.[49]

With this philosophy, I characterize the forest as a site of overlapping jurisdictions, a heterogeneous space, like any other, of multiple sovereigns. Some compete with each other, but each is unaware of or indifferent to most. And no sovereignty ever rules its subjects entirely. All these sovereigns—who are subjects to others—might exist in different sizes and operate according to different temporal scales, some faster or slower than others. Some of these sovereigns are human, but most not. Like any thing, humans believe themselves to be the center of the world, the measure of all things. As Ian Bogost points out, "When we welcome . . . things into scholarship, poetry, science and business, it is only to ask how they relate to human productivity, culture, and politics."[50] But humans may also be being used, more or less willingly or knowingly, by other sovereigns. Latour pointedly asks, "If you are mixed up with trees, how do you know they are not using you to achieve their dark designs?"[51] Nonliterary scholarship on the forest already knows this well: it has necessarily been ecological, zooarchaeological, and interested in climate and botany. In this work, humans do things with the forest, but the forest does its own thing too, before, around, and perhaps after humans. We might recognize how the king must work with his hounds and his horse, how during the hunt he is with the trees, with scavengers and other critters, with the climate, with the forest law, with the obligations of his own royal position and of being human, with his appetite, with an enormous array of things, desires, forces, and impediments.

Hunters had to cooperate, and this cooperation, too, I should stress, was as structured by fantasy, or perhaps more precisely, by its own

49. Harman, *Towards Speculative Realism*, 124.

50. Ian Bogost, *Alien Phenomenology, or, What It's Like to Be a Thing* (Minneapolis: University of Minnesota Press, 2012), 3.

51. Bruno Latour, *The Pasteurization of France*, trans. Alan Sheridan (Boston: Harvard University Press, 1993), 199, quoted in Bogost, *Alien Phenomenology*, 38.

umwelt, as any activity or self-maintenance.[52] I will stress again that I do not mean to contrast the fact of practical forest activities to the fantasy of sovereignty, for hunters know their world according to their own sensory and psychological schema, and, in various ways, the elements of their world knew the hunters likewise and each other as well. Hunters of course relied on and communicated with their horses and their dogs and even their equipment, which in their aggregate formed what Jeffrey Cohen called a "chivalric circuit"[53] and Susan Crane an "enigma, a reaching out beyond the human into interspecies [and technological] relationship."[54] As Crane remarks, hunting manuals are often language textbooks. They train hunters how to communicate with their dogs and with each other through horn calls or through the distribution of butchered prey.[55] In other words, the guides do more than train hunters in an elite jargon incomprehensible to their inferiors; they train hunters to join with the hunt, to become part of a forest machine. They train hunters to get along with their animals, with equipment, and their prey, which, to be worth hunting, must be able to resist.[56] That very resistance, in fact, at once enables and frustrates the hunt's theater of sovereignty.

The Lively Carcass

For elite hunting to remain, as Simon Schama termed it, "the most important blood ritual through which the hierarchy of status and honour around the king was ordered,"[57] and in which, as Crane writes, "nobility mimes its own myth of itself,"[58] the prey needed to die properly, through the efforts of authorized and properly skilled huntsmen. Since this death

---

52. On the extension of Jakob von Uexküll's concept of "umwelt" from animal environment relations to any interobject relation, see Levi Bryant, "The Interior of Objects," *Larval Subjects,* December 16, 2010, http://larvalsubjects.wordpress.com/2010/12/16/the-interior-of-objects/.

53. Jeffrey Jerome Cohen, *Medieval Identity Machines* (Minneapolis: University of Minnesota Press, 2003), 35–77.

54. Susan Crane, *Animal Encounters: Contacts and Concepts in Medieval Britain* (Philadelphia: University of Pennsylvania Press, 2013), 151.

55. Ibid., 112–14.

56. For cooperation, see Ryan Judkins, "The Game of the Courtly Hunt: Chasing and Breaking Deer in Late Medieval English Literature," *Journal of English and Germanic Philology* 112 (2013): 70–92, and Naomi Sykes, "Taking Sides: The Social Life of Venison in Medieval England," in *Breaking and Shaping Beastly Bodies: Animals as Material Culture in the Middle Ages,* ed. Aleksander Pluskowski (Oxford: Oxbow, 2007), 155.

57. Simon Schama, *Landscape and Memory* (New York: A. A. Knopf, 1995), 144.

58. Crane, *Animal Encounters,* 107.

was the forest's primary purpose, as S. A. Mileson has recently stressed,[59] an improper death frustrated the forest system as a whole. A living deer might always come to its right end; but once dead, it was past coercion or help, or, as this section will show, very *nearly* past such help.

Obviously, to ensure the forest functions, most hunters had to be defined as poachers, and as a result, implicitly encouraged to hunt a meat that had become so luxurious; then they would be caught and punished, acts that were as effective a means of upholding the status of the forest as elite hunting itself. However, because of the deer's own vitality, antipoaching laws might go awry, attesting to the disharmony between a deer's desires and those of the king. For example, in 1209, in Bridge Castle, in the Forest of Shrewsbury, Shropshire, when a deer wandered through a postern gate, the castellans took possession of it, perhaps to poach it, or perhaps—since nothing is said about its being killed—to try to keep it safe long enough to be transported out of the castle before an inquiry took place. The deer was in the wrong place, and someone would have to answer for it. The castellans were not fast or secret enough, as the verderer, hearing of the deer's presence, had the sheriff and his retinue questioned and punished.[60] In other cases, people in the forest who stumbled across a deer's carcass panicked and hid to avoid any accusation, and eluded unfair punishment for poaching only by luck. Remarking on this case and others like it, Charles Young observed how antipoaching laws could create a "climate of fear," less because of worry over being caught out than because of the independence of deer, whose wandering and constant vulnerability to disease, injury, and poaching could cast a whole community into suspicion if the animal met its death at the wrong time and in the wrong manner.[61]

Like poachers, deer carcasses had to be dealt with to ensure that the forest continued to appear to be a zone in which its stock died only as the forest system intended. In 1251, the bailiff was commanded to clear Havering Park, located in Essex, of large, rotting carcasses,[62] which suggests an effort not only at practical cleaning but also at hygiene for the sake of the forest system as a whole, as this cleaning preserved the appearance that no game animal had died in the forest without royal approval

---

59. S. A. Mileson, *Parks in Medieval England* (Oxford: Oxford University Press, 2009).

60. G. J. Turner, ed., *Select Pleas of the Forest* (London: Selden Society, 1901), 8.

61. Young, "English Royal Forests," 13–14.

62. I. D. Rotherham, "The Ecology and Economics of Medieval Deer Parks," *Landscape Archaeology and Ecology* 6 (2007): 96.

or punishment. Forest keepers in Northamptonshire did still more, as they were instructed to suspend deer that had died of disease from trees. Exmoor Forest, in Southwest England, maintained forked trees, somewhat like gallows, for this same purpose.[63] This may have kept the carcasses from poachers or predators, but this seems an inefficient method. As strange as it sounds, the activity makes more sense if understood as a kind of punishment of the deer for dying improperly.

The most elaborate, and apparently most widely enforced, forest carcass law dates from at least 1238 and appears, among other places, in the "Customs and Assizes of the Forests" found in Cotton Vespasian B. vii. Other laws in this collection concern escaped and game-harassing dogs, the pasturing of pigs, and capture of rabbits (which should be sent to the king), as well as resource management, concentrating on the preservation of trees (particularly oak). The carcass law, one of its longer entries, runs as follows:

> If any dead or wounded beast should be found and it does not belong to
> a herdsman. First, there should be an inquiry in the four closest towns,
> which should be recorded; and the finder should be put by six pledges; its
> flesh should be sent to the nearest house of lepers, if there is one nearby
> in those parts, and this by the witness of the forester and the jury. If
> however there is no such house nearby, the flesh should be given to the
> sick and the poor. The head and skin should be given to the freemen of
> the nearest town; and the arrow, if one was found, should be given to the
> forester, and this should be recorded with his oath.[64]

> Si aliqua fera inveniatur mortua vel vulnerata, et non fuerit bercatorum.
> Ad prima placita debet fieri inquisitio per quatuor villas proprinquiores,
> que debet irrotulari; et inventor debet poni per sex plegios caro autem
> debet mitti ad proximam domum leprosorum, si que propre fuerit in par-
> tibus illis, et hoc per testimonium viridariorum et patrie. Si autem nulla
> talis domus fuerit propre, caro debet dari infirmos et pauperibus. Caput

---

63. Birrell, "Deer and Deer Farming," 121.

64. *The Statutes at Large from the Second Year of the Reign of King George the Third to the End of the Last Session of Parliament*, vol. 9, ed. Owen Ruffhead (London: M. Basket, 1765) Appendix, 25–26. For versions of the carrion law from fourteenth- and fifteenth-century Scotland, see John M. Gilbert, *Hunting and Hunting Reserves in Medieval Scotland* (Edinburgh: J. Donald, 1979), 297. Cf. the similar sequence for coroners' investigations into human deaths in Henry de Bracton, *On the Laws and Customs of England*, ed. George E. Woodbine, trans. Samuel E. Thorne, 4 vols. (Cambridge: Harvard University Press, 1968) 2:342.

et cutis debent dari liberis hominibus proxime ville; et sagitta, si que inventa fuerit, debet presentari viridariis, et in veredicto suo irrotulabitur.

To discourage poaching, it makes good sense not to let the neighboring folk or even the forester eat animals that had died "accidentally." The law does not, however, just prevent eating; rather, it sees to it that the meat will be eaten, or at least that it be given away, but only to certain disdained or marginal recipients. The requirement makes a good symbolic sense by correcting the deer's fault of dying improperly, since the masters of the forest once again determine how the animal can be eaten, a point that becomes clearer by reading the law alongside the *morticina* [carrion] laws of the early medieval penitentials.[65] The penitentials' often-repeated carrion laws forbid humans from eating the meat of any animal they did not intend to kill and sometimes require that the meat be distributed to pigs, dogs or, significantly, to *homines bestiales,* bestial men. To be sure, only slim evidence survives for the continued observance of these laws in the later Middle Ages: for example, William of Canterbury's late twelfth-century *Life and Miracles of Thomas Becket* includes a miracle about an injured pet sheep whose owner stabs it in the throat to kill it himself "ne morticinum fieret" [lest it become carrion].[66] Given that penitentials were a moribund genre by the thirteenth century, I am not arguing for direct influence but rather only that the penitentials and this law share an analogous logic and technique. Each set of laws demands that the death be understood as an assault either, in the case of the penitentials, on human dominance in general or on royal control over life and death in the forest. In neither case could the meat simply be discarded. To compensate for the assault, more had to be done. The meat had to be eaten, but not by dominant humans, who needed to avoid the appearance of giving their consent to what had happened by, as it were, sharing a meal with illicit animals or with poachers. To retain the appearance as masters, either in the human/animal sphere or just in the forest, these dominant humans had to rein in the unfortunate independence of the dead animal by a threefold action: recapturing it, distributing it, and disdaining it. The failure of human dominance was transmuted to a kind of paired generosity (toward dogs and bestial men, or, in later periods,

---

65. Maria Moisà, "The Rotten Gift: Caro Data Fuit Pauperibus," *Medieval Yorkshire* 26 (1997): 6–10.

66. James Craigie Robertson, ed., *Materials for the History of Thomas Becket, Archbishop of Canterbury,* vol. 1 (London: Longman & Co., 1875), 343. The saint resurrects the sheep.

people with leprosy or the poor) and, in a way, punishment, as the flesh was now given to despised eaters.[67]

The logic of punishment is particularly clear in, for instance, the forest's carrion law, given its similarities with the English law in which animals and nonhuman objects, "horses, boats, carts, &c., whereby any are slain, that properly are called *Deodands*," are to be sold by the king's agents, or have their assessed value levied as a fine, and the proceeds *deodandum* [given over to God], which is to say, to charity.[68] England's precociously centralized government meant it uniquely had a deodand law, and also, notably, meant England held no animal trials, since the deodand law covered all nonhumans.[69] The deodand law provides for a resolution in cases of death by misfortune; though no human agent could be blamed for the death, the violation to the king's peace could be repaired by blaming a nonhuman party. In this particular case, the victim of the crime, the deer, is also the guilty party, because it has broken the king's peace by dying improperly, claiming a kind of ownership over a life that ultimately was meant to be the king's property. Like other deodanda, it became an accursed object, but the peculiar characteristics of venison as compared to other deodanda required a peculiar solution. The ideological function of the deer for elite hunters barred it from the open market, as the deer would have lost its social value if it could be bought rather than only gifted.[70] Since the deer's carcass could not be sold legally, nor for the same reason translated into monetary terms (except in the thriving black market for venison), the actual flesh needed to be given away. Thus the same act at once accomplished charity and punished the guilty nonhuman agent.

---

67. For a longer version of this argument, attending particularly to the early medieval carrion laws, see my *How to Make a Human: Animals and Violence in the Middle Ages* (Columbus: The Ohio State University Press, 2011), 67–91.

68. From the "De Officio Coronatoris" of 1276; *The Statutes at Large, from "Magna charta" to the End of the Last Parliament, 1761,* vol. 1, ed. and trans. Owen Ruffhead (London: M. Basket, 1763), 61. For more on the deodand, see Teresa Sutton, "The Deodand and Responsibility for Death," *The Journal of Legal History* 18 (1997): 44–55, and Anna Pervukhin, "Deodands: A Study in the Creation of Common Law Rules," *The American Journal of Legal History* 47 (2005): 237–56.

69. J. J. Finkelstein, "The Ox That Gored," *Transactions of the American Philosophical Society* 71 (1981): 64–81; Piers Beirne, "A Note on the Facticity of Animal Trials in Early Modern Britain; or, The Curious Prosecution of Farmer Carter's Dog for Murder," *Crime, Law and Social Change* 55 (2011): 359–74.

70. Birrell, "Deer and Deer Farming," 113–14; Mileson, *Parks in Medieval England,* 80 and 106; Sykes, "Taking Sides," 155.

The confiscation of the skin and head, and their eventual return to the king's possession,[71] also corrects the deer's fault by ensuring that the king regains control over the parts most potently symbolizing forest mastery. Skin matters for elite hunting: the *Master of Game* instructs hunters to give it to whoever slew the deer.[72] Many medieval works attest to the symbolic importance of the deer's head: the courtly Tristan requires that it be carried in front of the hunting party as it returns to court; in a late twelfth-century Anglo-Norman romance, Ipomedon seduces the queen of Calabria by presenting her with "treis testes de cerfs ke il porterent, / granz esteient e asez beles" (4319–20) [the three deer heads that his men were carrying, which were large and quite lovely];[73] in 1334, Nicholas Meynell, a lord of the North Riding of Yorkshire, concluded a poaching expedition by affixing the heads of nine stolen harts to stakes planted in his enemy's park.[74]

The king's sovereign control begins to break down, however, in records of the law's actual enforcement. It works as intended when it deals with deer killed by poachers, like the men discovered in Northamptonshire in 1246 carrying a flayed doe in a sack[75]: the flesh was distributed to a hospital, the poachers' snare and doe's skin confiscated, and the poachers punished. But some records also show the law applied to deer that had died of hunger,[76] from wolves,[77] or from drowning after fleeing

---

71. Turner, *Select Pleas*, xxxix.

72. Edward of Norwich, *Master of Game*, 197.

73. Hue de Rotelande, *Ipomedon*, ed. A. J. Holden (Paris: Klincksieck, 1979). For the tendency of hunting societies to value heads highly, see Naomi Sykes, "Deer, Land, Knives and Halls: Social Change in Early Medieval England," *Antiquaries Journal* 90 (2010): 180. I have drawn further details from Marvin, *Hunting Law*, 76–80 and 109.

74. Derek Rivard, "The Poachers of Pickering Forest 1282–1338," *Medieval Prosopography* 17 (1996): 97. For a similar incident in 1272, see Turner, *Select Pleas*, 39, and for several more, Andrew G. Miller, "Knights, Bishops, and Deer Parks: Episcopal Identity, Emasculation, and Clerical Space in Medieval England," in *Negotiating Clerical Identities: Priests, Monks, and Masculinity in the Middle Ages*, ed. Jennifer D. Thibodeaux (New York: Palgrave MacMillan, 2010), 220. After a typical successful hunt, the head often was distributed to the poor (Sykes, "Taking Sides," 154).

75. Turner, *Select Pleas*, 84; also see 87 for a similar case the following year, and 106–7 for one in 1253. See also Stagg, *New Forest Documents*, 171, and F. H. M. Parker, "I. Inglewood Forest, Part III: Some Stories of Deer-Stealers," *Transactions of the Cumberland & Westmorland Antiquarian & Archaeological Society* n. s. 7 (1907): 7, on the flesh of an abandoned poached deer sent to the leper hospital of Saint Nicholas in Carlisle, and the skin to the Priory of Carlisle.

76. Parker, "Deer-Stealers," 7.

77. George Wrottesley, ed. and trans. "The Pleas of the Forest, *Temp.* Henry III and Edward I," in *Collections for a History of Staffordshire*, vol. 5 (London: Harrison and Sons, 1884), 162.

poachers,[78] as well as to a deer that had gone mad and died[79] and to a hart killed by another hart,[80] presumably during mating season. In all these cases, the flesh was still distributed to the leper hospitals or to the poor, and the skin and head, if not to the king, then at least to men in power. These are doomed attempts to empark the biological energy of the deer, their fear, their diseases, their predators. Poachers might be punished and controlled, but as Nietzsche, Foucault, and especially Wolfe observe, life itself will always do more than what the law or the state wants. We can finally observe, as well, that since English emparked deer tended to be smaller than truly wild animals,[81] the practices of forest management themselves produced animals that might have been increasingly less satisfying to hunt and certainly less laden with flesh.

Even after the death, the deer's carcass could continue to impede efforts at managing its life. The ideal temperature for curing a dead deer is no warmer than four degrees Celsius (forty degrees Fahrenheit), perhaps even cooler.[82] Since thirteenth-century England was warm compared to the following centuries,[83] during the time of the law's first promulgation, carcasses of deer left in the woods might have bloated or putrefied quickly. More to the point, while some modern people believe that game "n'est bon que faisandé"[84] [is good only if gamey], medieval people seem to have thought deer carcasses had an unusual tendency to go bad. Echoing Pliny, Albert the Great, Thomas of Cantimpré, and Vincent of Beauvais explain that "innards of a deer are very malodorous"; Thomas and Vincent both declare that dogs hate these innards; and all add that "twenty worms reside in the deer's cervical spine."[85] This belief was not

---

78. Ibid., 163.

79. Turner, *Select Pleas,* 182.

80. Ibid.

81. Naomi Sykes, Ruth Carden, and Kerry Harris, "Changes in the Size and Shape of Fallow Deer—Evidence for the Movement and Management of a Species," *International Journal of Osteoarchaeology* 23, no. 1 (2011): 66.

82. Field and Stream Editors, "Deer: Hang Time," *Field and Stream,* January 11, 2006, http://www.fieldandstream.com/articles/other/recipes/2006/01/deer-hang-time; Chris Raines, "Aging of Venison—Sounds Like a Good Idea, Right?" *Meatblogger,* November 29, 2010, http://meatblogger.org/2010/11/29/aging-of-venison-sounds-like-a-good-idea-right/.

83. For example, Brian M. Fagan, *The Little Ice Age: How Climate Made History, 1300–1850* (New York: Basic Books, 2000), 17–18.

84. Gustave Flaubert, *Dictionnaire des idées reçues,* ed. Étienne-Louis Ferrère (Paris: Louis Conard, 1913), s.v. "gibier."

85. Albert the Great, *Man and the Beasts (De Animalibus, Books 22–26),* trans. James J. Scanlan (Binghamton, NY: Medieval & Renaissance Texts & Studies, 1987), 97; Thomas of Cantimpré, *Liber De Natura Rerum: Editio Princeps Secundum Codices Manuscriptos,*

confined only to scholars technically barred from eating venison by their clerical vows. One manuscript of a widespread fourteenth-century Middle English recipe book requires a multiday sequence of covering, washing, hanging, salting, and boiling to keep venison from rotting, which it follows with another somewhat less complicated recipe for salvaging rancid venison.[86] Few of the other ninety-two recipes in the manuscript are as detailed as the recipe for keeping venison from rotting, and none of the others concerns preserving or rescuing food that had gone bad. The different treatment of venison makes practical sense: other meats tended to be eaten soon after and close to the site of butchery, while venison might be sent away as a gift or used to provision the king when he was traveling far from his own forests, as when Edward I was in Flanders in the late 1290s.[87] But the worry about putrefaction also suggests that even those who worked practically with venison considered venison a flesh that, if unsupervised, went off quickly. Deer left dead in the forest, or those that had died from disease or their own violence, those abandoned, partially consumed, or furtively hidden and rediscovered, would have been subject to other appetites, those of microbes, insects, and other critters, by the time the forester intervened to try to reestablish the king's sovereignty.

In other words, by the time the forester redistributed the flesh, the deer's carcass may have been inedible, at least for human eaters. Yet punishing the deer by feeding it to the poor or to people with leprosy works perfectly only if the meat is actually eaten. By the thirteenth century, these people were at once among the most reviled members of Christendom and its most suitable recipients of charity. In the more negative medieval portrayals, they were walking corpses,[88] "figure[s] for disfiguration itself,"[89] while more favorable portrayals understood lepers as endur-

---

ed. Helmut Boese (Berlin: W. De Gruyter, 1973), 123; Vincent of Beauvais, *Speculum Quadruplex Sive Speculum Maius: Naturale, Doctrinale, Morale, Historiale* (1624; reprint, Graz: Akademische Druck- u. Verlagsanstalt, 1964–65), 1345.

86. From "Diuersa Servicia," in Constance B. Hieatt and Sharon Butler, eds., *Curye on Inglysch : English Culinary Manuscripts of the Fourteenth Century (including the Forme of Cury)* (London: Oxford University Press, Early English Text Society, Special Series 8, 1985), 73.

87. Bruce Lyon, "Coup d'oeil sur l'infrastructure de la chasse au moyen âge," *Le Moyen Âge* 104 (1998): 224–25. For more on salting venison, see Birrell, "Procuring, Preparing, and Serving Venison," 180–92.

88. Cathérine Peyroux, "The Leper's Kiss," in *Monks and Nuns, Saints and Outcasts: Religion in Medieval Society,* ed. Sharon Farmer and Barbara H. Rosenwein (Ithaca, NY: Cornell University Press, 2000), 179.

89. Julie Orlemanski, "How to Kiss a Leper," *Postmedieval* 3 (2012): 148. See also, Carole Rawcliffe, "Learning to Love the Leper: Aspects of Institutional Charity in Anglo-

ing purgatory on earth. A gift of carrion might have been either garbage or a pious donation or both, simultaneously. Thus at times, they received charity designed to mark them as abject, as with this hunting law or as at York, which in 1301 mandated that they receive diseased pork.[90] But people with leprosy of course have their own ways of life, inaccessible to the symbolic needs of the dominant and temporarily able human community. Here we see the ethical value of object-oriented ontology's insistence on the withdrawn and inaccessible particularity of any given being, of whatever size, duration, or type. People with leprosy might not have needed or wanted the charity offered them by the forester, even if the carcass was in good condition: some leprosaria kept dairy cattle, hens, and pigs, and had fishing rights,[91] which ensured they had a medicinally appropriate diet. Game, on the other hand, was a medically unsuitable meal for the sick.[92] A mangled, perhaps rotten carcass, made of the wrong kind of meat, might have been an especially unwanted or unnecessary gift.[93] Here again, the forest law might fail in its encounter with the independence of the life it tried to enlist in its projects.

Further forest sovereigns could be enumerated, each with its own interests and frustrations. There is virtually no limit to this kind of analysis of the forest's vast nonhuman worlds. The forest's multiple interests are not only those of elite hunters, nor even only of the poachers that have received so much attention in the "history from below." In a kind of political "history of the surround," we should try to account for the perspectives and interests of prey, microbes, climate, and the law itself, for whom human hunters of whatever sort are *their* surround. Sovereign pretension is best combated not by eliminating the sovereign but by remembering how many more sovereigns inhabit any given realm, each behaving as if it itself were the sole master.

Norman England," in *Anglo-Norman Studies XXIII: Proceedings of the Battle Conference 2000*, ed. John Gillingham (Woodbridge, UK: Boydell & Brewer, 2001), 238–39.

90. Patricia H. Cullum, "Leperhouses and Borough Status in the Thirteenth Century," in *Thirteenth-Century England III: Proceedings of the Newcastle Upon Tyne Conference, 1989*, ed. Peter R. Coss and Simon D. Lloyd (Woodbridge, UK: Boydell & Brewer, 1991), 43.

91. Carole Rawcliffe, *Leprosy in Medieval England* (Woodbridge, UK: Boydell & Brewer, 2006), 214, 327–28.

92. Ibid., 213.

93. Hanawalt, "Men's Games, King's Deer," 182, makes a similar point in passing.

2

Sovereign Meat

REASSEMBLING THE HUNTER KING
FROM MEDIEVAL FOREST LAW TO
THE WEDDING OF SIR GAWAIN AND
DAME RAGNELLE

Jeanne Provost

The Kyng was sett att his trestylle-tree
With hys bowe to sle the wylde veneré
And hys lordes were sett hym besyde.
As the king stode, then was he ware,
Where a greatt hartt was and a fayre,
And forthe fast dyd he glyde.

The hartt was in a braken ferne,
And hard the houndes, and stode fulle derne,
Alle that sawe the kyng.
"Hold you stylle, every man,
And I wolle goo myself, yf I can,
With crafte of stalkyng."
The Kyng in his hand toke a bowe
And wodmanly he stowpyd lowe
To stalk unto that dere.
When that he cam the dere ful nere,
The dere lept forthe into a brere,
And evere the Kyng went nere and nere.

So Kyng Arthure went a whyle
After the dere, I trowe, half a myle,
And no man with hym went.

And att the last to the dere he lett flye
And smote hym sore and sewerly—
Such grace God hym sent.
Doun the dere tumblyd so theron,
And felle into a greatt brake of feron;
The kyng folowyd fulle fast.
Anon the Kyng bothe ferce and felle
Was with the dere and dyd hym serve welle,
And after the grasse he taste.
—The Wedding of Sir Gawain and Dame Ragnelle, *19–48*[1]

AT THE START of the mid-fifteenth-century English romance *The Wedding of Sir Gawain and Dame Ragnelle,* King Arthur leaves behind his entourage while shadowing a deer in Inglewood forest. Arthur and the deer take turns anticipating and answering each other's movements, and through this parallel image, the poem highlights the king and the beast's shared physicality as both of their bodies become enmeshed in the ritual of the hunt. Arthur, standing still, perceives the deer nearby and, ordering his grooms not to follow him, creeps toward it; the hart hears him and freezes, replicating Arthur's initial stillness. As the king draws closer, the deer retreats. When Arthur shoots the deer and it falls wounded amidst the ferns, he sinks down after it to kill it, test its fat, and butcher it. The deer mirrors the king as it flees, and Arthur ultimately prepares its body for even nearer connection with his own by butchering it to be consumed as meat, a goal hinted at by the gastronomic puns on his ability to "serve welle" the deer and "taste" its fat (47–48). During the special ritual of the hunt, the bodies of deer and king reflect each other until, if the hunt is successful, the body of the human sovereign subsumes that of the deer.

If the romance's opening hints at the potential slippage between royal subjects and cervine objects by comparing the movements of Arthur and his prey, this slippage grows explicit in the remainder of the episode. Arthur's privileged position as a hunter of vulnerable beasts shifts when the Baron Gromer Somer Joure surprises him beside the felled venison. The poem highlights Arthur's isolation, his being helpless as the animal

---

1. All quotations of *The Wedding of Sir Gawain and Dame Ragnelle* are taken from Thomas Hahn, ed. *Sir Gawain: Eleven Romances and Tales, Middle English Texts* (Kalamazoo, MI: Medieval Institute Publications, 1995), 41–80. Online. http://d.lib.rochester.edu/teams.

he has just killed, by pointing out that, having ordered his grooms to
stay behind (29), "no man with hym went" (39), and he is "with the
dere alone" (49). As Arthur kneels by his prey, Somer Joure proclaims
that Arthur has wrongly deeded away the Baron's lands. Fully armed, the
Baron means to decapitate the king, who wears only his forest-colored
hunting gear, to requite this slight. Arthur describes his situation thus:
"Shame thou shalt have to sle me in *veneré*, / Thou armyd and I clothyd
butt in grene, perde" (82–83; my italics). "Veneré," meaning at once
"the chase" and "wild animals hunted as game" (*OED* "venery," n.1 def.
1a, 2) suggests equally that Arthur has just slaughtered a deer and that
Somer Joure intends to slaughter Arthur. If the episode opens on a king
slaying a deer, then Somer Joure's threat intimates that a king's flesh,
too, can be butchered.

The romance's opening passage introduces Arthur engaged in a sport
in which matter and metaphor mutually construct each other. From one
perspective, the trope of the deer hunt that proves sovereignty pervades
medieval literature and culture.[2] From another, this connection between
kings and deer, recursively written down in a manuscript, takes the literary shape it does in part due to the material transformation of nonhuman
into human flesh. Karl Steel illustrates that medieval philosophical and literary discussions of such exchanges between human and animal flesh had
to deal with the seemingly contradictory ideas that humans were superior
to other animals, that humans and other animals sometimes consumed
each other's flesh, and that human flesh was also meat. They typically did
so by asserting the superiority of human flesh, whether by suggesting that
it enjoyed a special capacity for resurrection or that it was more delicious
than the flesh of other animals.[3]

In this chapter, I argue that *The Wedding of Sir Gawain and Dame
Ragnelle* unsettles this orthodox medieval belief in human ascendancy

---

2. Robert Pogue Harrison, *Forests: The Shadow of Civilization* (Chicago: University of
Chicago Press, 1993), 69–75; Barbara Hanawalt, "Men's Games, King's Deer: Poaching
in Medieval England," *Journal of Medieval and Renaissance Studies* 18.2 (1988): 191. On
venison in general as a prestige meat, see Hanawalt, "Men's Games, King's Deer," 180–
82; John Cummins, *The Art of Medieval Hunting: The Hound and the Hawk* (Edison, NJ:
Castle, 1988), 7. The opening hunt that introduces a challenge or test of sovereign identity is a common theme in several Gawain romances, such as *Sir Gawain and the Carle of
Carlisle, The Awyntyrs off Arthur,* and *The Avowyng of Arthur,* though the hunted animal
in the last is not a deer but the other beast of the forest stipulated by medieval English
forest law, a boar.

3. Karl Steel, *How to Make a Human: Animals and Violence in the Middle Ages* (Columbus: The Ohio State University Press, 2011), 108–35.

over other animals. Reading the romance alongside cases from the forest courts of post-Conquest Britain, I explore ways that human sovereignty was mediated by venison. I suggest that, in the romance, venison's role in Arthur's hunt subtly reminds the audience that kings are made of meat, in more ways than one. After foregrounding venison's role in sovereignty, the romance depicts the loathly lady, whose blazon references many, varied components of forest communities and ecosystems, as an alternative vehicle of kingship that replaces the violence of the hunt with sympathy for many forms of life.

A binary focus in scholarship on medieval forests and the hunt echoes their double role in marking Arthur as a king and hinting at his potential dismemberment. On one hand, the hunt constructs hierarchies.[4] Susan Crane explains, for example, "The formalities of the hunt *à force* construct a microcosmic model of creation, in which creation's hierarchy of humans over animals reinforces the human social hierarchy."[5] Conversely, medieval people often regarded forests as legally confusing places where criminals escaped punishment. In this vein, Jacques Le Goff describes the forests as "a place where . . . the hierarchy of feudal society broke down."[6] Other readings synthesize these two views, suggesting that forest savagery substantiated royal power. Robert Pogue Harrison argues that the king's power to hunt the beasts of the forest made him the top predator in a food chain ordained by natural law, affirming his status as the head of the social order.[7]

---

4. On the use of the hunt to ground aristocratic identity, see Richard Almond, *Medieval Hunting* (Stroud, UK: History Press, 2003), 29–60; Cummins, *Art of Medieval Hunting*, 5; and William Perry Marvin, *Hunting Law and Ritual in Medieval English Literature* (Cambridge: Brewer, 2006), 105–30. On the construction of the deer park at Hesdin that interwove a natural setting with human artifice for aristocratic entertainment and display, see Sharon Farmer, "Aristocratic Power and the 'Natural' Landscape: The Garden Park at Hesdin, ca. 1291–1302," *Speculum* 88.3 (2013): 647. On the hunt as demarcating both gender and class identity, see Hanawalt "Men's Games, King's Deer." On the hunt as constructing the hierarchy between Arthur as metaphorical boar and the animal he degrades by mastering it, see Steel, *How to Make a Human*, 189–203.

5. Susan Crane, *Animal Encounters: Contacts and Concepts in Medieval Britain* (Philadelphia: University of Pennsylvania Press, 2013), 72.

6. Jacques LeGoff, *The Medieval Imagination* (Chicago: University of Chicago Press, 1985), 110. See also Corinne J. Saunders, *The Forest of Medieval Romance: Avernus, Broceliande, Arden* (Cambridge: Brewer, 1993), 115–22; and Sylvia Huot, *Madness in Medieval French Literature: Identities Lost and Found* (Oxford: Oxford University Press, 2003), 182–93.

7. On the Robin Hood legend as emerging out of the specific situation of the peasantry and the gentry with respect to forest law, see the *Past and Present* studies reprinted in R. H. Hilton, ed., *Peasants, Knights, and Heretics: Studies in Medieval English Social*

The bifurcated role of forests as places that created hierarchy and sub-verted it results partly from the place of slaughtered animal flesh at the heart of the hunt's ritual of sovereignty. Alongside the growing critical understanding that we construct human identity by subjugating nonhu-man animals comes the recognition that human bodies and many human assemblages are made, in myriad senses, out of meat.[8] In this sense, medi-eval sovereignty performed via the hunt was a group formation mediated by cervine bodies. Bruno Latour explains:

> If a social difference is expressed in or "projected upon" a detail of fash-ion, but this detail—let's say a shine of silk instead of nylon—is taken as an intermediary transporting some higher meaning—"silk is for high-brow," "nylon for lowbrow"—then it is in vain that an appeal has been made to the detail of the fabric. It has been mobilized purely for illustra-tive purposes. . . . If, on the contrary, the chemicals and manufacturing differences are treated as so many mediators, then it may happen that without the many indefinite material nuances between the feel, the touch, the color, the sparkling of silk and nylon, this social difference might not exist at all.[9]

History (Cambridge: Cambridge University Press, 1976). See also Marvin, *Hunting Law,* 131–57, on the symbolic connections between the brutality of the hunt and aristocratic courtliness; and Randy P. Schiff, "The Loneness of the Stalker: Poaching and Subjectivity in *The Parlement of the Thre Ages,*" *Texas Studies in Language and Literature* 51.3 (2009): 263–93, on the interplay between poaching and aristocratic privilege, where in some cases poaching was less subversive than in others.

8. For medievalist forays into posthumanism, see Donna Beth Ellard on the influence of bird behaviors on Old English poetry, in "Going Interspecies, Going Interlingual, and Flying Away with the Phoenix," *Exemplaria* 23, no. 3 (2011): 268–92, Jeffrey Jerome Cohen's use of actor-network-theory in "An Abecedarium for the Elements," *Postme-dieval* 2, no. 3 (2011): 291–303, and Steel's Derridean discussion of the predication of "the human" on the subjection of the animal in *How to Make a Human* (29–60); for the latter, see also Jacques Derrida, *The Animal That Therefore I Am,* ed. Marie-Louise Mallet, trans. David Wills (New York: Fordham University Press, 2008). On the act of reading as an encounter with the inhuman body of a literary text, and the interactions between human readers and "fictional figures who never were and never can be hu-man" (32), see Eileen Joy, "Like Two Autistic Moonbeams Entering the Window of My Asylum: Chaucer's Griselda and Lars von Trier's Bess McNeill," *Postmedieval* 2 (2011): 316–28. On the affective interpermeability of extimate experience—exterior objects and the images from others' rhetoric—and internal experience, see L. O. Aranye Fradenburg, "Beauty and Boredom in *The Legend of Good Women.*" *Exemplaria* 22, no. 1 (2010): 65–83. On human imbrication with other species, see Donna Haraway's discussion of the ways that human and canine species have mutually constructed each other in *When Species Meet* (Minneapolis: University of Minnesota Press, 2004). On recent directions in posthumanist literary study more broadly, see Cary Wolfe, "Human, All Too Human," *PMLA* 124, no. 2 (2009): 564–75.

9. Bruno Latour, *Reassembling the Social: An Introduction to Actor-Network-Theory*

In Latour's analysis, humans are not the only actors in social phenomena; the nonhuman animals and objects we "use" in our social aggregates fundamentally influence the shape these aggregates take.[10] Viewed through a Latourian lens, medieval English kings performed their sovereignty via a hunt mediated by the butchered bodies of deer. This particular mediator, deer meat, influenced kingship in complex senses. For one thing, deer meat signified not only the king's power over animal bodies *but also* the physical vulnerability of his own suffering, animal flesh.

In a reading of Ernst Kantorowicz's *The King's Two Bodies,* Eric Santner defines the royal flesh's suffering as fundamental to the late medieval discourse of sovereignty. The split between the king's mortal body and the immortal body of sovereignty eventually evolves to cast the king's human body as suffering and dying incessantly.[11] L. O. Aranye Fradenburg describes a similar split between sublime and material chivalric bodies from a psychoanalytic vantage point, saying, "By entombing/keeping death inside oneself as standing for the incorrigible thing, the knight of faith triumphs over the flesh, thereby acquiring the excoriating, militant, angelic body," and yet such shimmering sublimations always also point back toward the mortal flesh that enables them.[12] Santner says, "The immortality of the King reveals itself to be a kind of death drive immanent to the very station of kingship."[13] In light of Santner's argument, the performance of the hunt might be interpreted in one sense as a ritual exchange of the many dying bodies of the hunter-king's prey for his "always dying" human body. Yet the performance of kingship via the butchered body of the deer also, to borrow Santner's phrase, "fattens the creaturely life" of the king's body.[14] The hunt both asserts the sovereign's

---

(Oxford: Oxford University Press, 2005), 40.

10. For an influential application of actor-network-theory to literary criticism, see Jane Bennett, *Vibrant Matter: A Political Ecology of Things* (Durham, NC: Duke University Press, 2010). Recent enthusiasm for actor-network-theory can be seen in the special issue of *postmedieval 4, no. 1 (2013),* coedited by Jeffrey Cohen and Lowell Duckert.

11. Ernst H. Kantorowicz, *The King's Two Bodies: A Study in Mediaeval Political Theology,* rev. ed. (Princeton: Princeton University Press, 1957; rev. 1997); Eric L. Santner, *The Royal Remains: The People's Two Bodies and the Endgames of Sovereignty* (Chicago: University of Chicago Press, 2011). Santner's analysis focuses on Kantorowicz's reading of *Richard II.* Kantorowicz says that the conjunction of mortal and immortal sovereign bodies emerges into a notion of sovereignty in which "kingship itself comes to mean death, and nothing but death," and "the king that never dies has been replaced by the king that always dies and suffers death more cruelly than other mortals" (30). See also Joseph Taylor's discussion of Santner in chapter 7 of this volume, 181–83.

12. L. O. Aranye Fradenburg, *Sacrifice Your Love: Psychoanalysis, Historicism, Chaucer* (Minneapolis: University of Minnesota Press, 2002), 158.

13. Santner, *Royal Remains,* 48.

14. Ibid., 6.

power over animal bodies and casts his own body as just more meat to be consumed by kingship.

In the hunt, deer meat is thus an actor in the king's sovereignty and also a subtle reminder that his body, too, is composed of vulnerable flesh like that of the animals he butchers, and it is with this double role of deer meat in sovereignty in mind that I use the words "body," "flesh," and "meat" somewhat interchangeably in this chapter. From one perspective, my semantic alignment of human bodies, nonhuman meat, and the flesh that human and nonhuman species share is a tacit acknowledgement of the political lives of animal bodies. In this line of thought, Robert Mills draws from Agamben's concept of "bare life"—the state outside of political existence that allows animals and exiled humans to be killed without legal consequences—to consider the political lives of nonhuman animals: "Drawing attention to the fact that human victims of violence are aligned, by analogy, with animal existence brings into focus one side of the problem. Yet we also need to confront sovereignty's ability to capture within the domain of the political nonhuman animal life."[15] In his chapter in this volume, Karl Steel speaks similarly of the political lives of deer, noting how the king obviously binds his love of forest animals to the love of killing them, paradoxically including them within his political order by exposing them to "an unconditional capacity to be killed."[16] As Mills and Steel suggest, deer are also agents in the ritual of sovereignty that is the hunt; there are, as Steel puts it, many sovereigns, and "some of these sovereigns are human, but most not."[17] Conversely, my choice to align bodies, flesh, and meat acknowledges the capacity of all human bodies, including the king's, to be killed violently, butchered, and even— as I will suggest *The Wedding of Sir Gawain and Dame Ragnelle* hints— eaten. When we notice the role of the cervine body in sovereignty, we can also observe that the body of the human king has much in common with the bodies of the deer he fells, dismembers, and consumes. All animal bodies, human and nonhuman, are flesh, and all flesh is potentially meat.

I begin this analysis of the king-making power of deer meat with a reading of medieval forest court records. These laws that used meat to elevate kings, I show, also bespoke deep instability within the lines separating the bodies of humans, especially human kings, from those of deer. Next, I argue that *The Wedding of Sir Gawain and Dame Ragnelle* offers

---

15. Robert Mills, "Judicial Violence, Biopolitics, and the Bare Life of Animals," *New Medieval Literatures* 12 (2010): 127.

16. Georgio Agamben, *Homo Sacer: Sovereign Power and Bare Life,* trans. Daniel Heller-Roazen (Stanford: Stanford University Press, 1998), 85.

17. See chapter 1 of this volume, 46.

an alternative to the forest laws' vexed linkage of sovereign human and slaughtered cervine species. Instead, its author uses the copiously untidy figure of Dame Ragnelle to foreground the diverse forest inhabitants coexisting, sometimes contentiously, with the king's hunt.[18] Many kinds of power interacted in medieval English forests; many more things than deer lived there. The romance reassembles Arthur's sovereignty, shifting kingship's mediators from the slaughtered meat of the sovereign/ cervine body to an alternative: the fluctuating food chain of a complex ecosystem.[19]

The Vulnerable Forest

The hunter/hunted theme opening the romance, by situating the action within the ritual of the hunt, also places it within the forest laws governing the king's hunting grounds.[20] Somer Joure resents the deeding

---

18. On various historical uses of English forest land, see Jean Birrell, "The Medieval English Forest," *Journal of Forest History* 24, no. 2 (1980): 78–85; and for a history that emphasizes beneficial legal management of English forests as complex ecosystems, see Birrell, "Deer and Deer Farming in Medieval England," *The Agricultural History Review* 40.2 (1992): 112–26.

19. This analysis of the loathly lady as a figure from romance who tests and reworks the ideals expressed in forest law recalls the historicist literary analyses of early Robin Hood legends and ballads in Hilton's collection, *Peasants, Knights, and Heretics*, in which the legend is seen as a vehicle for peasants and gentry to reimagine their place and privilege with respect to late medieval forest law; see the essays by R. H. Hilton, J. C. Holt, Maurice Keen, and T. H. Aston in the collection. As Stephen Knight argues, "The good outlaw: the idea attracts attention in every culture. Someone rejects conventional law, breaks it, is exiled from it, is pursued and imprisoned by its agents, but somehow always escapes and somehow always represents a system of social and personal order that is better than what we have now. The good outlaw tests contemporary law, reveals its weakness, and holds out the promise of greater freedom" (*Robin Hood: A Mythic Biography* [Ithaca, NY: Cornell University Press, 2003,] xi). If Robin Hood and his men often break the law in order to change it, the loathly lady makes her way into the law by marrying Gawain and then offers a new way to understand the law through this ingress. She represents sovereign authority as power grounded in a diverse ecosystem, rather than in subjugation of one privileged creature within a monoculture. This view of the loathly lady story echoes Fradenburg's argument that medieval fantasies have "the power to remake the social realities in which we live and desire. This means not that fantasies are inevitably benign, but that, far from being illusions powerless to affect reality, fantasies affect the ways that people and societies feel and act" ("'Fulfild of fairye': The Social Meaning of Fantasy in the *Wife of Bath's Prologue* and *Tale*," in *The Wife of Bath: Complete, Authoritative Text with Biographical and Historical Contexts, Critical History, and Essays from Five Contemporary Critical Perspectives*, ed. Peter G. Biedler [Boston: Bedford, 1996], 206).

20. On Somer Joure's complaint as foregrounding the issue of land law and *seisin* more generally, though not in the specific context of forest law, see Sheryl L. Forste-

away of his land, and Inglewood, the scene's setting, had by the thirteenth century been appropriated as royal forest.[21] Though the romance's author does not say that Somer Joure objects to losing land in Inglewood specifically, the Baron seems to preside over this forest, ordering Arthur to return in a year to meet him. He treats Arthur as an interloper, granting him a twelvemonth before judgement, the forest court's legal limit for detaining a poacher unable to make bail.[22]

Under forest law, the death of a deer or a boar commanded at least as much official scrutiny as the death of a human tenant of the same land. The 1217 Charter of the Forests required an inquisition into the death of any protected animal in the king's demesne forests.[23] At this inquisition, representatives from four neighboring townships bore witness to whatever they knew regarding weapons or snares found at the scene or eyewitness accounts of the creature's demise. If someone without license to hunt venison there had killed the beast, he either paid a fine or went to prison for a year and a day (though during Henry III's reign this period of imprisonment before trial became less stringently enforced). After a year and a day, he could produce pledges guaranteeing his appearance at the Forest Eyre, the traveling, central forest court.[24] Intervals among eyres varied from county to county, but G. J. Turner says, "Seven years was supposed to be the proper interval between one eyre and another," and through the reign of Henry III the eyres more or less kept to this standard.[25] At the eyre, a convicted poacher either paid a fine or, if he fled the courts, suffered outlawry.[26]

This elaborate, juridical attention to nonhuman deaths made some people uneasy. In part, the forest records' grisly reports of the wounds dealt to illicitly hunted animals hint at this anxiety. The eyre records meticulously enumerate the wounds of poached deer. For example, the 1209 Northamptonshire Eyre describes one hart's corpse not only as dead but as "wounded," its antler "fractured as far as the brain."[27] A

Grupp, "A Woman Circumvents the Laws of Primogeniture in 'The Weddynge of Sir Gawen and Dame Ragnell,'" *Studies in Philology* 99, no. 2 (2002): 105–22.

21. Charles R. Young, *The Royal Forests of Medieval England* (Pittsburgh: University of Pennsylvania Press, 1979), 62–63.

22. Young, *Royal Forests*, 102.

23. G. J. Turner, *Select Pleas of the Forest* (London: Selden Society, 1901), xxxviii–l.

24. Ibid., xl.

25. Ibid., lvii.

26. Young, *Royal Forests*, 41–43, 97–98.

27. Turner, *Select Pleas*, 4.

1246 Northamptonshire Inquisition Roll tells of a deer with "one gash in the left side and another gash in the left part of the neck."[28] By graphically describing slaughtered deer's bodies, court-appointed coroners foregrounded the capacity of all animal bodies, deer and human alike, to be reduced to dead flesh. In so doing, they also highlighted the ability of living deer to feel pain just like living humans.

An account of a man's murder by poachers in Northamptonshire similarly enumerates his wounds. In 1246, poachers shot at two foresters and killed one, Matthew of Brigstock:

> And the aforesaid malefactors standing at their trees turned in defense and shot arrows at the foresters so that they wounded Mathew, the forester of the park of Brigstock, with two Welsh arrows, to wit with one arrow under the left breast, to the depth of one hand slantwise, and with the second arrow in the left arm to the depth of two fingers, so that it was despaired of the life of the said Matthew.[29]

> Et predicti malefactores ad fusta sua stantes turnaverunt in defensum et in forestarios sagittas suas direxerunt, ita quod vulneraverunt Matheum forestarium de parco de Bricstok' cum duabus sagittis waliscis, scilicet cum una sagitta sub mamilla sinistra ad profunditatem unius palme de belongo et cum alia sagitta in brachio sinistro ad profunditatem duorum digitatuum, ita quod de vita dicti Mathei desperabatur.

Certainly, this record of Matthew's death differs significantly from those of poached deer. The records do not tell of *desperabatur*, of anyone having "despaired" for the lives of deer (although the bureaucratic nightmare that discovering a dead deer entailed certainly might inspire despair). Yet the Northamptonshire Roll enumerates Matthew's injuries much as it does those of deer. The wounds to his body, an arrow under his left breast and a second under his left arm, bring to mind the gashes in the left side and neck of the slain animal in the 1246 Inquisition Roll. Both accounts of poachers' violence catalog similar bodily hurts to humans and to forest beasts.

Another set of records conveys explicit discomfort by eliding the difference between slain deer and slain humans. The 1255 Huntingdon Roll catalogues the death of a stranger this way:

---

28. Ibid., 82.
29. Ibid., 80.

A certain fawn was found dead in the field of Ellington. An inquisition was made by the townships of Ellington, Woolley, Alconbury, and Brampton who know nothing thereof.

A certain stranger was found slain in Sapley. Richard Lenveyse the walking forester, who was the first finder, does not come, nor was he attached.[30]

Quidam feto inventus fuit in campo de Elynton' mortuus. Inquisicio facta per villatas Elinton' (alibi), Wlfley (alibi), Alcumbir' (alibi) et Brampton' (ij marce) qui nichil inde sciunt.

Quidam homo extraneus occisus inventus fuit in Sappel.' Ricardus Lenveyse forestarius pedes primus inventor non venit nec fuit attachiatus.

The sentences introducing the discoveries of the two corpses match syntactically. This structural similarity and the episodes' adjacency, the human stranger's death described directly after the deer's, implicitly compare human and cervine fatalities. The record says officials attached neither the forester who found the man's body nor any of the neighboring townships, requiring no one to appear before the forest courts to testify regarding the stranger's death.[31] The eyre roll adds that the case must now belatedly be investigated and judged under the common law, since "the law of the land concerning the death of a man ought not to be abated on account of the assize of the forest."[32] Though the eyre judges eventually referred the dead stranger's case to common law court, the fawn's death received more immediate attention than the human's.

Besides the similarities between human and cervine deaths suggested by the forest inquisition's language and structure, the death of a royal deer also really could threaten human bodies, whether those of poachers, of foresters, or even of innocent commoners who happened to be found near slain deer. For example, the 1209 Northamptonshire Eyre rolls tell the following story:

The foresters found in the wood of Nassington a doe with its throat cut, and hard by they found Henry the son of Benselin lying under a certain

30. Ibid., 19.
31. The term "attachment" is complicated and sometimes contradictory as applied to forest law. Young suggests that it could mean several things: the forest court of attachment that attended to ministerial matters about every forty-two days, a fine or punishment or appearance before that court, or a requirement issued within that court to appear later before the more powerful forest eyre (*Royal Forests*, 88–92).
32. Ibid.

bush. And they took him and put him in prison. He comes before the justices and denies that he ever knew anything of that doe, except only that he went into that wood to seek his horse. The foresters took him and led him to that doe. The foresters and verderers, being asked if he were guilty thereof or not, say that they do not think that he was guilty, but they believe rather that Richard Gelee the reaper of Newton is guilty thereof, because he fled as soon as he heard that the aforesaid Henry was taken. And because Henry himself has taken the cross, and is not suspected, and has lain for a long time in prison, it is granted to him that he may make his pilgrimage.[33]

Forestarii invenerunt in bosco de Nassinton' unam damam habentem gorgiam abscisam et prope inde invenerunt Henricum filium Bence latentem sub quodam bussone; et ipsum ceperunt et in prisonam posuerunt. Idem venit coram iusticiariis et defendit quod de dama illa nunquam aliquid scivit, nisi tantum quod ibat in bosco illo ad querendum equm suum. Forestarii illum ceperunt et duxerunt asque ad damam illam. Forestarii et viridarii, requisiti si ipse culpabilis sit inde vel non, dicunt quod non credunt quod ipse culpabilis sit, set credunt melius quod Ricardus Gelee messarius de Neweton' sit inde culpabilis quoniam fugiit quamcito audivit quod predictus Henricus captus fuit. Et quoniam ipse Henricus cruce signatus fuit et non malcreditur et diu iacuit in carcere, concessum est ei quod ipse faciat peregrinacionem suam.

The confusion clouding this account underscores forest law's reputation for being harsh and arbitrary. Apparently, Henry remained imprisoned even after the foresters who had arrested him became convinced of his innocence. The eyre roll does not specify how long he lay there, not even genuinely suspected of the crime, only that it was "a long time." The roll also does not make clear whether a lower court first convicted Henry before the eyre retried him or the foresters simply imprisoned him on suspicion of guilt to await the eyre. Regardless of the nebulousness of Henry's legal situation, his discovery near a dead deer in the royal forest clearly put his life in peril. Given that the eyre's somewhat erratic interval was seven years and suspected poachers were sometimes held to await the eyre longer than the prescribed year and a day,[34] he may have been in prison for a very long time. Not everyone survived lengthy incarceration

---

33. Ibid., 3.
34. Ibid., ivii.

in thirteenth-century prisons, where conditions could be grim.[35] Henry's potentially deadly imprisonment simply for being discovered near a slaughtered doe illustrates the butchered animal's contagious power to harm humans unlucky enough to come too close.

Comparisons between humans and deer structure the written rules of the inquest and pervade the language of the court records. The rhetorical comparison materialized in the contagious power of violence against deer to cause violence to human flesh. In turn, this transferrable violence slipped back into the court's uneasy descriptions of real forest cases, reinscribing the legal similarities between the wounded bodies of two sentient species. This was the embodied metaphor at forest law's heart: violence to deer was violence to humans.

The King's Venison

Especially, harming the king's deer meant harming the king. The forest records represent this harm most apparently as metaphorical and economic, but accounts of the king's economic loss from poaching often shift registers to describe lost limbs, heads, and innards. If the rolls for the 1269 Rutland Eyre speak of deer beset and killed "to the loss of the lord king and the detriment of his forest,"[36] figuring deer's bodies as commodities, the records of the 1272 Northamptonshire Eyre map this economic injury back onto the physical body of the king:

> And they were shooting in the same forest during the whole day aforesaid and killed three deer without warrant, and they cut off the head of a buck and put it on a stake in the middle of a certain clearing . . . placing in the mouth of the aforesaid head a certain spindle; and they made the mouth gape toward the sun, in great contempt of the king and his foresters.[37]

> Et fuerunt bersantes in eadem foresta per totum diem predictum, et occiderunt tres feras sine warento, et abciderunt capud unius dami et posuerunt illud super unum pelum in medio cuiusdam trenchie, que vocatur Harleruding', inponendo in os predicti capitis quendam fucellum; et fecerunt illud iniare contra solem in magnum contemptum domini regis et forestariorum suorum.

---

35. Young, *Royal Forests*, 102–3.
36. Turner, *Select Pleas*, 44.
37. Ibid., 38–39.

By labeling the poachers' satirical staking of the deer's head as "in great contempt of the king," the court affirms this grisly display to be about more than desecrating the king's property. It parses the severed head not as contempt for the king's law or for the king's special prey, but as contempt for the king himself. He has been duped along with his buck, and hence the poachers' display of the buck's head symbolizes violence not only against the law of the body politic but also against the body politic's head, the king.[38]

The forest vert, too, stood for the king's flesh in the material metaphors of the eyre rolls. The 1269 Rutland Eyre describes the damage to the undergrowth in the park of Ridlington this way:

> Peter de Neville agisted very many animals in that place after it was enclosed, which ate the shoots of the stumps of the oaks which had been sold and of the underwood which had been felled; and he caused a great part of those stumps to be uprooted and made into charcoal, so that it can never grow again, to the loss of the lord king and his heirs of one hundred pounds.[39]

> Petrus de Nevill' fecit agistare quam plurima animalia in placea illa postquam inclusa fuit que corroserunt sciones cipporum quercuum venditarum et subbosci prostrati et magnam partem eorum cipporum fecit eradicare et carbonare ita quod nunquam recrescet ad dampnum domini regis et heredum suorum de centum libris.

Vividly physical damage to the vert intrudes into this account of the king's economic loss. Peter de Neville's animals devoured foliage and left the stumps *prostrati,* literally, "overthrown" or "prostrated," corpselike, "ita quod nunquam recrescet," unable ever to grow again for the king or his heirs. This rampant overthrowing of the king's living trees, now unable to reseed themselves, renders the king unable to produce saplings from this ravaged piece of land for his progeny. The episode thus links the king's reproduction of heirs and the forest's reproduction of timber.

Although waste to the vert diminishes the king's wealth in individual

---

38. Barbara Hanawalt interprets this incident as conveying a sexual insult in that the spindle, a woman's tool here acting symbolically as a phallic symbol, invades the mouth of the male buck, representing sexual inversion, "with the buck's throat representing the vagina and the female implement the penis" (*Of Good and Ill Repute: Gender and Social Control in Medieval England* (Oxford: Oxford University Press, 1998), 153). In this reading, too, the metaphorical gelding of the king implies violence against his person.

39. Turner, *Select Pleas,* 48.

locations, connecting his body with the forest greenery signifies sovereignty's power to rejuvenate through the king's scions. The forest leaves'
and branches' capacity to renew themselves contrasts with the finality
of the death of each individual royal deer. Sharing the vert's enormous
capacity to suffer harm, the king's reign seems almost infinitely sus-
tainable in light of this metaphor. When the Park of Ridlington suffers
detericio, a word with senses as both "injury" and "economic loss," the
eyre roll says the damage to the undergrowth "cannot in any way be
estimated."[40] Yet this damage's very immensity bespeaks the vastness of
sovereign authority. His reign can sustain unquantifiable damage because
the sovereignty itself is inestimably valuable. However, within this mode
of accounting, the individual life and body of the king are expendable as
long as the renewable royal line marches on. Thus, even the reference to
the king's potential to engender heirs is a reminder that he must sire sons
because he will die. If the renewability of the vert describes sovereignty
as vast and immortal, like Kantorowicz's supernatural sovereign body,
the performance of sovereign power by killing individuals of a species
gestures back to the vulnerability of the king's unique, human body.

The reminder of sovereign suffering was far from hypothetical. Forests
could indeed be dangerous for human kings, whose hunting privileges
the barons and the commoners often resented and whose sovereignty
often tempted other aspirants to the throne, as witnessed in William
Rufus's suspect death while hunting in the New Forest.[41] While William
Rufus's death actually took place during a hunt, in a less direct sense, the
hunt may have occasioned other violent deaths of kings in which medieval sovereignty was steeped. Edward of Norwich's *The Master of Game,*
an early fifteenth-century translation of Gaston Phoebus's famous fourteenth-century hunting treatise, *Livre de Chasse,* describes the hunt as
training for battle. According to Edward, the hunt taught noblemen the
skill and self-discipline to fight in the wars that often killed English sovereigns and royal scions, to maintain the lands that proved their worth.[42]

---

40. Ibid., 45.

41. Young, *Royal Forests,* 7–8. See also Taylor's extensive discussion of Rufus' death
throughout chapter 7 of this volume.

42. See Edward of Norwich, *The Master of Game,* eds. William A and F. N. Baillie
Grohman (Philadelphia: University of Pennsylvania Press, 2005). On the hunt as training
for war and the lord's value as equivalent to the lands he could maintain through his skill
on the field, Edward of Norwich says that his predecessor Gaston Fébus "saw never a
good man that had not pleasure in some of these things, were he ever so great and rich.
For if he had need to go to war he would not know what war is, for he would not be ac-
customed to travail, and so another man would have to do that which he should. For men

From Edward the Black Prince and Richard III's deaths on the battlefield to Richard I's death by gangrenous crossbow wound to Edward II and Richard II's likely murders, the deaths of English kings often did not come peacefully in old age. Those not killed in battle often died at the will of battle-ready barons with the military strength to bid for control of the realm. In this sense, the performance of sovereignty in the hunt not only figured metaphorically the violence to sovereign bodies through which many English kings gained and lost the throne. It also, at least in Edward and Gaston's view, coached the top echelons of society to be military experts, well-skilled to engage in deadly struggles for power and land.

Forest law's equation between the death of a deer and the death of the king was an ethically and materially problematic formula for tabulating the value of living, embodied individuals. On one level, this comparison between deer and kings depicted the sovereign's value as immeasurable, like the vast worth of his lands. Yet the same comparison also drew much of its impact from eliding the differences among individuals and species—every slaughtered deer stands for every dying king—in a metaphor that depicts the forest as a monoculture, filled only with deer and human kings. This metaphor between deer and kings fails to account for the diverse biota and abiota that composed forest ecosystems. Even as the hunt represented kingship as vast and renewable, like the land, it also performed sovereignty through ritual slaughter of the individual, living bodies in that land. It implied, too, that only very specific kinds of lives had worth, and even those only in rigidly limited ways. Even kings became a uniformly expendable resource, "always dying," and often in violence, like each creature butchered in the hunt.

The Copious Land

In *The Wedding,* the beastly equivalence shadowing the king's body in the forest records is brought into the light via the grotesque figure of the loathly lady, Dame Ragnelle. Upon capturing Arthur in Inglewood, the Baron Somer Joure demands to be told "whate wemen love best in feld and town" (91) in exchange for the king's life. After much searching with help from his knight Gawain, Arthur meets the horrible Dame Ragnelle

say in old saws: 'The lord is worth what his lands are worth'" (12–13). For the broader context of this notion of the hunt as proof of martial prowess, see Eric J. Goldberg, "Louis the Pious and the Hunt," *Speculum* 88 (2013): 618.

who offers him the answer, "To have the sovereynté, withoute lesyng, / Of alle, bothe hyghe and lowe" (423–24), in exchange for Gawain's hand in marriage. Gawain marries this purulent bride and, upon gallantly offering to make love to her, finds her transformed into an exquisite young woman. When she asks him to choose between her beauty lasting at night or in the daytime the baffled knight returns the choice to her, and she then promises to be lovely for him both day and night.

Many critics associate the loathly lady in the variants of this tale with forests and the land.[43] Other readings describe her as a figure of abundance or, to borrow Patricia Parker's term, *copia:* the land's fertile yields, or various kinds of human liberality.[44] If the hunt opening the romance enacts political control of the land via a neat equation between human and cervine bodies, then the copious Ragnelle represents an alternative understanding of sovereignty. Her blazon, her ambiguous answer to Somer Joure's question, and her vast and varied appetites at the wedding banquet all represent the range of other vehicles, besides deer meat, that participate in sovereignty. Her body metaphorically replaces the symmetrical equation between human hunters and living prey, acted out in the hunt, with the messy interlacement of many kinds of lives within a shifting ecosystem, lives that sometimes conflicted with the king's exclu-

---

43. See Susan Carter, "Coupling the Beastly Bride and the Hunter Hunted: What Lies behind Chaucer's *Wife of Bath's Tale*," *Chaucer Review* 37, no. 4 (2003): 329–45. Carter notes the typical presence of the forest as a motif in loathly lady tales, suggesting that "the generic loathly lady's beastliness signals that she belongs in the wilderness; her unstable flesh is chaotic like the forest," as opposed to the orderliness of the city (330–31). Early twentieth-century critics traced the loathly lady's Celtic antecedents as an earth goddess or a mythic avatar of the land. In some of these variants, she tells the rightful king that she is "the sovereignty" itself. The king, in the Irish tale, shows himself worthy to rule this mutable land by his courteous willingness to make love to a hideous crone, and the loathly lady, through her unappealing erotic demands, represents the many-shaped needs of the land that the king's exceptional generosity must satisfy. The king's selflessness in turn awakens the fertility of the countryside. See R. S. Loomis, "The Hag Transformed," *Celtic Myth and Arthurian Romance* (Chicago: Academy Chicago, 1997), 296–301; Alfred Nutt, *Studies on the Legend of the Holy Grail* (London: Strand, 1888), 167; Manuel Aguirre, "The Riddle of Sovereignty," *Modern Language Review* 88 (1993): 273–82; and John Bugge, "Fertility Myth and Female Sovereignty in *The Wedding of Sir Gawen and Dame Ragnelle*," *Chaucer Review* 39, no. 2 (2004): 198–218.

44. On *copia* and Chaucer's Wife of Bath, who tells the most famous loathly lady story, see Patricia Parker, *Literary Fat Ladies: Rhetoric, Gender, Property* (London: Methuen, 1987), 16. On the loathly lady's connections to liberality, see Alcuin Blamires, *Chaucer, Ethics, and Gender* (Oxford: Oxford University Press, 2006), especially 136–38, 143–46; and Rebecca A. Davis, "More Evidence for Intertextuality and Humorous Intent in *The Weddynge of Syr Gawen and Dame Ragnelle*," *Chaucer Review* 35, no. 4 (2001): 433.

sive hunting rights.[45] In her incongruent figure, we find revealed the impossibility of forest law's attempt to draw up human and animal flesh into neat accounts of the land. Her untidy *copia* also suggests that the law, once its instability is understood, can be seen to work best when generosity and gentleness replace neat valuations of human and nonhuman life.

Arthur's vow to Somer Joure represents just such an impossible attempt to assign a neat, legal valuation to life. His oath binds him to return in a year with an answer true enough to trade for his head (97–105).[46] Yet, as his day approaches, Arthur becomes increasingly unwilling to put any such price on his own flesh. Culling every corner of the countryside, the king and Gawain record each answer they are offered in a book, as if making a tax survey of the land. But upon comparing notes they discover a dilemma:

> Syr Gawen had goten answerys so many
> That had made a boke greatt, wytterly.
> To the courte he cam agayn.
> By that was the Kyng comyn with hys boke,
> And eyther on others pamplett dyd loke.
> "Thys may nott faylle," sayd Gawen.
> "By God," sayd the Kyng, "I drede me sore;
> I cast me to seke a lytelle more
> In Yngleswod Forest.
> (207–15)

After all their careful recording of accounts, they remain uncertain how to add up the data about what women love best. Will their two big books full of answers be enough? And, regardless of the number, will

---

45. On medieval art and literature as often representing the land's change and unpredictability as well as reconciling seasonal uncertainty with the order of creation, see Derek Pearsall and Elizabeth Salter, "The Landscape of the Seasons," in *Landscapes and Seasons of the Medieval World* (Toronto: University of Toronto Press, 1973), 119–24. On the conflict between forest and agricultural land see Young, *Royal Forests,* especially on assarts: 109, 117, 121–23, and 134.

46. On the legally binding power, and waning status, of the oath by the fifteenth century, see Richard Firth Green, *A Crisis of Truth: Literature and Law in Ricardian England* (Philadelphia: University of Pennsylvania Press, 2002), 41–78, 121–64. On the central importance of oaths in the *Wedding* specifically, see Colleen Donnelly, "Aristocratic Veneer and the Substance of Verbal Bonds in 'The Weddynge of Sir Gawen and Dame Ragnell' and 'Gamelyn,'" *Studies in Philology* 94, no. 3 (1997): 323–31.

any be right? Whereas Gawain calls these answers "many" and "enough," Arthur, fearing he may need more, anxiously counts up the time until his next meeting with Somer Joure. The problem Arthur faces is the lack of any single right way to signify the value of his living body. Everyone in his land believes this crucial knowledge to be traded for the safety of his skin is something different, and no one else's answer rings quite true to him.

In this state of painful uncertainty about the entanglement of his flesh in a legally binding equation, Arthur rides back into the forest on a different kind of chase than the one that got him into trouble. There, he encounters the quarry he now pursues in place of the deer he hunted at the romance's beginning: someone with an answer worth his life. This person, the loathly lady, denies rational attempts to render an account of her body or indeed of any body made of meat. Instead, her unruly form brims with many things that fatten and flow from living flesh:

> Her face was red, her nose snotyd withalle,
> Her mowithe wyde, her tethe yalowe overe alle,
> With bleryd eyen gretter then a balle.
> Her mowithe was nott to lak:
> Her tethe hyng overe her lyppes,
> Her chekys syde as wemens hippes.
> A lute she bare upon her bak;
> Her nek long and therto greatt;
> Her here cloteryd on an hepe;
> In the sholders she was a yard brode.
> Hangyng pappys to be an hors lode,
> And lyke a barelle she was made.
> And to reherse the fowlnesse of that Lady,
> Ther is no tung may telle, securly;
> Of lothynesse inowghe she had.
> (231–45)

The loathly lady's body gives a material reply to Arthur's uncertainty that his book of answers will be enough. Somer Joure wants the king's life as forfeit for his lost lands, but Ragnelle's body proves the impossibility of assessing the value of land or of the lives in it. Her antiblazon contains a panoply of terms invoking the land as source of copious things—some with apparent value, and some without it. Her eyes like a "balle," for instance, suggest not only rounded hills and an emblem of sovereignty itself (*OED* "ball," n.1) but also livestock or agricultural produce, in the sense of "balle" as a woolly sheep (*MED* "bale," n.5) or a bale of hay (*MED* "bale," n.3). Her shoulders are measured in yards, like a plot of land. She is made like a barrel for commodities such as grain or wine, and her breasts resemble the udders of the horses that sometimes plowed the land.[47] These sack-like paps "to be an hors lode" also suggest the quantity of goods a given horse could transport (*OED* "load," n. def. 2), a varying amount, as medieval workhorses had hard lives, and individual animals bore up differently under the strain.[48]

While the land does produce wealth in all these ways, however, its harvests are never certain, and her body also expresses the instability of the land's yields. For instance, though her eyes may look like sheep or bales of flax, they also potentially resemble the wildfires that burned up crops and livestock (*MED* "bale," n.2), as well as the human corpses that resulted from the famines and plagues due to crop failures of the previous century. They might even evoke the grief such disasters occasioned (*MED* "balle," n.1); numberless eyes have shed tears over such unaccountable losses. Besides things that have value and things that destroy value, the land also produces unpurchaseable waste: the lady's copious effusions of snot, rheum, and hair like an abundant "cloteryd hepe" that suggests massive piles of objects both useful and worthless: grain, stones, dirt, armies, anthills, corpses, dust, or dung (*MED* "hep," n.). Heaps of such things can be treasure or rubbish, depending on their context. Stones, for example, built walls and cluttered fields. Dung polluted cities and provided manure for overworked agricultural land.[49] With a back fashioned like a lute and an enormous mouth, she is even a reminder that the land is a source not only of the produce that satiates hunger but also of music, as a source of the wood and sheep gut used to make lutes that human voices accompanied. She embodies seemingly innumerable connections between the ecosystem's value and human lives. The last line of Dame Ragnelle's horrible blazon, "Of lothlynesse inowghe she had" (245), sums up neatly the diverse abundance she represents. Luckily for Arthur, his land contains more things, both precious and worthless, than venison and kings.

Like Dame Ragnelle's unshapely ugliness, the answer she gives Arthur, "sovereignty in marriage," is enough. If one has sovereignty, one

---

47. Ester Pascua, "From Forest to Farm to Town: Domestic Animals from ca. 1000 to ca. 1450," in *A Cultural History of Animals in the Medieval Age,* ed. Brigitte Resl (New York: Berg, 2007), 90–92.

48. Ibid., 91–92, 99.

49. Ibid., 89–90, 99–100.

can ask for many different things, as she explains, "For where we have
sovereynté, alle is ourys" (425). The potential answers to the question,
"What is a King's life worth?" are as sundry as the responses to the question, "What do women want?" that Dame Ragnelle says range from the
desire to be fair to the wish to be flattered and thought young, the wish
to be satisfied in bed to the wish to be married (408–19). Both a sovereign's value and a woman's desires are accounted differently by different
people. Equations that add up sentient variables do not yield neatly mea-
surable outcomes.

At the wedding feast, the plenitude of Dame Ragnelle's appalling
body manifests itself in her myriad appetites. After the ladies of the court
dress her for her wedding, she approaches the dais looking "fulle foulle"
(602). She is full of foulness, no lack of ugliness there, and in fact "foul,"
along with its alternate spellings of "fowl" and fowlle," becomes her epithet, used to describe her twenty-two times in the course of the poem.
She is also full of fowl, with a pun on the many domestic and game
birds she consumes to the horror of the reluctant guests at her wedding
reception. She voraciously eats her fill of three capons and three curlews.
Although her seemingly bottomless hunger provokes the disgust of all
around her, it compliments Arthur, for, as the author tells us, she can eat
all she wants because his land produces plenty:

> Ther was no mete cam her before
> Butt she ete itt up, lesse and more,
> That praty, fowlle dameselle.
> Alle men then that evere her sawe
> Bad the deville her bonys gnawe,
> Bothe knyght and squyre.
> So she ete tylle mete was done,
> Tylle they drewe clothes and had wasshen,
> As is the gyse and maner.
>
> Meny men wold speke of diverse service;
> I trowe ye may wete inowghe ther was,
> Bothe of tame and wylde.
> In Kyng Arthours courte ther was no wontt
> That myghte be gotten with mannys hond,
> Noder in Forest ne in feld.
> (613–27)

The passage brims with references to unquantifiable variety. No meat comes before her that she doesn't devour, both "less and more," until the meal ends, and then a bit longer, since she hangs around eating, like a rude latecomer at a closing restaurant, until the waitstaff collect and wash the table linens. Just as she has "enough" loathliness, so too Arthur's "diverse" board, supplied by forest game and the animal and vegetable produce from his agricultural lands (both "tame and wild," from "forest" and "feld") furnishes "enough" for her to eat and leaves the king's guests with "no want." The abundant meal served at Arthur's table takes the measure of his holdings and shows them to be copious, so it is aptly described by the word "mete," denoting animal meat or any kind of food (*MED* "mete," n.1 def. 2), a property boundary (*MED* "mete," n.2), or a unit of measurement, especially for agricultural goods (*MED* "met," n. def. 1).

Perhaps Dame Ragnelle repulses Arthur's dinner guests partly because her eating habits reflect those of peasants, which sometimes vied with courtly diets for the same territory. Not long before the feast, the author describes Dame Ragnelle as a "fowlle sowe" (597) with "two tethe on every side / As borys tuskes" (548–49), associating her with chickens and pigs, animals whose meat was not exclusively aristocratic fare.[50] In fact, pigs in particular represented dining habits competing with the king's hunt. Pigs were part-time forest-dwellers for much of the Middle Ages. They foraged among the trees for oak mast, a habit only permitted seasonally as pig pannage damaged new woodland shoots before they matured to provide cover for the king's deer.[51] Her piggishness attests that neither royal appetites nor even human ones, given the porcine taste for acorns, are the only kinds of hunger in the land.

Some of Dame Ragnelle's animal features suggest she might not even turn down a nice piece of human flesh. Though she consumes pigs and fowl, the writer also compares her directly to these omnivorous animals. When the wedding guests watch Dame Ragnelle eating all kinds of meat, part of their horror may arise from the resemblance between her semihuman form and the bodies of the game animals on the table, a reminder that humans, too, have flesh some species find edible. In particular, pigs in medieval communities were sometimes tried in court and punished for consuming human flesh, and some medieval fiction also associated

---

50. Ibid., 85, 97–99.

51. Ibid., 82–86; Birrell "Forest," 80–81, Young, *Royal Forests*, 57, Turner, *Select Pleas*, xxvi.

them with human cannibalism.[52] When the wedding guests, watching in disgust as Ragnelle gnaws to the bone enough cooked birds to feed six men, pray for the devil to gnaw *her* bones, too (617), the symmetry of their curse is suggestive. Ragnelle's epithet as "fowl" referred not only to game birds but also to birds of prey (*MED* "foul," n.1), and in fact, the only kind of fowl she compares herself with explicitly is the owl (310, 316). The undertone of cannibalistic predation in her affinity with this raptor that sometimes dines on other birds is heightened by her long, claw-like nails (607), her big boar's teeth, and her enormous mouth, which, surrounded by grey hairs (551–53), seems faintly lupine. In view of her predatory physiognomy and her enormous appetite, when Arthur and Guinevere come to collect Gawain after his wedding night, perhaps they fear they may not find him alive (724–26, 754–55) because they expect she has eaten him up like another tasty capon.

Dame Ragnelle's voracious hunger for every kind of meat at the king's table harks back to Arthur's parsimonious "taste" of the fat of a single slain deer at the opening of the poem (48). "Taste" in this context can mean both that Arthur carefully measures the deer's fat and that he tastes it in a carefully measured way, by taking a mere morsel to discover the flavor of this particular deer (*MED* "tasten," v.1, 4). Arthur's frugal, intimate test/taste of the deer's body makes his connection to its flesh seem palpable and signifies that its body is precious to him like his own finite life. Dame Ragnelle's copious consumption multiplies this moment of scrumptious embodiment to show that Arthur's lands contain countless different ways of enjoying one's own flesh and consuming the flesh of another. She represents the horrifying and delectable knowledge that inhabiting a body, even a human one, always means: hungering and inspiring hunger, eating and being eaten.

If every body incorporates myriad kinds of desire and delectation, then an individual's life and death will always mean different things from different points of view. Furthermore, how life plays out and death comes depends considerably on others' tastes. Arthur must recognize the contingency of his own life when Dame Ragnelle demands Gawain's

---

52. On pig attacks on human children in medieval communities, see E. P. Evans, *The Criminal Prosecution and Capital Punishment of Animals* (London: William Heineman, 1906), 140–46, especially his account of the trial of a French sow for eating a human child; on various subtle and explicit connections between pigs, pork, and cannibalism, see Claudine Fabre-Vassas, *The Singular Beast: Jews, Christians, and the Pig* (New York: Columbia University Press, 1997), 126–27; Geraldine Heng, *Empire of Magic: Medieval Romance and the Politics of Cultural Fantasy* (New York: Columbia University Press, 2003), 63–65; Steel, *How to Make a Human*, 108–35.

hand (and, as far as Arthur knows, Gawain's life with it) in exchange for the answer he needs. By law, only Gawain can make this choice; Arthur cannot force his consent (291–93). Gawain's generous willingness to risk his life for Arthur's and marry Ragnelle sight unseen, even if she turns out to be a fiend (343–44), his connubial determination to make love to his horrible bride "for Arthours sake," even though all she actually asks him for is a kiss (635–39), and his ultimate ceding of everything he has, "body and goodes, hartt, and every dele" (682), to her all evince his copious love for Arthur. Arthur's life is thus saved by the unaccountable tastes of others: Gawain's overabundant enthusiasm to give his possessions, his body, and even his desire, because Arthur means that much to him, his profusely generous sacrifice set in motion by the monstrous Ragnelle's incongruous choice of Gawain for her mate. Arthur's discovery that his life depends on the unpredictable preferences of others proleptically illustrates John Muir's famous comment on ecological interconnection: "When we try to pick out anything by itself, we find it hitched to everything else in the universe."[53]

Sovereign Gentilness

The most famous loathly lady, the hag from Chaucer's *Wife of Bath's Tale*, gives a pillow-sermon on *gentillesse*, a word that meant both kindness and membership in the ruling class. The word's semantic linkage of the two concepts sometimes was used to represent kindness as an essential characteristic of true nobility (*MED* "gentil," adj. def. 1). Indeed, this is how Chaucer's loathly lady sees it, describing *gentillesse* as a trait that varies from one person to another and cannot be annexed as property; kindness, not simply inheriting royal blood, is what makes one worthy of one's place at the top of the social order.[54] In her own prayer for "gentilnes" (811), Dame Ragnelle asks Arthur to remake his right to rule as a kindness contrasting with Somer Joure's willingness to slay Arthur without "mercy" (165). Now, Somer Joure's life depends on Arthur's choice, and, fortunately for him, the king decides to be kind. This choice shifts the grounds of Arthur's sovereignty from a grindingly repeated performance of his power to kill to a recognition that his own embodiment connects him with countless other kinds of life.

---

53. John Muir, *My First Summer in the Sierra* (Boston: Houghton Mifflin, 1911), 157.

54. See III.1146–64 of Chaucer's "Wife of Bath's Tale" in Geoffrey Chaucer, *The Riverside Chaucer*, 3rd ed., ed. Larry D. Benson (Boston: Houghton Mifflin, 1987), 120.

From an ecological perspective, Arthur's struggle to account for his own life has shifted his role as top predator of his favorite sort of meat. William, Eugene, and Howard Odum describe such fluctuations in predators' dietary choices in terms of a "pulsing paradigm."[55] Even in stable systems, they explain, where predator-prey, parasite-host, and plant-herbivore cycles maintain an overall balance, population levels shift back and forth between consumer and food species. Consumer species, to avoid wiping out their food and thus themselves, sometimes self-limit their own populations through behaviors that help sustain both groups and can even promote biodiversity. Some of these behaviors can include diet-switching by the consumer, behaviors and mechanisms by the consumer to improve the resilience and health of the food, and even cannibalism within predator populations. In other words, in healthy ecosystems, it is in the eater's best interest to promote the diversity of the community and the health of the eaten.

From an ethical point of view, Arthur's pardon reflects his recognition that each life, and each individual's suffering, is unaccountable. The pleasure or pain that any living creature feels in inhabiting its body evades fixed legal value. Thus, the spectacle of the hunt that neatly translates suffering bodies into political authority is an illusion. The king pardons Somer Joure because the Baron "may nott amendes make" (815) for his treason. From an accounting standpoint, Arthur's decision seems not to add up. Legal penalties for treason in medieval England could include disemboweling, drawing and quartering, disinheritance of heirs, and beheading,[56] the last of which might seem like fair compensation for Somer Joure's threat to take Arthur's head. On the other hand, in practice, punishment for treason was at the king's discretion, and medieval kings pardoned traitors for many reasons.[57] Arthur's assertion that the baron's crime cannot be compensated for, coming after the king's struggle to account the value of his own life, acknowledges that torturing or dismembering Somer Joure cannot possibly recompense what happened to Arthur or translate into evidence for his kingship. As Elaine Scarry says about the use of torture to support unstable political regimes, the body's

---

55. On the various ways that eaters sometimes promote the welfare of the eaten, see W. E. Odum et al., "Nature's Pulsing Paradigm," *Estuaries* 18, no. 4 (1995): 550. On cannibalism as promoting biodiversity in certain kinds of predator-prey cycles, see J. Ohlberger et al., "Community-Level Consequences of Cannibalism," *American Naturalist* 180, no. 6 (2012): 791–801.

56. J. G. Bellamy, *The Law of Treason in England in the Later Middle Ages* (Cambridge: Cambridge University Press, 2004), 8–23.

57. Ibid., 173.

pain can create the "spectacle of power," but torture's dramatic conversion of that pain into legitimate authority is "wholly illusory."[58]

The Wedding closes on the author's appeal to God for pity that echoes Ragnelle's plea for Somer Joure. He laments that he has been imprisoned for a long time, in "paynes strong" (852). He begs for "pety," (850), "sympathy for another's suffering" (*MED* "pitie," adj. 2a), from his embodied God, a "Kyng Royalle" (847) "born of a virgyn" (841) and thus alive in suffering flesh.[59] In a similar vein, Derrida says that in order to deconstruct the human/animal false binary, the right question to ask is not, "Can [animals] reason?" but rather Bentham's question, "Can they suffer?"[60] Though kingship was assembled through the wounded bodies of animals, suffering animal flesh also contained the potential realignment of sovereignty as sympathy. Sympathy with the pain of another. Sympathy with the hunger of another, perhaps even another profoundly different from oneself, for life.

---

58. Elaine Scarry, *The Body in Pain: The Making and Unmaking of the World* (New York: Oxford University Press, 1985), 27.

59. On the medieval emphasis on Mary's body as having provided the material, human flesh of Christ that suffered on the cross, see Caroline Walker Bynum, *Fragmentation and Redemption: Essays on Gender and the Human Body in Medieval Religion* (New York: Zone, 1991), especially 151–80.

60. Derrida, *Animal*, 27–28. I am indebted to Steel's similar arguments on this point, apropos of *Sidrak and Bokkus,* in "How to Make a Human," *Exemplaria* 20 (2008): 8–9, 19. On similar depictions of cross-species sympathy, see Susan Crane's discussion of Canacee's compassion for the falcon in Chaucer's *Squire's Tale* (*Animal Encounters,* 130–36).

3

The Physician and the Forester

VIRGINIA, VENISON, AND THE BIOPOLITICS OF VITAL PROPERTY

Randy P. Schiff

WHILE MAKING a series of asides that amplify his critique of the negligent care of young and vulnerable youths, Chaucer's Physician observes that a former "theef of venysoun" [deer poacher] can best "kepe" [police] a "forest" (VI.85).[1] Despite the presumption that a university-trained doctor would have systematically studied rhetoric and thus been expected to be a tolerable orator,[2] many critics see this poaching reference as merely one of a number of awkward digressions from the story of doomed Virginia. Critical condemnation of the Physician's wandering style can be seen most clearly in reaction to the comment's immediate context—the assertion that "olde" governesses are best suited to watch over lords' daughters, either because they have always been honest or because they have experienced sexuality's "olde daunce" and then "forsaken" it (VI.82–85).[3] Lee Patterson's reading of the Physician's turn to

---

1. All citations from Chaucer are from *The Riverside Chaucer,* 3rd edition. ed. Larry D. Benson (Boston: Houghton Mifflin, 1987). Glosses are my own.

2. On the standard training of physicians such as Chaucer's pilgrim, including in rhetoric, see Carole Rawcliffe, *Medicine and Society in Later Medieval England* (Phoenix Mill, UK: Sutton, 1995), 106–16.

3. For influential readings of the Physician's asides as both locally digressive and as indicative of more general disorganization within his performance, see Trevor Whittock, *A Reading of the Canterbury Tales* (Cambridge: Cambridge University Press, 1968),

the forest typifies the critical reception of Chaucer's medical professional as discursively clumsy: he sees the poaching reference as an "irrelevant" flourish in a tale prone to superficial rhetorical excess.[4]

I will maintain that the Physician's juxtaposition of foresters, governesses, and parents is part of a clear—and biopolitical—design. Considering that the Physician tells a well-known story about a father who kills his daughter rather than let her live as a slave, it is surprising that his cautionary commentary on the dangers of poor parenting and childcare has not been typically read as apropos. That the Physician's interest in the supervision of children is not simply about the fear of their becoming socially corrupted (VI.63–71) but also concerns their becoming victim to such horrors as the honor-killing at the center of the tale is clear from the doctor's discursive movement from governesses to parents: he exhorts "fadres" and "modres" to be wary that their own behavior may make their children "perisse" [perish], much as "many a sheepe and lomb" have been "torent" [slain] by wolves due to a "shepherde softe and necligent" (IV.93–104). By endorsing a relativist approach to life that embraces formerly criminal hunters who, like previously promiscuous governesses, have "forlaft" [forsaken] their "olde craft" (VI.83–84),[5] the Physician's poaching reference clearly condemns the parental absolutism that leads to Virginius's murder of his daughter. Far from random observations maladroitly strung together by a discursively careless Physician, his digressions on poaching, parents, and governesses thus converge in a methodical condemnation of Virginius and his wife alongside the more obvious villains, Apius and Claudius. The Physician, whose vocation involves him fundamentally in the fostering of life, cannily blends his asides on proper guardianship and forest regulation to paint both lawful parents and foresters as agents of death.

This juxtaposition of fatal poaching and parenting serves a deeper aspect of the Physician's agenda—the cultivation of a politics of life

---

179–83; Anne Middleton, "*The Physician's Tale* and Love's Martyrs: 'Ensamples Mo Than Ten' as a Method in *The Canterbury Tales*," *Chaucer Review* 8 (1973): 19–20; and Lee Patterson, *Chaucer and the Subject of History* (Madison: University of Wisconsin Press, 1991), 369–70.

4. Patterson, *Chaucer and the Subject*, 370.

5. For another medieval example of a poacher whose very poaching recommends him for participation in enforcing the law, see my "Sovereign Exception: Pre-National Consolidation in *The Taill of Rauf Coilyear*," in *The Anglo-Scottish Border and the Shaping of Identity, 1300–1600*, ed. Mark Bruce and Katherine Terrell (Palgrave Macmillan, 2012), 37–41. See also Jeanne Provost's discussion of Robin Hood in chapter 2 of this volume (59n7).

aimed at replacing the thanatopolitical status quo.[6] I will argue that
the Physician's appeal to forest law in the midst of discussing the surveillance of adolescents—far from being either simply comparative or
digressive—catalyzes the doctor's systematic effort to present himself as
a biopolitical authority. The invocation of the legal regulation of deer
killing, launched alongside a critique of Virginius's transformation of
his daughter into disposable property, highlights the fatal nature of the
current legal regulation of bodies. Moving beyond a mere hostility to
the legal profession, which is often associated with medieval medical
practitioners,[7] Chaucer's Physician portrays the grisly poles of a legal
system grounded in thanatopolitics. Both Apius, the corrupt judge who
concocts a conspiracy to rape Virginia by having her legally defined as a
false plaintiff's stolen slave, and Virginius, the *paterfamilias* who commits an honor-killing that would be justifiable under the rubric of the
Roman law of *vita necisque potestas*—the right of any free Roman father
to treat children as bare life, killing them without any state interfer-
ence—reveal a legal world that, the Physician suggests, leads only to
death.[8]

Far from being anachronistic, biopolitical concepts illuminate the
Physician's cultivation of a politics of life that encompasses the legal bodies of both children and deer. If biopower has to do, as Michel Foucault
argues, with the management and production, rather than the mere dis-

---

6. I adopt this political terminology from Roberto Esposito, who links "thanatopolitics" with the explicit turn to a biological agenda in Nazi nation-building; see *Bíos:
Biopolitics and Philosophy*, 2004; trans. Timothy Campbell (Minneapolis: University of
Minnesota Press, 2008), 110–45, and from Giorgio Agamben, who reads biopolitics as
stretching back at least into Roman antiquity, see *Homo Sacer: Sovereign Power and Bare
Life*, trans. Daniel Heller-Roazen [Stanford: Stanford University Press, 1998], 72–90. I
understand the Physician as fashioning a biopolitical version of Virginia's victimization—
not as a personal history but as an allegory intended to figure any population member's
subjection to law.

7. See Beryl Rowland, "The Physician's 'Historial Thyng Notable' and the Man of
Law,'" *ELH* 40 (1973): 165–78.

8. On the legal rights of the *paterfamilias* (the male head of the household in Roman law), see Andrew Borkowski and Paul du Plessis, *Textbook on Roman Law*, 3rd ed.
(Oxford: Oxford University Press, 2005), 113–20. On *vita necisque potestas*, the Roman
legal tradition that afforded fathers the exceptional right to slaughter children without
incurring murder charges, see Borkowski and du Plessis, *Textbook on Roman Law*, 114–16;
Brent D. Shaw, "Raising and Killing Children: Two Roman Myths," *Mnemosyne* 54, no.
1 (2001), 31–77; and Agamben, *Homo Sacer*, 87–90. This law is particularly relevant to
Livy's story of Virginia, insofar as crimes related to daughters' sexual proclivities were
prominent in the (rare) exercise of this power; see Borkowski and du Plessis, *Textbook on
Roman Law*, 115.

ciplining and repression, of subject bodies,[9] then the Physician's injection of forest life and law into Virginia's story foregrounds premodern biopower concerns. The medieval forest reveals the thoroughgoing interpenetration of law and life: within a precisely bounded space, all life-forms, whether animal or vegetable, are actively and explicitly subject to the law. Life and law are not simply inseparable within a medieval forest but rather interrelated in a biopolitical system aimed at managing rather than merely policing resources. Laws against taking wood without permission are not, for example, based on territorialism or revenue protection but on the need to ensure that ample vegetative cover enables the pleasures of the hunt; laws against taking deer without license are not primarily repressive but rather caught up in a sophisticated system of resource protection grounded both in classifying types of forest animals and the proper times in which certain beasts can be hunted.[10] In regulating hunting seasons and the types and ages of killable deer, forest law is fundamentally biopolitical, insofar as it is designed to organize organic elements necessary for populations to thrive and, hence, produce a sufficient number of deer to be both enjoyably and profitably slain. Forest law, then, is not aimed at restricting wood gathering or hunting in a zero-sum game but rather at ecological management that, in providing forest nourishment and cover, also offers the deadly pleasures pursued by hunters.[11] The forest law is not, in the end, a law of death, but of life—aimed at maintaining deer communities and the vegetative environments that they require.

A number of authors restaged the legal drama of Virginia's murder that was launched by the Roman historian Livy, including Jean de Meun, whose version in the thirteenth-century *Romance of the Rose* almost cer-

---

9. For Foucault's argument that biopolitics (which he dates from the eighteenth century) involves the management and multiplication of life through attention to the population, whereas the sovereignty it displaces was confined to the discipline and taking or repression of life, see Michel Foucault, *Society Must Be Defended: Lectures at the Collège de France, 1975–1976*, ed. Mauro Bertani and Alessandro Fontana, trans. David Macey (New York: Picador, 2003), 243–47.

10. For an analytical survey of medieval English forest law, see Charles R. Young, *The Royal Forests of Medieval England* (Philadelphia: University of Pennsylvania Press, 1979), 74–113. On the extension of English forest law into late medieval parks, see S. A. Mileson, *Parks in Medieval England* (Oxford: Oxford University Press), 121–45.

11. For analysis of the pleasures linked with hunting in medieval forests, whether the highly ritualized (and often eroticized) traditions of aristocrats or of the lower classmen who often literally poached such joys from aristocratically managed forests, see Matt Cartmill, *A View to a Death in the Morning: Hunting and Nature though History* (Cambridge, MA: Harvard University Press, 1993), 59–75.

tainly influenced both Chaucer's version and that of his contemporary, John Gower.[12] Virginia's tragic story clearly appealed to poets striving to visualize the limits of the law: Virginia's innocence frames both individual and public responses to institutional corruption. By invoking forest law in the network of asides devoted to the regulation of children, Chaucer's Physician develops a biopolitical analysis of premodern power that interrelates game animals and Virginia as entities whose lives and bodies are each fatally subject to legal force. By linking venison as the material goal of illicit hunters with Virginia as the target of a judge's venereal desires, the Physician imagines law as transcending human-animal difference. Exploring what Giorgio Agamben analyzes as exceptional modes of asserting legal privilege, I will show that the classificatory management of bodies in the *Physician's Tale* is enforced precisely through lawful violation of the law.[13] Much as kings assert exceptional authority over the lives of deer who inhabit the legal limits of forests or parks, so does Virginius wield, due to his status as a free Roman father, an exceptional legal power to transform his daughter into a *homo sacer*—a human who can be killed with impunity.[14]

Chaucer's key innovation, I will maintain, is to emphasize the sociospatial dimension of the competing jurisdictions in Virginia's story. At the heart of the *Physician's Tale* lies a collision between the power of state authority, represented by the corrupt judge Apius and his various co-conspirators, and that of the domestic sphere whose borders are defended by Virginia's father. Whereas Virginius wields the basic Roman legal power of *vitae necisque potestas,* the location of his jurisdiction differs significantly in renderings of this legal drama. In Livy's and Gower's versions, Virginius opportunistically slaughters his daughter in the same site where the legal decision declaring his daughter a stolen slave

12. For analysis of Livy, Jean de Meun, and other possible sources of Chaucer's *Physician's Tale,* as well as of Gower's version, see Helen Storm Corsa's edition in *A Variorum Edition of the Works of Geoffrey Chaucer, vol. 2, The Canterbury Tales, Part 17, The Physician's Tale* (Norman, OK: University of Oklahoma Press, 1987), 4–10.

13. On exceptional modes of asserting legal privilege, in which juridical authority is established by extrajudicial means, see Carl Schmitt, *Political Theology: Four Chapters on the Concept of Sovereignty,* 1st. rev. ed., 1934; trans. George Schwab (Chicago: University of Chicago Press, 2005), 5–35; Giorgio Agamben, *State of Exception,* trans. Kevin Attell (Chicago: University of Chicago Press, 2005), 1–40; Timothy C. Campbell, *"Bíos,* Immunity, Life: The Thought of Roberto Esposito," *Diacritics* 36, no. 2 (2006): 12–14; and Achille Mbembe, "Necropolitics," *Public Culture* 15, no. 1 (2003): 11–40.

14. On the *homo sacer* as a foundational figure in Western political discourse, see Agamben, *Homo Sacer,* 72–86; Lemke, *Biopolitics,* 54–56; and Mbembe, "Necropolitics," 22–30.

is rendered,[15] while Jean de Meun does not make clear where Virginia is killed.[16] Chaucer's Physician has Virginius deliberately take his daughter back home before he commits the exceptional act of what he understands to be a mercy killing. Chaucer's Virginius thereby calls attention to parental rights as a spatial phenomenon—not just a privilege based on class but one tied to absolute control over an area. In tune with the trends of domestic spaces becoming "more private" in the depressed housing markets of the Physician's fourteenth-century London, and with what Vanessa Harding identifies as the "more conditional" nature of private space in Chaucer's times, Virginius's actions would resonate as a legal means of staking a territorial claim.[17]

In critiquing both the corrupt judge and the homicidal father, the Physician uses Virginia's story to carve out a position of authority by co-opting the voice and moral authority of Nature early in his tale. In the end, the Physician's performance suggests an answer to what seems, in the context of a brutal Roman world dominated by corrupt clerks and brutal warrior-fathers, merely rhetorical—namely, Virginia's question, "Is ther no remedye?" (VI.236). Choosing a word that can in Middle (as in modern) English refer both to a legal or medical mechanism, the Physician suggests that in an older world guided only by the laws of fathers, Virginia has no "remedye"—but in a modern world guided by medical practitioners who can transform the law of the state into a biopolitics focused on managing and multiplying life rather than merely policing it, Virginia could perhaps thrive. The Physician's vivid dramatization of the horror of an honor killing—that is, a sacrifice aimed at communicating

---

15. In Livy, Virginius tactically asks for a moment to mourn with his daughter, and then kills her in the Forum marketplace where Apius's judgment was made; see *History of Rome,* vol. 2, ed. T. E. Page et al., trans. B. O. Foster, Loeb Classical Library (Cambridge, MA: Harvard University Press, 1922), III.xvlviii.2–8. In Gower, an enraged Virginius impulsively kills his daughter shortly after he appears at the site of the legal decision; see *The Complete English Works of John Gower, vol. 2,* ed. G. C. Macaulay, Early English Text Society e.s. 82 (1901), VI.5210–52.

16. In Guillaume de Lorris and Jean de Meun, *Le Roman de la Rose,* ed. and trans. Armand Strubel (Paris: Librairie Générale Française, 1992), 5620–35, the precise time or location of Virginia's killing is not made clear, though the presentation of her decapitated head is localized in Apius's court (5635).

17. See Vanessa Harding, "Space, Property, and Propriety in Urban England," *Journal of Interdisciplinary History* 32, no. 4 (2002): 562, 567. For analysis of the evolution of space in medieval Europe, and of the increasing importance of private space in the postplague era, see Georges Duby, *A History of Private Life: Revelations of the Medieval World,* ed. Georges Duby, trans. Arthur Goldhammer (Cambridge, MA: Belknap Press, 1988), especially ix–xiii and 3–33.

patriarchal values, triggered by a male authority's shame over the potential loss of control over a female relative—sets his own medically trained voice against a world of corrupt judges and atavistic aristocrats.[18] Virginia's use of a word that recalls both medical and legal discourses moves the narrative from mere pathos to the imagination of an alternative social model. Speaking as a member of an elite, trans-European class of doctors whose expertise was valued all the more in a plague-rattled late medieval era, the Physician uses the story of Virginia's class-based, territorialist slaughter to imagine an affirmative politics of life that could replace the thanatopolitics of a world run by those such as Jephthah, Apius, or Virginius.[19]

Prosopopoeia and the
Medical Appropriation of Natural Authority

As we have seen in the reception of his poaching reference, critics often portray the Physician as a clumsy, sensationalist, and ultimately ineffective storyteller. In discussing the lengthiest section of original material inserted into the tale—an instance of prosopopoeia, in which the Physician takes up the voice of Nature in order to praise Virginia's virtue and beauty—Patterson imagines a doctor who only superficially controls his tale. Arguing that the *Physician's Tale* is "fraudulent or 'counterfeit' hagiography," Patterson asserts that the Physician is not "aware" of the irony that his attempt to voice Nature runs counter to Nature's rhetorical query, "Who kan me counterfete?" (VI.13) [Who could counterfeit

---

18. For analyses of honor killings that highlight patriarchy and shame as the key elements in such murders, see Nancy V. Baker, Peter R. Gregware, and Margaret A. Cassidy, "Family Killing Fields: Honor Rationales in the Murder of Women," in *Violence against Women* 5, no. 2 (1999): 164–84; and Necla Mora, "Violence as a Communicative Action: Customary and Honor Killings," *International Journal of Human Sciences* 6, no. 2 (2009): 499–510. On the sacrificial vocation of knighthood with which Chaucerian characters would associate soldiers such as Virginius, see L. O. Aranye Fradenburg, *Sacrifice Your Love: Psychoanalysis, Historicism, Chaucer* (Minneapolis: University of Minnesota Press, 2002), 205–6.

19. On critical movement away from focus on political domination toward arguments about the proliferation of life and positive political resistance as "affirmative biopolitics," see Campbell, "*Bíos,* Immunity, Life," 3. Prominent examples of affirmative biopolitics include Esposito's *Bíos*; Melinda Cooper, *Life as Surplus: Biotechnology in the Neoliberal Era* (Seattle: University of Washington Press, 2008); and Michael Hardt and Antonio Negri, *Multitude: War and Democracy in the Age of Empire* (London: Penguin, 2004).

me?].[20] Patterson thus sees the ornate, extended treatment of Nature's voice as merely an act of deceptive craftsmanship, communicating only the tale-teller's "fallen historicity."[21]

I would argue that not only is the Physician laser-focused throughout his discursive co-option of Nature's voice but also that the fallenness in his emphatically "historial" (155) [historical] tale is directed toward a clear, underlying political objective: taking up Nature's voice is a key step in a biopolitical program. In assessing traditional critical frustration with the Physician's failure to deliver a clear exemplary meaning, Daniel Kempton illuminates the tale-teller's motivations: "The *Physician's Tale* is *about* nothing other than the Physician, a medieval doctor of physic."[22] Through prosopopoeia, the Physician performs the subject of his vocational expertise. In speaking as Nature, the Physician is not doing something new or particularly medical; as George Economou shows, medieval philosophers and theologians, ranging from Bernardus Silvestris to Alan of Lille, famously channeled Nature's voice in order to work out such issues as the coexistence of corporal and spiritual elements.[23] However, the Physician, *as* a medical practitioner, co-opts the authority of nature as he ventriloquizes her discourse, demonstrating precisely the biological mastery to which his profession pretends. If the Physician's story of Virginia is, as I maintain, aimed at undermining the authority of traditional state government (embodied by Apius and his co-conspirators) and that of the military aristocracy (figured by Virginius), then the Physician's voicing of Nature serves a jurisdictional function—namely, to mark the biological realm of life as the territory of medical professionals.

With Nature's assertion that no human could produce a person as beautiful as Virginia (VI.11–18), the Physician strategically opens his presentation by establishing Nature's transcendent authority. The Physician's argument that art cannot counterfeit nature not only privileges his own area of expertise as the best suited to analyzing nature's most beauteous productions but also communicates a basic assumption of medieval medical practitioners—that the healer does not do something new but rather works to restore the balance of humors that, according to standard

---

20. Patterson, *Chaucer and the Subject,* 370.

21. Ibid.

22. Daniel Kempton, *"The Physician's Tale:* The Doctor of Physic's Diplomatic 'Cure,'" *Chaucer Review* 19, no. 1 (1984): 26 (author's emphasis).

23. See George D. Economou, *The Goddess Natura in Medieval Literature* (Cambridge, MA: Harvard University Press, 1972), 58–92. For further analysis of medieval intellectuals' philosophical and political use of allegorized Nature, see Stephanie Batkie's essay in this volume.

premodern medical theory, should be returned to natural harmony.[24] Moreover, the Physician's Nature performs an explicitly literary service. In having Nature argue that nature is a horizon within which any artist must work, the Physician channels his medical training into aesthetic reflections on working within received limits—a subject that is eminently appropriate for the Canterbury tale-telling contest. Finally, the rhetorical co-option of nature's authority advances the Physician's biopolitical agenda as he moves from the description of Virginia as an ideal biological specimen to describing her doom within the death-saturated world of archaic Rome.[25]

In returning repeatedly to questions of surface and form, the Physician deploys his professional medical perspective to develop his jurisdictional, biopolitical claim. If ecology builds, as Timothy Morton argues, on the insight that "everything is connected,"[26] then the Physician performs political ecological work in showing the causal links between macrocosm and microcosm that were central to medieval medical discourse. As an expert in the medieval science of astrology (which encompassed what modernity distinguishes as astronomy), the Physician clearly holds a materialist worldview, in which any physical phenomenon depends upon the relative positions and temporal contexts of the planets and stars.[27] The Physician uses his description of Nature and her perfect creation, Virginia, to blur the boundaries between the objective and the subjective, as both physical appearance and moral predisposition fall under the expert eye of the medical doctor. Building on

---

24. On medieval humoral theory, see Rawcliffe, *Medicine and Society*, 43–44, and Nancy G. Siraisi, *Medieval and Early Renaissance Medicine: An Introduction to Knowledge and Practice* (Chicago: University of Chicago Press, 1990), 104–6. Given the common medieval association of internal humoral balance with social order (Rawcliffe, *Medicine and Society*, 44), humoral theory supports the view that the Physician systematically integrates biological and political issues.

25. On the range of definitions of biopolitics, with the primary sense of natural life as the basis of all politics, see Thomas Lemke, *Biopolitics* (New York: New York University Press, 2012), 3–4.

26. Timothy Morton, *The Ecological Thought* (Cambridge, MA: Harvard University Press, 2010), 1.

27. Any medieval Physician's connection of astronomical and terrestrial phenomena implies a thorough materialism that refuses to see a fundamental gap between the material and moral life of humans and the most distant external worlds. On the fundamental role of astrological theory in medieval medicine, see Siraisi, *Medieval and Early Renaissance Medicine*, 67–69, 135–36; Rawcliffe, *Medicine and Society*, 82–94; and Sophie Page, *Astrology in Medieval Manuscripts* (London: British Library Press, 2002), 52–57.

the argument—common in medieval universities since the twelfth century—that Nature supervises all activities in the sublunary sphere,[28] the
Physician discursively blends the political and the biological. He uses
recognizably political language in framing his tale: Nature made Virginia with "sovereyn diligence" (VI.10) [sovereign care], the Physician
explains, injecting the language of political authority into his portrait
of Virginia's creation. Much as, according to Chartrist doctrine, Nature
as the vicar of God represents in more circumscribed form the glory
of God, so does Virginia also microcosmically reflect not just the glory
but the authority of Nature: she possesses "soverayn bountee" (VI.138)
[sovereign grace].

The Physician's Nature makes no distinction between Virginia's body
and behavior, situating each within Nature's own domain of biological
life. By subtly coordinating his speaking *as* Nature with his speaking as
an expert *on* nature, he claims jurisdiction over the ethical assessment
of Virginia. In the Physician's materialist, medical vision, all is focused
on surfaces: there is no interior soul or self that escapes the shaping
hand either of Nature or of the physician whose expertise consists of
understanding her handiwork.[29] The Physician highlights the manner in
which the anatomical and the moral exist on the same spectrum by moving immediately from identifying physical details, such as Virginia's hair
"dyed" the color of sunbeams by "Phebus," to the description of her
being utterly "virtuous" in all aspects of her "condicioun" (VI.37–42).
The Physician thus blurs the boundaries between Nature's and his own
voice in order to establish medical discourse as expert "as wel in goost as
body" (VI.43) [regarding both spirit and body]: Virginia's physical and
moral beauty are thus each medicalized.

---

28. See Economou, *Goddess Natura*, 6.

29. It is crucial to recognize that late medieval individuals clearly had a sense of an
interior self that transcended the physical world, as has been demonstrated by Masha
Raskolnikov's study of body-soul debates, in *Body against Soul: Gender and 'Sowlhele' in
Middle English Allegory* (Columbus: The Ohio State University Press, 2009), 1–8. That
the notion of an interior self is not a modern invention escapes the attention of much
current ecocriticism, which uncritically accepts assertions that current environmental
crises arise from an epochal shift effected by René Descartes splitting all matter into
the realms of *res cogitans* (the ideational) and *res extensa* (the material); see Nandita Batra, "Dominion, Empathy, and Symbiosis: Gender and Anthropocentrism in Romanticism," *Interdisciplinary Studies in Literature and Environment* 3, no. 2 (1996): 103–4,
and Morton, *Ecological Thought*, 25–26. Medievalist studies such as Raskolnikov's belie
such an epochal understanding of dualism and suggest that a materialist outlook such
as the Physician's would have been a remarkable intervention.

The Physician's politicization of the biological does not simply consist of praise of Virginia as an ideal specimen but also involves the condemnation of the earthly agents who martyr her.[30] Continuing to blend
the language of political and biological authority, the Physician asserts
that it is "sovereyn pestilence" to "bitrayseth innocence" (VI.91–92). The
Physician here makes a clear reference to his most striking and lucrative
mode of social service—his efforts to heal those touched by the various outbreaks of plague that ravaged Europe since the Black Death of
1347–49.[31] That the Physician is an ambitious and successful practitioner
is clear from the *General Prologue:* Chaucer stresses that the "Doctour
of Phisik" (I.411) has "wan" [earned] considerable money "in pestilence" [during plague-time] (442), and also follows well-trodden satirical
paths by discussing the Physician's lucrative cultivation of his professional
"frendshipe" with "apothecaries" (425–28).[32] The Physician's materialist ambition does not, however, preclude him from having an ambitious
moral and political agenda. In describing the tale's moral crises as literally
pestilent (VI.91), the Physician weaves his medical jurisdiction over bodies into the tale's ethical fabric. The Physician's pointed deployment of
the discourse of infectious disease not only reinforces his alignment with
Nature's authority but it also points to death as his discursive means of
entering into politics. The very power that the emergency conditions of
plague opened up for medical professionals also creates an entryway into
the broader political sphere, and the Physician uses this opportunity to
systematically expose both parents and legal authorities as incapable of
bringing Virginia anything but death.

---

30. The portrait of Virginia's suffering clearly partakes of traditional martyrdom imagery. For analysis of hagiographic aspects of the *Physician's Tale,* see Catherine Sanok,
"The Geography of Genre in the *Physician's Tale* and *Pearl,*" *New Medieval Literatures* 5
(2002): 177–201.

31. For thoroughgoing surveys of the social and economic effects of plague in late
medieval England, see Colin Platt, *The Black Death and Its Aftermath in Late Medieval
England* (Toronto: University of Toronto Press, 1997); and Stuart J. Borsch, *The Black
Death in Egypt and England: A Comparative Study* (Austin: University of Texas Press,
2009).

32. On satirical treatment of collusion between medical doctors and apothecaries, see
Jill Mann, *Chaucer and Medieval Estates Satire* (Cambridge: Cambridge University Press,
1973), 95–96. See also Julie Orlemanski, who notes that frequent satire of medical greed
led to physicians being associated with the "brute materiality" (400) in which they were
expert, in "Jargon and the Matter of Medicine in Middle English," *Journal of Medieval
and Early Modern Studies* 42, no. 2 (2012): 395–420. On apothecaries in medieval England, see Rawcliffe, *Medicine and Society,* 148–65.

Of Daughters and Deer:
Imagining Competing Jurisdictions over Bodies

In both strategically using prosopopoeia to co-opt the authority of Nature and in structuring his asides so as to interrelate the potentially fatal regulation of deer with that of children, the Physician thus lays the groundwork for a distinctly biopolitical deployment of Virginia's story. The Physician's biopolitics emerges through twinned emphases: the regulation of bodies (whether deer subject to forest law or children subject to the strictures of guardianship) and the contestation of juridical territories. The central conflict of the *Physician's Tale* consists of a clash between state and familial jurisdictions, in which all male players share the assumption that Virginia is human property. Class and space are essential ingredients in the Physician's political ecology. Whether Virginia is a free daughter subject to legal execution by her father or a slave subject to physical abuse by her owner depends not only on the Roman law of persons, which identified humans as slave or free, but also on the assertion of territorial boundaries. Much as the deer that are key to the Physician's poaching reference acquire their legal status based on their spatial location in the forest, so do the bodies inhabiting the *Physician's Tale* meet their destinies only at the intersections of law and space.

The Physician is acutely concerned with the issue of territoriality— the juridical control of space that occurs at multiple scales within any society.[33] As we have seen, one of Chaucer's chief interventions is the moving of Virginia's death to a clearly domestic locale: whereas Virginia is killed in earlier versions either in the immediate vicinity of Apius's court or in an undisclosed location, Chaucer's Physician has Virginius take his daughter home. It is within the confines of their private, familial space that Chaucer's Virginius and Virginia discuss her imminent beheading. That the Physician condemns Virginius's act is clear. Given his professional concern with caring for bodies and his preceding critique of negligent parents and guardians, we can safely assume that the Physician aims to condemn Virginius's choice of death rather than a life of slavery for his daughter. Concern with legal location compli-

---

33. For analytical surveys of geographical, sociological, and anthropological views of territoriality, see Robert Sack, *Human Territoriality: Its Theory and History* (Cambridge: Cambridge University Press, 1986), 5–27; and Saskia Sassen, *Territory, Authority, Rights: From Medieval to Global Assemblages*, rev. ed. (Princeton, NJ: Princeton University Press, 2008), 1–23.

cates this scene, enabling separate indictments of Virginius's knightly, and Apius's state-based, thanatopolitics. As Harding maintains, private space in the Physician's postplague England was highly unstable: it was more "conditional" than in previous eras, with a depressed land market enabling numerous efforts to convert private areas into public spaces.[34] The Physician's decision to have Virginius choose to return home to kill his daughter creates a political moment that is exclusive to his Canterbury pilgrimage audience: only those listening to his tale experience the private domain being delimited, as Virginius asserts his legal space by exercising his paternal authority to slaughter his child. Virginia's death marks the space of precisely those private domains that required strenuous identification in the Physician's unstable age. Juridical territory thus proves to be key to the Physician's critique, as both judge and knight assert authority from their respective loci of power. The Physician's preceding reference to forest law aids in clarifying such spatialization of the law by reminding his audience how bodies can be lethally subject to the legal imposition of boundaries.

At the heart of the *Physician's Tale* lies the contestation of such legal borders, as Virginius uses his overdetermined killing of Virginia to highlight his defense of domestic sovereignty. For Romans, the authority of the state was not continuous but punctuated with free male citizens maintaining certain sovereign rights that the state could not undermine. While most, if not all, premodern societies featured such fragmented modes of sovereignty,[35] the Roman case transmitted from Livy to Chaucer's Physician involves the quintessential example of the defense of domestic territorial sovereignty. The right of *vitae necisque potestas* was one of the clearest and most ancient examples of domestic sovereignty respected by the Roman state. As with many archaic laws, the rationale of this paternal power over life and death required ritual performance to maintain, and the Physician's Virginius asserts his individual rights by

---

34. See Harding, "Space, Property, and Propriety," 562–67.

35. For a powerful survey of the fragmented nature of premodern sovereignty, see Sassen, *Territory, Authority, Rights,* 32–53; and see also Stuart Elden's analyses of land practices postdating the "fracturing of the West" after Rome's fall (99) to feudal-era Europe, which lacked a "territorial system" (156), in *The Birth of Territory* (Chicago: University of Chicago Press, 2013), 99–156. For a call to medievalists to avoid projecting modern notions of national sovereignty into medieval political contexts that involved more dynamic and discontinuous notions of territory, see my *Revivalist Fantasy: Alliterative Verse and Nationalist Literary History* (Columbus: The Ohio State University Press, 2011), 5–9; 157–62.

providing a domestic "frame" for the sacrifice of his daughter.[36] Virginia's apparent consent to her father's decision to execute her, which has been analyzed in terms of politics and privacy by Lianna Farber,[37] contributes to the legal framing of the scene: that both daughter and father play a role in the sacrifice highlights the defense of specifically familial rights. Virginius's action bears with it the overdetermined energy of ritual coupled with its pragmatic use—Virginius's sense that his daughter's death is preferable to her living in shame as a sexual object who would also be subject to death at the hands of her legal owner, Claudius—overshadowed by the excessive, emphatically performative nature of the killing.

The overdetermined nature of Virginius's honor-killing of his daughter is twofold—both in its being unnecessary and in its being excessively dramatic. It becomes clearly and thus cruelly evident that there was no need for Virginius to kill his daughter: just moments after Virginius shocks all by displaying his daughter's severed head in Apius's court, the mob is quickly convinced that a morally questionable Apius rigged the trial. It is thus clear that Virginius simply could have waited for mob justice to defend his cause—and the force of a suspicious "peple" [citizenry] becomes crystal clear when at the close of the tale a "thousand" of them "thraste" [burst] into the courtroom and assert their will (VI.260–73). Indeed, Apius's earlier fear that he could not possibly take Virginius's daughter by physical "force" because Virginia's family is too "strong of freendes" (VI.133–35) impels him instead to turn to institutional force to seize this knight's daughter.[38] This further suggests that Virginius might simply have used military force to alleviate his situation.[39] Both the mob and Virginius's strong military position show that his slaughter of Virginia was not desperate but rather a deliberate, political act: he defends his own power by shaming Apius's court, with his daughter's body little more than a legal prop. Much as in *Sir Gawain and the Green Knight* the

---

36. In referring to the framing function of ritual, I rely on Mary Douglas's emphasis on the symbolic work performed by ritual acts that produce a "marked off time or place" in which the power of the act can resonate beyond its immediate effects; see *Purity and Danger,* rev. ed. (London: Routledge, 2002), 78.

37. See Lianna Farber, "The Creation of Consent in the *Physician's Tale,*" *Chaucer Review* 39, no. 2 (2004): 159–62.

38. That Virginius, hearing of the legal finding, assumes that he must give his daughter up by "force" (VI.205) shows that the Physician, blurring the boundaries between text and body, places military and legal force on the same plane.

39. Jerome Mandel discusses Virginius's failure to use military force to overrun Apius's court as the sign of his own political failure, which links him with the corrupt judge, in "Governance in the *Physician's Tale,*" *Chaucer Review* 10, no. 4 (1976): 323.

head of a hunted animal becomes a prop in a performance of Bertilak's power over the living entities within his domain,[40] so does the presentation of Virginia's severed head in court ritually communicate Virginius's power over a being who lived within the legal limits of his domain.

If his actions indeed defend paternal jurisdiction by invoking the law of *vitae necisque potestas*, then Virginius expresses his sovereign power in an overdetermined mode that Agamben sees as key to Western power—in a state of exception. In the time following Apius's biopolitical decision to alter Virginia's status according to the law of persons, determining that she is Claudius's stolen slave rather than Virginius's free daughter,[41] Virginius stands in a profoundly ambivalent legal zone. He is no longer legally her father, though his actions and thoughts, such as his "fadres pitee" (VI.210) [father's pity], show that he continues to act paternally. That the Physician produces a domestic setting for Virginius's honor-killing highlights the exceptional nature of the archaic law that haunts the tale: despite the state's effort to strip him of the parental power he has assumed since Virginia's birth, he will nevertheless perform what is ultimately determined to be a legal killing of his daughter—the legality of which only emerges after he brings her head to the site of the law. Much as the forest invoked by the Physician in his aside on poaching imagines a strictly delimited space in which deer are rendered killable by their spatial location, but only by particular agents, so does a domestic stage elaborate Virginius's exceptional defense of paternal legal privilege. As we shall see, the Physician portrays such an instance of exceptional sovereignty in order to condemn it, with the tragic figure of Jephthah highlighting the equally self-destructive realms of state and domestic power, whose thanatopolitics might be dislodged only by a future state dominated by medical, rather than military, logic.

---

40. In *Sir Gawain and the Green Knight,* Bertilak shows through the display of animal parts his mastery over the life-forms on the grounds of his castle, Hautdesert. In a striking instance of such deployment of animals to assert his lordly power, he ceremoniously has a "bores hed" [boar's head] carried before him while entering the castle (1616), ed. and trans. William Vantuono, rev. ed. (Notre Dame, IN: University of Notre Dame Press, 1999).

41. The Roman law of persons stands as a particularly clear example of the *bios-zoē* distinction: the same physical bodies (*zoē*) emerge as distinctly relative to whatever legal decisions and customs define a particular individual (*bios*). On the Roman law of persons, see Borkowski and du Plessis, *Textbook on Roman Law,* 86–90; on the distinction between *zoē* as merely material and *bios* as the cultured "way of life proper to men," see Agamben, *Homo Sacer,* 66.

Countering Thanatopolitics:
Imagining a New Juridico-Medical Remedy

In observing that the Physician's "studie was but litel on the Bibel" (I.438) [study rarely included the Bible], Chaucer's narrator does more than simply suggest that this pilgrim is a secular-minded professional whose practice required familiarity with intellectual materials that often derived from pagan Greco-Roman and Islamic Arabic sources.[42] The *General Prologue* portrait of a highly educated practitioner who clearly scorns biblical learning situates the Physician in direct competition with many of those clerks on the Canterbury pilgrimage whose social authority is explicitly tied to Christian textual culture. Indeed, the Physician's being paired in Fragment VI with the Pardoner, whom H. Marshall Leicester Jr. has influentially read as a "disenchanted" subject whose ironic attitude toward biblical texts reveals a Weberian sense that Christianity's power is primarily invested in institutional rather than spiritual structures,[43] suggests that the narrative moment occupied by the Physician features an explicit attack on ecclesiastical sources of power. Even as the Physician's disinterest in the Bible registers the internal challenge to exclusively Judeo-Christian training in an English profession "dominated" by the clergy until the mid-fifteenth century,[44] such highlighting of his relation to biblical textuality intensifies the significance of his sole sustained biblical reference—to the bloody tale of Jephthah who, as both judge and father, ritually slaughters his daughter.[45]

While Patterson rightly reads the Physician's reference to Jephthah as suggesting Virginius's "culpability" in the killing of his daughter, he prefers to read the comparison of the murderous biblical judge with the self-righteous Roman father as coming from a Chaucer safely behind the scenes: for Patterson, the Physician seems "unaware" of the "pre-

42. On the pivotal influence of Greco-Roman and Muslim Arabic traditions in the rise of medieval Christian medicine, see Siraisi, *Medieval and Early Renaissance Medicine,* 1–16.

43. See H. Marshall Leicester Jr., *The Disenchanted Self: Representing the Subject in the Canterbury Tales* (Berkeley: University of California Press, 1990), 35–64.

44. Rawcliffe, *Medicine and Society,* 110.

45. For valuable surveys of commentary on Chaucer's use of the Jephthah narrative, see Sonya Brockman, "The Legacy of Jephthah's Daughter: Chastity, Sacrifice, and Feminine Complaint in Chaucer's *Franklin's* and *Physician's Tales,*" *Medieval Feminist Forum* 46, no. 2 (2010): 68–84; and Richard Hoffman, "Jephtha's Daughter and Chaucer's Virginia," *Chaucer Review* 2 (1967–68): 20–31.

sumption" of his characters to enact a divinely authorized violence.[46] I would argue differently—that the Physician is fully aware of the conflation of divine and human authority in the actions of his tale and that this single biblical reference made by the decidedly non-biblical Physician is intended to resonate, saturating the tale with its pointed commentary on Virginia's tormentors. Much as the earlier reference to poaching law illuminates the conjunction of legal limits and living bodies as central to the tale, so does the Physician's striking deployment of the biblical story of Jephthah expose the tale's interest in a destructive biopolitics shared by state and domestic patriarchs.

At the heart of the Physician's turn to biblical discourse is the technique of analysis—in its literal sense of breaking up a whole into its constituent parts. Jephthah's appearance forces us to see the seemingly opposed and autonomous individuals, Apius and Virginius, as possessors of the same unitary, patriarchal agency. Having analyzed the Hebraic Judge Jephthah, who fulfilled the dual roles of judge and warrior-king, as the judge Apius and the domestic lord Virginius, the Physician portrays a tragic recurrence of the same. As long as societies remain trapped in a system that conflates the roles of father, judge, and warrior, innocent daughters will die. Spoken by a figure whose profession is fundamentally geared toward preserving life, the Physician's narrative condemns the thanatopolitics of both the biblical and classical worlds that ground medieval Western culture.

Parallels between Apius and Virginius, who each stand in judgment over Virginia, are in one respect quite transparent. As many scholars have noted, explicit legal language links Apius's fiendish manipulation of Roman law to declare Virginia a slave and therefore transferrable property to Virginius's own assumption of the role of a legal judge over Virginia: responding in kind to Apius's act of judgment, Virginius pronounces to his daughter, "Take thou thy deeth, for this is my sentence" (VI.224).[47] In addition, a more subtle parallel runs through the Physician's compact but highly wrought tale, signaling a patriarchal mode of authority that thrives on exceptional rule. The key word "pitous," which becomes a force that passes through bodies as readily as laws in a fundamentally biopolitical circuit connecting judge, knight, and audience,

---

46. Patterson, *Chaucer and the Subject*, 369n; see also 370.

47. Angus Fletcher makes the fascinating suggestion that such legal sentences are continuous with the larger literary "sentence" of the Physician's tale, with Virginia alone unable to participate in passing any sentence; see "The Sentencing of Virginia in the *Physician's Tale*," *Chaucer Review* 34, no. 3 (2000): 304–6.

clarifies the Physician's aggressive attack on the clerical and knightly elites who dominated the thanatopolitical medieval state. Referring both to deadly legal force and to the prerogative of surviving males to express sorrow concerning innocent females' utter subjection, the term "pitous" points to the Host's active participation in a critique of patriarchy's self-maintenance via female bodies.[48]

In the grotesque moment of justifying his barbarous honor-killing of his daughter, the Physician's Virginius deploys the language of exception, framing his fatal action as due to the love and mercy usually linked with life: "For love, and nat for hate, thou most be deed [dead]; / My pitous hand mot smyten of thyn heed" [must cut off your head] (VI.225–26). While Virginius's description of his familial decapitation as merciful—as "pitous"—might initially seem merely a local critique from a horrified Physician-storyteller, the echo of this word both within and beyond the tale shows that the uncertainty embedded in the concept of the pitiable illuminates the politicotheological core of the *Physician's Tale*. It is not just the resolute enforcer of domestic legal rights who is "pitous" but also the law of the state: blurring, in biopolitical fashion, the boundaries of texts and bodies, the "bille" [legal bill] that dooms Virginia is also described as "pitous" (VI.166). With the judge's bill and Virginius's sword each characterized by a sense of pity that defies the distinction between pity and horror, the verbal echo of these two judges points to a thanatopolitics according to which the innocent must die because of the arbitrary power of force-bearing male judges. This kind of pity is, after all, the same sort of language key to the discourse of *vitae necisque potestas*, which assumes that he who holds legal force maintains an essentially exceptional power to save by killing—that what is seemingly destructive actually fulfills the positive function of enabling citizens to enjoy their own domestic sphere of power. Pity connects the murderous hand of the legalistic father with that of the legal judge, in a language whose sacralization of paternal power was central to maintaining the exceptional status of free Roman males.

---

48. For seminal analyses of female subjection to patriarchy in the *Canterbury Tales,* see Carolyn Dinshaw, *Chaucer's Sexual Poetics* (Madison: University of Wisconsin Press, 1989), 88–155; Fradenburg, *Sacrifice Your Love,* 155–75; and Elaine Tuttle Hansen, *Chaucer and the Fictions of Gender* (Berkeley: University of California Press, 1992), 26–57, 208–92. For a sustained argument that Virginia as virgin serves allegorically to reveal a patriarchal system that simultaneously desires and destroys femininity, see R. Howard Bloch, *Medieval Misogyny and the Invention of Western Romantic Love* (Chicago: University of Chicago Press, 1991), 93–111.

By having a key audience member of the Canterbury pilgrimage echo this crucial word, the Physician links ancient Rome and his contemporary England as each in exclusively male possession of an exceptional legal power that precedes any distinction between horror and mercy as regards the pitiable. In a move that exposes the Physician's careful restriction of pity to the male holders of only superficially distinct modes of legal force,[49] Chaucer has his Host participate in this circuit of pity. Reflecting on the tale that has just been presented before the Canterbury pilgrims, the Host refers to Virginia's death as having "pitously" transpired (VI.298), and expresses his great "pitee" (VI. 317) [pity] for her. The Host, whom Carolyn Dinshaw analyzes as a figure for the patriarchal center grounding the *Canterbury Tales*,[50] highlights through his double repetition of the verbal link between Apius and Virginius the exceptional nature of clerical and knightly assertions of juridical right. Both the knight-father and the judge deal in a legal power that precedes the distinction between horror and mercy, in a world where all political might is as saturated with pitiable death as the alternately joyous and murderous Jephthah who haunts the tale. The "fadres pitee" (VI.211) [paternal pity] running through Virginius's heart is shown to be just as deadly as the "cursed" (VI.176) and "pitous bille" (VI.166): both lead to Virginia's suffering and death. Pity often serves as a biological marker of nobility in the *Canterbury Tales*; indeed, the Riverside edition describes the verse, "For pitee renneth soone in gentil herte" (I.1761) [For pity comes easily to a noble person's heart], as "Chaucer's favorite line."[51] The prominence of this term as a physical sign of class throughout the *Canterbury Tales* suggests that the Physician's audience would readily note both its identifying purpose and its restriction to male power-players, who themselves become interrelated with the deadly laws and judgments that they enact.

Voiced by an agent of life—a medical practitioner who, however much he profits from sickness and death is, nevertheless, essentially committed

---

49. In discussing that "favorite Chaucerian theme of pity," Thomas B. Hanson calls attention to the Physician's refusal to refer to Virginia's death proper as pitiable and attributes this to the Physician's cold-heartedness, rather than, as I assert, to the doctor's careful deployment of pity as a key to patriarchal circuits of power; see "Chaucer's Physician as Storyteller and Moralizer," *Chaucer Review* 7, no. 2 (1972): 135–37.

50. On the Host as a figure for patriarchy within the *Canterbury Tales,* see Dinshaw, *Chaucer's Sexual Poetics,* 93–94.

51. Benson, *Riverside Chaucer,* 834n. In the first of the *Canterbury Tales,* the *Knight's Tale,* "pitee" [pity] is equated with "gentillesse" [nobility] by the Theban widows (I.920), and we get the first rendition of Chaucer's oft-repeated linkage of nobility with pity, "For pitee renneth soone in gentil herte" (I.1761).

to bringing back humoral balance and restoring health—the *Physician's Tale* portrays Apius's legal state and Virginius's knightly domain as autocratic spheres of death. That both institutions—the patriarchal state and the patriarchal family—lead only to the death of the innocent suggests the need for a new, affirmative, life-based politics, in which physicians have more authority than tyrannical, atavistic judges like Apius, Virginius, or Harry.[52] Exposed as equally arbitrary through their participation in the deadly circuit of pity, Apius, Virginius, and Harry serve negatively to argue for some other mode of governance that could provide an alternative, life-based outcome for Virginia. The Physician is not, as Patterson assumes, simply an empty conduit for Chaucer, with his Jephthah reference merely a sign of his biblical clumsiness: rather, he explicitly indicts the death-saturated worlds of both pagan Rome and the Christian Middle Ages, in which corrupt clergymen, aristocrats, and businessmen can all present themselves as—to take up the language describing Harry in the *General Prologue*—both "governour" and "juge" (I.812–13). The Physician's use of the figure of Jephthah to conflate secular and clerical authorities who take up the separate yet related roles of warrior-king and judge and his jarring use of an exclusively male pity over legitimized acts of legal violence might at first seem entirely negative. However, by listening to Virginia's own questioning voice, we can identify another, more hopeful vision generated by the grisly scenes of a Roman world chillingly similar to the clergy-dominated medieval West.

In the build-up toward Virginius's grisly honor-killing, Virginia asks a poignant question that moves beyond mere pathos to stand as the clearest indication of the Physician's political agenda: she asks, "Is ther no remedye?" (VI.236). In this moment of intense deliberation—in an interval of time absent due to the opportunistic nature of the act in Livy's original recounting or in the rushed narrative of Jean de Meun's version—Virginia uses a word that blurs the boundaries between the legal world at the heart of the tale and the medical profession informing the Physician's worldview. Suggesting either a legal mechanism or a medicinal treatment, Virginia's "remedye" directs itself not only to either of the aggressive lawmen in the tale (whether the corrupt judge Apius or the merciless father Virginius) but also to the future Physician voicing her "historial" (VI.156) [historical] tale. If told by another pilgrim, Virginia's

---

52. Even as he professionally dwells in and profits from plague and death, the Physician would, much like Esposito in his vision of an "affirmative biopolitics," bring to the intersections of biology and law a sense that the informing "norm" must be "life" (*Bíos,* 194).

query might seem merely rhetorical, but both the Physician's professional occupation as a preserver of life and the preceding series of asides that stress the importance of parenting and guardianship clarify that there *is* an answer to Virginia's dilemma—namely, to choose life rather than give in to the inevitability of the thanatopolitics of the day. Virginia's question reveals a futural, biopolitical vision: the Physician, through his very voicing of this legal tragedy, bids us imagine a medicojuridical establishment that will challenge and displace the thanatopolitical limits of either pagan Rome or his own, Christ-centered age.

If the Physician is indeed engaging in what Patterson calls "'counterfeit' hagiography,"[53] then it is crucial that we read his choice of a Roman historical example as a sign that his martyrdom narrative has nothing to do with the soul that is central to Christocentric literary manifestations. Virginia suffers within the jurisdiction in which the Physician has already sketched out his authority—that of the body, with any moral dimensions of Virginia's person conceived as every bit as material as her corporeal self. In a political statement that links Virginia's death with the limitations of a legal world dominated by men such as Apius and Virginius, the Physician implies that a future state informed by medical practitioners such as he could provide a positive answer to Virginia's query: rather than delivering the gruesome legal mechanism of Virginia's head, the Physician—or any judge or government guided by his biopolitical expertise—might offer life. Sketched out through the blood and violence of a thanatopolitical past that, with its patriarchal sharing of clerical and militarist power looks hauntingly like the pilgrims' late medieval England, the Physician's rendition of Virginia's honor-killing allows her own, tragically rhetorical question to suggest a future age that might be able to answer *yes*, that life itself is a remedy that can transcend the death-dealing politics of patriarchal modes of juridical exception.

If ecology is, as Morton suggests, centrally concerned with the fact that all things are connected, then the medieval Physician—who was trained to see everything, from one's moral life to the sores on one's skin, as connected to an environment that stretched from the terrestrial air we breathe to the heavenly spheres whose movements affect every aspect of our lives—is eminently suited for ecological analysis. Chaucer's Physician, whose study of the destructive force of pity moves across the bodies of father and daughter to the legal text of a fateful bill and the literary criticism of a hypermasculine Host, presents a picture of political power that

---

53. Patterson, *Chaucer and the Subject*, 370.

is profoundly skeptical of male exception. The Physician's diagnosis of power depends for its force on a reference that is readily linked with contemporary ecological discourse—namely, to poached deer, whose status as killable bodies chillingly links up with Virginia's own transformation into lively property, subject to legal slaughter by her owner. By transforming Virginia's story into one structured by legal jurisdictions—whether the strictly delimited world of the medieval forest upon which any law against poaching depends, or the domestic sphere to which Virginius retreats in order to assert his possession of an exceptional jurisdiction that trumps Apius's equally horrible authority—the Physician demonstrates how physical life is thoroughly defined by the limits imposed by law. The Physician depicts this nightmarish vision of juridically outlined death in order to invite us to imagine, along with Virginia, a world where a Physician's politics of life can bring remedy where now exists only the clerkly and knightly law of death.

PART II

Objects, Networks, and Land

4

On the Line of the Law

THE LONDON SKINNERS AND
THE BIOPOLITICS OF FUR

Michelle R. Warren

FRANZ KAFKA'S "Before the Law" ("Vor dem Gesetz") presents a provocative enigma. A man asks a doorkeeper for admittance to the Law; he is told that he cannot pass and that, in any case, there will be more doorkeepers, each more terrifying than the last. Although the doorkeeper does not bar the way, the man decides to wait. After many years, with his eyesight failing, he wonders why no one else has come before the Law; the doorkeeper replies that the door was meant only for him and slams it shut.[1] This story has attracted many disparate interpretations. For the politics of ecology, it functions most readily as an allegory of sovereignty. The nature of this sovereign authority, however, remains opaque, and so the tale serves as a flexible cipher for contradictory political philosophies.

If we approach the Law with Michel Foucault in mind, Kafka's tale traces the historical shift from sovereignty (embodied in a law-giving monarch) to biopower (structural state control of the body). The man arrives from the country thinking, "The Law . . . should be accessible to

---

1. Franz Kafka, "Before the Law," *Parables and Paradoxes: Bilingual Edition*, ed. Nahum M. Glatzer (New York: Schocken Books, 1975), 60–65; all citations of Kafka are from this edition. I would like to thank the students in my 2009 literary theory seminar for making me read Kafka, and the students of 2011 for a pivotal conversation about fur. Some new ideas crystallized in conversation with the theory seminar at the Fannie and Alan Leslie Center for the Humanities, Dartmouth College (September 2012). Sources referenced in this article were handled in paper form, except where otherwise noted; subscription-based electronic archives paid for by Dartmouth College.

every man and at all times" (61). In the end, though, he discovers the opposite: he alone did not pass through his own door. The Law is thus individualized, adapted to police only one subject and to do so to perfection. In this Foucauldian vein, Jeffrey Nealon argues that Kafka's text demonstrates the costs of "mistaking" biopower for sovereignty:

> The man from the country performs his relation to power on a sovereign model of law, which is to say he sees power as centralized, housed in a specific place or person. . . . The parable, however, shows us that biopower is already at work everywhere (inside and outside the law, as well as Kafka's text). Unlike sovereign power, biopower is wholly immanent to the socius rather than organizing it from above, or from some central location that's hidden behind everyday social structures (behind Kafka's series of doors). . . . Mistaking biopower for sovereign power is a matter of misdiagnosing a back-and-forth relation of force as the one-way street of sovereignty.[2]

The tale's jarring conclusion turns sovereignty into a narrative fiction that has masked a far more insidious situation, in which the man has all along been a particularized body rather than a generic citizen. In this model, which Foucault elaborated as the modern social condition, power does not eliminate alterity but instead names it with ever increasing particularity so as to contain it more fully.[3] There is nothing more particular than an entire legal structure dedicated to only one man.

In this Foucauldian view, where biopower supersedes sovereignty, the man from the country "fails" because he misdiagnoses his situation. By contrast, Giorgio Agamben sees biopower as coextensive with sovereignty—and so finds in Kafka's tale a story of success. For Agamben, the tale serves as an "exemplary abbreviation" of the sovereign ban—"the law that *is in force* but does not *signify*."[4] Since the doorkeeper tells the

---

2. Jeffrey T. Nealon, *Foucault beyond Foucault: Power and Its Intensification Since 1984* (Stanford, CA: Stanford University Press, 2008), 52.

3. Michel Foucault, *The Birth of Biopolitics: Lectures at the Collège de France, 1978–79*, ed. Michel Senellart, trans. Graham Burchell (New York: Palgrave Macmillan, 2008).

4. Giorgio Agamben, *Homo Sacer: Sovereign Power and Bare Life,* trans. Daniel Heller-Roazen (Stanford: Stanford University Press, 1998), 49, 51; author's emphases. Commentary on philosophical relations between Agamben and Foucault is voluminous; some especially compelling entry points are Malcolm Bull, "Vectors of the Biopolitical," *New Left Review* 45 (2007): 7–25 (www.newleftreview.org), and Rey Chow, *Entanglements, or Transmedial Thinking about Capture* (Durham, NC: Duke University Press, 2012), 81–105 (accessed via e-Duke Books Scholarly Collection).

man that it *is* possible to enter, "but not at this moment," Agamben concludes: "Kafka's legend presents the pure form in which law affirms itself with the greatest force precisely at the point in which it no longer prescribes anything."[5] In other words, the injunction not to enter is not a law but a deferral. The man from the country is excluded from the law, by the law itself—and so paradoxically also subject to it: "According to the schema of the sovereign exception, the law applies to him in no longer applying, and holds him in its ban in abandoning him outside itself."[6] In this view, there is "nothing" to enter because the Law isn't about anything except its own access, its power to exclude those it governs. As a result, the door's closure at the end signifies for Agamben the man's triumph: he has made the doorkeeper act at last, successfully disrupting the force of the Law: "We can imagine that all the behavior of the man from the country is nothing other than a complicated and patient strategy to have the door closed in order to interrupt the Law's being in force."[7] Contrary to the common feeling that "nothing happens" in the story (expressed most extensively in Jacques Derrida's commentary), Agamben concludes, "The story tells how something has really happened in seeming not to happen."[8] The terrifying *state of exception* that made life coincide with law is over. The man from the country becomes a hero who shows that it is indeed possible to outlive sovereignty and even to evade biopower. Pointedly, where many commentators assume that the man from the country dies, Agamben notes that the text says only that he is "close to the end."[9]

In the end, Agamben and Foucault both point toward an optimism latent in Kafka's tale, a view contrary to the critical tradition that equates "Kafkaesque" with crushing forms of oppression. Biopower casts a different light on Kafka. In Nealon's reading, the pervasiveness of biopower includes a promising vulnerability: "We might learn from a Foucauldian analysis of Kafka's parable that we needn't wait for admittance to the law through a central portal. Rather, the 'entrances' to the 'law' of biopower

---

5. Agamben, *Homo Sacer,* 61, 49.

6. Ibid., 50.

7. Ibid., 55.

8. Ibid., 57. The relation between Derrida and Agamben is explored more fully by Catherine Mills, "Agamben's Messianic Politics: Biopolitics, Abandonment and Happy Life," *Contretemps* 5 (2004): 42–62 (http://sydney.edu.au/contretemps, accessed 7 August 2012), and by Kalpana Rahita Seshadri, *HumAnimal: Race, Law, Language* (Minneapolis: University of Minnesota Press, 2012), especially 80–83 on Kafka (Seshadri begins the book with an epigraph from Agamben that refers to Kafka's tale).

9. Agamben, *Homo Sacer,* 55.

are actually and virtually everywhere, as ubiquitous as the form of power itself."[10] In this view, the man is not destroyed, but resistant; he belongs to a community of survivors. From a different line of reasoning, Agamben relishes the promise of the Law's suspension. Once the door closes, sovereignty—and the sovereign ban—are lifted, opening space for a messianic future.[11] The man from the country is the messenger who makes possible the messiah's arrival. Of course, following Agamben's observation about the man's survival, we must also note that the door likewise is only "going to shut": in the last narrative moment, it remains open. Even if the Law's suspension remains thus "in suspense" (perhaps nothing has happened after all), a space has been opened for thinking beyond the Law.

I would like to suggest that neither we as readers nor the man from the country have to wait until the tale's end to find ambiguities that disrupt the Law. No one, in other words, has to wait for the messiah to act. Two biopolitical details, all but overlooked in the tale's voluminous commentary, reorient the representation of sovereignty from almost the beginning: fur and fleas. Before anything else, the sight of the doorkeeper's fur coat sparks the man's decision to wait before trying to pass through the door: "When he looks more closely at the doorkeeper in his furred robe, with his huge pointed nose and long, thin, Tartar beard, he decides that he had better wait until he gets permission to enter" (61). The large nose may intimidate, but the fur draws the line of the Law. A few lines later, describing the doorkeeper's periodic questions to the man, the narrator notes that they are put "impersonally, as great men put questions" (63). Whether or not the doorkeeper is a man of high social standing, he speaks and dresses the part of the lawmaker. Seemingly, he can withhold or give permission.

Dead animal fur tells a different biopolitical story than living human flesh. A clue to its significance comes not from political theory but from Jacques Derrida's playful deconstruction of Kafka's tale. In a parenthetical comment—the only comment I've encountered that addresses the doorkeeper's *Pelzmantel*—Derrida characterizes the fur as "the artificial hair, that of the town and the law, which will be added to the natural hairiness [the beard]."[12] If the doorkeeper represents the urban in con-

10. Nealon, *Foucault*, 53.

11. Agamben, *Homo Sacer*, 56–57; Agamben, *Potentialities: Collected Essays in Philosophy*, ed. and trans. Daniel Heller-Roazen (Stanford, CA: Stanford University Press, 1999), 172–74.

12. Jacques Derrida, "Before the Law," trans. Avital Ronell and Christine Roulston, in *Acts of Literature*, ed. Derek Attridge (New York: Routledge, 1992), 195.

trast to the man from the country, this description of the coat deconstructs the city/country binary. Animal fur is itself perfectly "natural," and the animal likely hails from the country, just like the man. Yet tawed and worked into a coat, the animal becomes urban artifice. The pelted coat becomes the sign of its own opposite: country and town, nature and legislation. The fur, in other words, can reconcile history (Foucault's sovereignty), meaninglessness (Derrida's endless deferral), and transcendence (Agamben's coming messiah) by reorienting the tale's subjects around an object.

Further weakening the line of the law, the coat turns out to support its own ecosystem of animate life. The man from the country, having tried many times over the years to gain permission to enter, makes a final effort by addressing the fleas in the fur: "He grows childish, and since in his prolonged watch he has learned to know even the fleas in the doorkeeper's fur collar [*Pelzkragen*], he begs the very fleas to help him and to persuade the doorkeeper to change his mind" (63). Speaking to the bugs is the sign of the man's final desperation and, perhaps, the sign of another "mistake"—pests taken for subjects. On the flip side, the man's begging interpolates the fleas into the human community; the parasites of the (dead) animal become advocates before the Law. They inhabit, literally, the locus of social difference—the fur that signifies the doorkeeper's status—even as the doorkeeper's own function as mediator between the man and the Law has turned into a charade that forecloses communication ("questions put impersonally"). Language passes the time, not the meaning, under a law in force with no signification.

In this world, fleas are as a good a lawyer as any. When we take them into account, though, they shift the terms of sovereignty. On the one hand, the fleas support traditional "Kafkaesque" pessimism: when the man speaks to them, he reveals the depth of his hopelessness before the final failure. On the other hand, if we imagine with Agamben a time after the Law, the fleas become part of a liberated interspecies community. The group of survivors (the man, the former doorkeeper, the fleas) will soon have to forge new social relations outside the Law. Will they do so as equals or will new hierarchies emerge? Or will they invent some other form? Perhaps they will exemplify what Rey Chow and Julian Rohrhuber have called "intermedial capture"[13]—an aimless, anarchic existence that remains after the Law. Captivated in a "heteronomous affective assemblage,"[14] they are suspended in a whorl of time structured by the

13. Chow, *Entanglements*, 52.
14. Ibid., 55.

tale itself yet beyond its own bounds. Perhaps, in a more Foucauldian version of capture, nothing will change: they (and we, the readers) will remain captured within the labyrinth of doors, even when the doorkeepers have left.[15] In the end, we do not know if the fleas replied to the man's entreaties or if they exerted their power through silence, or even indifference.

The ordinary lifecycle of the flea is also relevant here. By any standard, these are exceptionally long-lived fleas if they have accompanied the man all these years. They have "passed time" in every sense, gradually teaching the man to know them. They cannot live, though, by dead animal fur alone—or actually at all. Fleas require warm blood. They have, then, made the doorkeeper their subject all these years, nipping at the skin at the edges of the collar. Notoriously good jumpers, they have perhaps also ventured the short distance to the warm body from the country. They have been sovereign all along, periodically infused with the blood of the men who think that their destinies lie elsewhere. In this sense, they resemble the tick that occupies Agamben in *The Open*—capable of intense relationships with their world; capable of extended periods of waiting.[16] Indeed, perhaps they have already brought the two men together, mingled their blood, dismantled the Law through organic insurrection. The door that seems to structure the relationship between the men does not exist in the fleas' world; instead, the fleas themselves make the men relate—as the man's question indicates. They are not merely mediators but sovereigns disposing of the bodies of their subjects. The fleas teach us to look "elsewhere" than the Law for fundamental social structures—not necessarily to look "deeper" but perhaps more closely, with magnifying glass in hand.

The more I've thought about Kafka's enigma, the more I've wondered what it might suggest to us about the intersection of law, fur,

---

15. The parable might illustrate "the subject's capture by a power entirely indifferent to him. . . . The force of this indifferent power is therefore entirely owing to the subject's own immobilizing fascination with its empty majesty" (Sergei Prozorov, *Foucault, Freedom and Sovereignty* [Aldershot, UK: Ashgate, 2007], 75–76).

16. Giorgio Agamben, *The Open: Man and Animal,* trans. Kevin Attell (Stanford, CA: Stanford University Press, 2004), 45–47. The fleas and ticks embody a model of interspecies relation quite different from Donna's Haraway's *When Species Meet* (Minneapolis: University of Minnesota Press, 2007)—where the emphasis on communion with dogs curiously never mentions fleas, except in passing as a modifier for their own destruction (poisonous "flea products" [50]). Parasites such as fleas provoke a rather different approach to biopolitics and animal studies—one that must reckon with Michel Serres's *Le parasite* (which begins, appropriately enough, with the fable of the country rat and the city rat [Paris: Grasset, 1980]).

and social status in the Middle Ages. How could the doorkeeper's flea-ridden coat shape an understanding of medieval fur? In London, regulations involving fur and other luxury clothing are frequently understood to have no force: they are periodically reissued because no one is abiding by them—or perhaps because everyone is. And yet they stand in force. Perhaps they are emphasized *especially* when they are not enforced. The moralized discourse of luxury, moreover, can turn fur into a matter of religious orthodoxy—and thus a matter of life and death in the late fourteenth and early fifteenth centuries. Meanwhile, the purveyors of furs—artisans and merchants of the Skinners' Guild—stand on the line of the law with every sale.

The doorkeeper's fur coat points toward the interplay of biopower and sovereignty, rather than to their periodization (Foucault) or transcendence (Agamben). Fur takes us right to the line that supposedly divides nature and culture, country and town, animal and human, *zoē* and *bios*. Practices around what might be called *necrofashion* address, head-on, the discourse of death in biopolitics, pushing its target from the powers of the *one true sovereign* to the distributed effects on every citizen's role in discontinuous yet effective circuits of sovereignty—perhaps even beyond the citizens to the parasites who also rule the city.

Sumptuary Discourse

Sumptuary regulation is the most literal place to explore the "line of the law" as it is shaped by fur. Sumptuary discourse, in the form of both petitions and the statutes that sometimes resulted from them, relies on both sovereignty and biopower. In late medieval England, we get a glimpse of how sovereignty gathered biopower to itself and even how "those below" begged the sovereign to extend the Law over their own bodies. In this way, sumptuary discourse reconfigures Foucault's periodization of power: instead of a slide from (medieval) sovereignty to (modern) biopower, we can perceive their entanglements and mutual dependencies.

Sumptuary regulation, especially when it targets fur, masks the multimodalities of the socius behind a simpler form of linear hierarchy that is more comfortable for everyone. The petitions and statutes are about keeping everyone in the place they actually most desire, freeing them from the daily negotiation of status and the oppressions of competitive fashion. Nealon's description of Foucauldian biopower describes well the development of sumptuary discourse: "Power increasingly comes to target the economic relations among bodies, rather than the bodies 'themselves.'"[17] Petitions are precisely a "tactic" in Foucault's sense:

> What enabled sovereignty to achieve its aim of obedience to the law, was the law itself; law and sovereignty were absolutely united. Here [with governmentality], on the contrary, it is not a matter of imposing law on men, but of the disposition of things, that is to say, of employing tactics rather than laws, or, of as far as possible employing laws as tactics; arranging things so that this or that end may be achieved through a certain number of means.[18]

Foucault may be addressing "modernity" here, but this description describes well the "tactic" of petitioners who endeavored to use the law to achieve specific social and economic goals. The petitionary process itself suggests that law and sovereignty were not always so united—or if they were, that the sovereign did not only enforce.

Sumptuary petitions and statutes stretching from 1337 to 1510 trace a process that converted biopower into sovereignty. The regulations define precisely how clothing materials, including fur, should match the social standing of the bodies that wear them. Kim Phillips describes how petitions arising from the House of Commons were, in fact, articulated in collaboration with the Lords and Council. Phillips concludes that despite their appearance of comprehensiveness, the regulations target primarily the upper-middle echelons of knights, gentry, and merchants (those with rank and/or money): this is where distinctions matter most because they are most prone to confusion.[19] In other words, those most subject to regulation were also those most involved in creating regulations. Phillips argues that petitions originated in large part with knights coming from the counties to join parliament, where they interacted more regularly with a broader swath of society and so experienced increased pressure to maintain their own status: "Knights who were used to occupying positions of hegemonic masculinities in their own communities found them-

---

17. Nealon, *Foucault*, 53.

18. Michel Foucault, *Security, Territory, Population: Lectures at the Collège de France, 1977–1978*, ed. Michel Senellart, trans. Graham Burchell (New York: Picador, 2007), 99.

19. Kim M. Phillips, "Masculinities and the Medieval English Sumptuary Laws," *Gender and History* 19 (2007): 24–25 (Wiley Online Library, accessed 13 April 2012). Phillips usefully summarizes the relevant petitions and statutes (33–37), with reference to *The Parliament Rolls of Medieval England, 1275–1504 [PROME]*, ed. Christopher Given-Wilson (Leicester: Scholarly Digital Editions, 2005) and *The Statutes of the Realm* (London: Dawsons, 1810–28). Citations here are from *PROME* (accessed throughout April 2012).

selves subordinate to the Lords."[20] And so they negotiated to occupy
a well-restricted middle ground: they acknowledged the superiority of
the titled lords while ensuring that they had their own inferiors; they
legitimized their own rank by appealing to the traditional aristocratic
hierarchy.[21] In drawing the lines of exclusion, they exercised a structural
sovereignty from outside the Law. By endeavoring to dispose of "things,"
they sought control of their own social and political standing. As both
"men from the country" and doorkeepers to the law, knights in parliament participated in defining their own sartorial and corporeal disposi-
tions. They simultaneously manipulated the Law and performed their
own subjection.

One of the key markers of rank defined in sumptuary is fur. Indeed,
in the earliest petition of 1363, the fact that everyone is wearing fur (*pel-
lure*) is identified as the one practice that most disturbs the desired categories of social difference and affordability.[22] In general, petitions divide
society into two groups: those who can wear ornamental fur and those
who cannot. Sometimes the line follows rank (the adorned must be at
least a lord), sometimes annual income (at least £100 in 1337, at least
£40 in 1379). Additionally, those designated as fur-wearers are subdivided
according to relative class of pelt. The distinction between knights/may-
ors and esquires/gentlemen/burgesses is made partly according to the
types of furs they wear—and all are restricted from the most luxurious
furs.[23] The matching of fur types to social ranking is most elaborated in
the petitions of 1402 and 1406.[24] In both cases, ermine and weasel (and
marten in 1402) are reserved for the highest ranks (banneret or higher).
Clerks, esquires, and their wives (and a host of other citizens in 1406)
are further excluded from miniver, greys, and browns. Finally, valets are
the lowest rank allowed to wear fur; in this case only, the petitions offer
a list of what is allowed (rather than what is excluded): lamb, fox, and
rabbit (and otter in 1402). The class unnamed here would be ordinary
knights—of sufficient standing to have their own esquires and valets, but
not themselves leaders or lords. Presumably they wore a lot of miniver,
grey, and brown (in 1399, new knights are described as wearing hoods

---

20. Phillips, "Masculinities," 29.

21. Ibid., 33.

22. *PROME,* Edward III, October 1363 (Membrane 3, Item 25).

23. Phillips, "Masculinities," 25.

24. *PROME,* Henry IV, September 1402 (Membrane 6, Item 76); March 1406 (Membrane 7, Item 110).

of miniver[25]). While society, then, had many ranks and income levels, the animals are divided into only three groups. Once this animal typology becomes the basis for a social typology, human society also appears to have three main levels. The petitions are at pains to establish clear parallels between the hierarchies of the natural world and those of the social order.

Every rule, however, is soon followed by various exceptions—for uniforms, social role (royal servant, civic official, scholar, etc.), citizenship, sometimes by individual name, and so on.[26] Sometimes the list of exceptions is longer than the rule. In both 1402 and 1406, after delineating what only bannerets and above can wear, the petitions recognize a blanket exception for men-at-arms when they are armed, who may dress however they wish ("q'ils purront user vesture ce qe lour plerra"). The petition of 1483 includes far fewer exemptions—and so the king adds a number of individual names in his reply (unranked members of his household, for the most part). In other words, certain public duties justified the display of high-status clothing as a sign of the social status of the office itself. Fur, quite specifically—in parallel with Kafka's doorkeeper—marks accession to new roles under and within the Law.

Skinners are not named in any of the sumptuary petitions or regulations. In passages on enforcement of the rules, however, other crafts are singled out as responsible for not making or selling items to those who should not have them—drapers will be punished according to the king's judgment (1363), tailors are threatened with imprisonment and fine (1406), and both tailors and cobblers may be fined (1406, 1463). Craftsmen and merchants are thus meant to function as enforcers alongside magistrates. As providers of furred ornament, skinners likewise occupied a potent "doorkeeping" position on the line between fashion and status. Indeed, when the 1406 petition envisions excommunication for those contravening the law, temporal infractions became spiritual—and craftsmen could turn transgressive fashionistas into full-fledged outlaws. Of course, as Phillips and others have pointed out, there is no evidence of either enforcement or prosecution before the mid-sixteenth century.[27] Evidence from wills suggests that people followed the rules,[28] by and

---

25. *Chronicles of London,* ed. Charles L. Kingsford (London: Oxford University Press, 1905), 49 (edited from Cotton Julius B II, f. 45).

26. Phillips, "Masculinities," 25–26.

27. Ibid., 23.

28. Kristen M. Burkholder, "Threads Bared: Dress and Textiles in Late Medieval English Wills," *Medieval Clothing and Textiles,* ed. Robin Netherton and Gale R. Owen-Crocker (Woodbridge, UK: Boydell & Brewer, 2005), 133–53.

large, which suggests in turn that the rules themselves were useful to everyone (despite complaints in the petitions about cross-class dressing and economic ruin, or moralizing condemnations of luxuriant excess[29]).

Through the lens of biopolitics, sumptuary legislation turns a forest-based clothing material into a tool of urban hierarchies. The producers and sellers of pelts sit on the line of the law. By legislating the social use of fur, the petitions and statutes turn nature into culture. Furs are supposed to clarify status distinctions, but the commodification of both constantly challenges social hierarchy. The distinction that is threatened is not so much one between man and animal, but between men. At stake in sumptuary concerns, according to Phillips, are "homosocial cartels of power."[30] The merchant furriers stand on this line—with all the power to break down social divisions in the urban space, where furs can be sold and seen. The portability of any status conferred by a garment—like the doorkeeper's fur coat—suggests both the emptiness and ubiquity of the Law. Anyone can wear the coat; in the absence of the coat, the man from the country might have just walked in.

Luxury Values

A particularly dramatic debate over the line of the law took place "before the law," when the *Twelve Conclusions of the Lollards* were posted on the doors of Westminster Hall while Parliament was in session (27 January–15 February 1395). Perhaps a doorkeeper kept them from entering; perhaps the Law is most vulnerable on the outside where all can see. Kafka's doorkeeper, in any case, draws our attention to the twelfth conclusion—an accusation against the immorality of luxury:

> Þe multitude of craftis nout nedful usid in our Chirche, norsschith michil synne in wast, curiosite and disgysing. þis schewith experience and resun provith, for nature with a fewe craftis sufficith to nede of man. þe correlari is, þat sytthin seynt Powel seyth, we havende oure bodili fode and hilling we schulde holde us apayed, us thinketh þat goldsmethis and armoreris

---

29. John Hardyng, *Chronicle,* ed. Henry Ellis (London, 1812), 346 (Google Books, accessed 6 August 2012), 347 (critique of Richard II's court); Matthew Giancarlo, "Dressing up a 'Galaunt': Traditional Piety and Fashionable Politics in Peter Idley's 'Translacions' of Mannyng and Lydgate," in *After Arundel: Religious Writing in Fifteenth-Century England,* ed. Vincent Gillespie and Kantik Ghosh (Turnhout, BE: Brepols, 2011), 429–47.

30. Phillips, "Masculinities," 24.

and alle manere craftis nout nedeful to man aftir þe apostle schulde ben distroyd for þe encres of uertu.[31]

The criterion of "need" raises much the same issue as Derrida's comment on artificial hair: what is the line between nature and artifice? "A few crafts" are natural to human needs, but others are not. Goldsmiths and armorers are specifically excluded as "not needful": they promote wasteful spending, lead the mind astray, and disguise the body. Furs can perform all three of these task as well, and yet they are also "needed" for warmth. They thus resist the strict categorization required to distinguish between orthodoxy and heterodoxy.

When Roger Dymmok sets out to answer this conclusion in his *Liber contra XII errores et hereses lollardorum,* he engages the spirit of sumptuary discourse. He defends orthodoxy by defending luxury. Dymmok posits the "needfulness" of lords to be better housed and better fed than those they rule. In parallel, he turns to the usefulness of dress in maintaining social order:

> Thus it is right that princes and nobles, in accord with the decency of their estate, be housed in greater magnificence and fed more sumptuously than the remainder of the population; just so, it is likewise right that such be also arrayed more lavishly than the rest, in raiment of varied ornament. For it is not accordant to reason that a servant be dressed as splendidly as a lord, or an ordinary knight as a prince, or a monk as a secular. Rather, just as persons are distinguished from one another in estate and dignity, so reason requires that they be distinguished from one another also by means of different styles of dress.[32]

> Et sic oportet principes ac nobiles secundum sui status decenciam in edificiis et esculentis excellencius ordinari delicaciusque nutriri quam residuum populi; ita eciam oportet ipsos amplius quam ceteros diuerso ornatu uestium decorari, non enim est congruum racioni, quod ita splen-

---

31. In Roger Dymmok, *Liber contra XII errores et hereses lollardorum,* ed. H. S. Cronin (London: Kegan Paul, Trench, Trübner, 1922), 292. The use of St. Paul here to define a heterodoxy of "basic needs" points us back to Agamben's reliance on St. Paul to theorize a messianic time beyond the Law.

32. Dymmok, "Against the Twelve Errors and Heresies of the Lollards, part twelve," in Richard Maidstone, *Concordia (The Reconciliation of Richard II with London),* ed. David R. Carlson, trans. A. G. Rigg (Kalamazoo MI: Medieval Institute Publications, 2003), 112; also online, http://www.lib.rochester.edu/camelot/teams/maidap3f.htm (accessed 5 June 2012).

dide uestiatur seruus ut dominus, simplex miles ut princeps, monachus
ut secularis, set sicud distinguntur homines in statibus et dignitatibus,
sic racio exigit, ut diuersorum uestimentorum apparatu distinguantur.
(Dymmok, *Liber* 294)

Clothing reflects here preexisting differences between persons, rendering
vibrantly visible an underlying biopolitical "truth" of social stratification.
Disguising or "decorating" the body ("corporis decorum," *Liber* 292) is
not a problem if done for proper rather than sinful purpose. Later on,
Dymmok makes explicit the inherent orthodoxy of the craft itself, as the
material practice that makes virtuous luxury possible:

> So, just as persons can lawfully array themselves sumptuously and art-
> fully, in keeping with the proprieties of their estates, by the same token
> the artisans who make such adornment possible can lawfully pursue their
> crafts; moreover, such artisans are to be not done away with, but allowed
> or even encouraged, as needful collaborators for persons who would carry
> out their social and civic duties. (Dymmok, "Against the Twelve Errors"
> 116–17)

> Et sicud homines se possunt licite secundum sui status congruenciam
> ornare sumptuose et artificiose, ita artifices talium ornamentorum licite
> possunt suas artes exercere, et tales artifices non sunt destruendi set per-
> mittendi et fouendi, ut necessarii coadiutores hominum in conuersacione
> eorum politica et ciuili. (Dymmok, *Liber* 297)

The logic here passes from the users to the makers, positing directly their
"needfulness" as "collaborators" in the political system. Without "sumptuous and artful ornament," the argument goes, kings and lords would
be unable to carry out their duties and society would fall apart. Dymmok
goes on to give a specific defense of the goldsmiths and armorers named
in the twelfth conclusion (302–4). Throughout these passages delineating
the propriety of social difference, Dymmok could be drafting a new petition for sumptuary regulation. Indeed, he probably did identify with the
gentry class most concerned with stabilizing social hierarchies.[33] For him,
luxury goods—and the craftsmen who make them—become the foundation of good governance.

---

33. Fiona Somerset, *Clerical Discourse and Lay Audience in Late Medieval England*
(Cambridge: Cambridge University Press, 1998), 108, 120–21.

The point is timely, given Richard II's reputation for excessive luxury and the many criticisms he attracted for it.[34] Following Dymmok's reasoning, this propensity was not only acceptable but the very illustration of the highest morality and a principled defense against heresy. As John Bowers characterizes Dymmok's response: "The art of spending lavishly fortified the social order by teaching people to respect their superiors while discouraging the prince's rivals for power."[35] The political value of Dymmok's argument is nicely symbolized by the copy of his treatise that opens with a portrait of Richard II—seated in majesty with an ermine-lined mantle.[36] The flecked fur cascades from the collar to the floor, with a generous swath across the lap. The image evokes other sumptuous images of Richard, including his Westminster Abbey portrait and the Wilton Diptych (where his royal ancestors get all the fur, and he stands out with an intricately wrought woven robe).[37] These ostentatious displays of "needless craft" clearly put luxury on the right side of the Law.

The Skinners' Guild

Dymmok's defense of luxury justified "all manner of crafts," utilitarian or not. As purveyors of both ordinary and extraordinary furs, the Skinners could benefit from just about any argument: their product kept people warm, maintained useful social distinctions, and fed the luxuriant appetites of extravagant royals. The Skinners' Guild occupied a nearly unique position in the London civic order by supporting social stratification while allying themselves across hierarchical divisions. Sumptuary restrictions might have served the conservative interests of urbanized knights, but the Skinners themselves benefited from the gradual blurring of boundaries between the aristocracy and the merchant elite. Sumptuary

---

34. His consumption of fur stands out: 1,594 skins of minever for several robes in 1377; 36 ermine skins for a "mantelett" that covers the shoulders, 702 pured minever skins for one robe, a "capp" of 16 pured minever skins, and 5 ermines (Kay Staniland, "Extravagance or Regal Necessity? The Clothing of Richard II," in *The Regal Image of Richard II and the Wilton Diptych*, ed. Dillian Gordon, Lisa Monnas, and Caroline Elam [London: Harvey Miller, 1997], 88–89).

35. John M. Bowers, *The Politics of Pearl: Court Poetry in the Age of Richard II* (Cambridge: Boydell & Brewer, 2001), 28.

36. Cambridge, Trinity Hall, MS 17, f. 1r; reproduced in *The Cambridge Illuminations: Ten Centuries of Book Production in the Medieval West*, ed. Paul Binski and Stella Panayotova (London: Harvey Miller, 2005), 284.

37. London, Westminster Abbey, c. 1395; reproduced in *The Regal Image*, fig. 1, pl. 2.

regulations place the *makers* of luxury goods on the "line of the law" in several ways: they are "on the line" to uphold distinctions, and they are "on the line" that makes those distinctions even possible. In this interplay, the craft market functions as sovereign in a circuit of law that operates without the sovereign at the center. The sovereign and sovereignty function as competing, mutually cancelling, yet codependent systems. The Skinners sit in the structural location of this "exception."

The Skinners were one of the oldest guild establishments in London.[38] They first petitioned for royal recognition in 1327, asking Edward III to define the types and sizes of furs, and also to authorize punishment of those who might try to deceive the public with used or mislabeled materials. The petitioners sagely targeted their request as beneficial to the "comun profit des grantz et del communaute du Reaume," an expression that combines deference for the powerful ("des grantz") with the general welfare.[39] In response to the petition, Edward III granted the Skinners their first charter, incorporating many of the provisions for craft regulation detailed in the petition.[40] Later records show that disputes over the lawful conduct of the craft followed more or less these terms into the sixteenth century: cases involve counterfeit furs, deceitful sales, and unlawful mixing of different types of fur.[41] The guild itself was explicitly authorized to enforce the provisions of trade and to elect officers to search and inspect all relevant work.

The company took advantage of the second mayoralty of the skinner Adam de Bury in 1365 to register a city ordinance reinforcing some of the practical terms of the Charter: the main concerns were the mixing of furs, monitoring of "wild work" brought from the country into the city, and overall honesty in trade.[42] In a world where you could get pillo-

---

38. Histories of the Skinners provide summaries and extracts of many of the records referenced below: William Herbert, *The History of the Twelve Great Livery Companies of London*, 1834–37 (New York: Kelley, 1968); Henry Thomas Riley, *Memorials of London and London Life* (London: Longmans, Green, and Co., 1868); James Wadmore, *Some Account of the Worshipful Company of Skinners of London, Being the Guild or Fraternity of Corpus Christi*, 1876 (London: Blades, East, and Blades, 1902) (Google Books, accessed 27 June 2010); John James Lambert, *Records of the Skinners of London, Edward I to James I* (London: George Allen and Unwin, 1934).

39. National Archives, SC 8/260/12977 (http://discovery.nationarchives.gov.uk). Several other companies also asked for ordinances this year, the first of Edward's reign.

40. *Calendar of the Patent Rolls*, Edward III (1:34, membrane 18) (London: H. M. Stationary Office, 1891–1916) (http://www.uiowa.edu/~acadtech/patentrolls/).

41. Lambert, *Records of the Skinners*, 110–14, 304–7, 313–16, 321–30.

42. *Calendar of the Letter-Books*, ed. Reginald Sharpe (London: Corporation of London, 1899–1912; *British History Online*, http://www.british-history.ac.uk), *Book G*, ff. clxii, clxiv.

ried for making "mengled butter" or "lyght bread,"[43] the line of the law
in honest sales was no minor matter. The Skinners protected their civic
privileges carefully, as evidenced by later confirmations and expansions of
their royal charter. The document issued by Richard II in 1392 confirms
the charter of Edward III in all of its details, while adding new provisions for a religious fraternity dedicated to Corpus Christi.[44] This charter, in other words, has little to say about business practices, addressing
instead the company's role in civic life. The last medieval charter, issued
by Henry VI in 1437, confirms the rights granted by Richard II, covering
both trade practices and the religious fraternity.[45]

Collectively, the charters enshrine corporate autonomy in law while
also affirming the king's right to define the limits of that autonomy. They
require certain expenditures and display—and also set maximums. Elspeth
Veale and Sylvia Thrupp both paint pictures of the skinners as bejew-
eled and robed merchants, socially ambitious and landowning, an urban
elite only narrowly distinguished from the nobility.[46] Members of the
guild made some of the highest contributions to Edward III's war efforts,
while paying taxes comparable to earls, knights, and esquires; trade with
the royal household made Robert Persone one of the three richest men in
London in the early fourteenth century.[47] Merchants' incomes, according
to sumptuary regulation, would have determined their own fur-wearing
practices.

Within the guild, members expressed various levels of concern for
clothing, mirroring the larger social situation depicted in sumptuary petitions. The general membership, for example, requested the masters and
wardens not to purchase expensive clothing too frequently—and certainly
not by using the collective funds. Types of clothing were also regulated,
with fines imposed for appearing in the wrong array at the wrong time—
from wearing cloaks instead of gowns to the proper colors for apprentices.
And, of course, guild regulations addressed which types of fur should be
worn by whom in the service of the mayor.[48] All of these measures, like
sumptuary discourse more broadly, arose from concern for "good gov-

43. *Chronicles of London,* 187.
44. *Calendar of the Patent Rolls,* Richard II (5:286).
45. *Calendar of the Patent Rolls,* Henry VI (3:190–1, membrane 10).
46. Sylvia Thrupp, *The Merchant Class of Medieval London* (Ann Arbor: University of Michigan Press, 1948); Elspeth Veale, *The English Fur Trade in the Later Middle Ages* (Oxford: Clarendon Press, 1966), 79.
47. Veale, *English Fur Trade,* 53–54.
48. Lambert, *Records of the Skinners,* 113, 213, 214, 242, 245, 269.

ernance," which was understood to encompass both cost controls and social value.

Social standing and wealth all contributed to the skinners' political influence in the city. Political activity was concentrated among the powerful heads of the twelve great companies, a grouping that had been more or less fixed by the beginning of the fifteenth century. The mayor was elected exclusively from among these twelve. Government of the city was thus essentially a commercial affair. The merchants, for example, returned a mayor for a second term in 1385, against the will of the body of common citizens. And the city and the crown occasionally tussled over jurisdiction of civic affairs.[49]

The establishment by Henry IV of the position of King's Skinner in 1405 added a new dimension to the politics of fur in medieval London. Already subject to regulation by statutes of both the guild and the realm, furs had a new symbolic and literal relation to the sovereign. The King's Skinner was responsible for all work related to skins and furs in the Great Wardrobe, such that the guild forged a strong attachment to royal as well as aristocratic and civic culture. The man first named to this post was Henry Barton, a grant for life in recognition of his service to Henry's father John of Gaunt (Barton had previously been a yeoman of the chamber for Richard II). The position required wealth and involved a certain amount of financial risk, as debts for materials and services went unpaid and "loans" to the crown accumulated. Barton, himself well dressed in white fur, attended to all the details of producing and caring for the royal furs—from acquiring skins to arranging for the transport of fur garments.[50]

Barton's appointment to the Wardrobe coincided with the beginning of his lengthy career in civic politics: he was elected sheriff the same year, alderman the year after (a position he held for the next three decades), mayor in 1416 (the first skinner to hold the post since Adam Bury in 1374), and served in parliament in 1419. In 1428, London's merchants elected Barton mayor for a second time, despite new regulations that

---

49. Charles Kingsford, *Prejudice and Promise in Fifteenth-Century England* (London: Oxford University Press, 1925), 107–45; Caroline Barron, *London in the Later Middle Ages: Government and the People, 1200–1500* (Oxford: Oxford University Press, 2004).

50. Veale, *English Fur Trade*, 52–55, 81, 206–7; *Calendar of the Close Rolls,* Henry IV (2:408) (London: H. M. Stationery Office, 1892–1963); *Calendar of the Patent Rolls,* Richard II (5:488), Henry IV (1:116, 323). I have written elsewhere on Barton's literary patronage ("Lydgate, Lovelich, and London Letters," in *Lydgate Matters: Poetry and Material Culture in the Fifteenth Century,* ed. Lisa Cooper and Andrea Denny-Brown [New York: Palgrave Macmillan, 2008], 113–38).

restricted reelection due to the great expenses that came with the office.[51] Barton also had key roles in city management, from the tonnage and poundage collection (1408–10) to the wool custom (1410–16); as mayor he once oversaw a case involving fraudulent furs.[52] Other merchants found that Barton's standing in the city made him "an ideal trustee" and a desirable "feoffee-to-uses."[53] Although his civic duties had their costs, he was a major London property holder and also held estates in Buckinghamshire, Hertfordshire, Berkshire, and East Anglia.[54] He exemplifies the slippage of social distinction between the urban elite and country gentry.

Barton's stints as mayor bring back to the fore the convolutions of biopolitics. London's mayors obviously exercised significant political, juridical, and social functions. Their unique position is identified in fifteenth-century sumptuary discourse by the types of furs they were authorized to wear. In 1402, mayors are exempted—for all time—from the limits on esquires:

> Nor that any esquire should use fur of grey, cristigray, miniver, or brown, except those who are, or have been, or shall be, mayors in the cities of London, York, or the town of Bristol.

> Ne qe nulle esquier use furre de gray, cristigray, menyvere, ne bice, forspris les mairs qe sount, ount estez, ou pur le temps serrount, en les citees de Loundres, Everwyk, ou en la ville de Bristuyt.[55]

So adorned, a mayor looks precisely like a well-dressed knight. His wife, meanwhile, reaches an even higher status, authorized to wear ermine and weasel, which is typically reserved for the highest ranks. The petition of 1406 frames nearly the same exemptions for both mayors and their wives. In 1463, the petition expresses a direct equivalency between mayors and knights bachelor.[56] A chronicle description of a 1503 procession before the queen makes explicit the chain of equivalences between social status and dress: aldermen wore hoods like knights, and the mayor a baron's

51. Anne Lancashire, "The Mayors and Sheriffs of London, 1190–1558," in Barron, *London,* 338, 340; *Chronicles of London* 64, 71, 96, 126, 131.

52. *The History of Parliament: The House of Commons, 1386–1421,* ed. J. S. Roskell, Linda Clark, and Carole Rawcliffe (Stroud: Alan Sutton Publishing, 1992), 136; *Calendar of the Letter-Books,* Book K, f. 56–57.

53. *History of Parliament,* 137.

54. Ibid.

55. *PROME,* Henry IV, September 1402 (Membrane 6, Item 76).

56. *PROME,* Edward IV, April 1463 (Membrane 6–7, Item 20).

hood.[57] Since the mayors are also explicitly made responsible for enforcing the terms of these (hoped for) regulations, they are also explicitly exempt from the law they would judge.

The fashion choices of mayors have specific implications in Dymmok's moralization of sumptuous dress. In a passage that replies to Conclusion Five, Dymmok uses mayors as an analogy for lordship in general. Here, he explains how temporal and spiritual officeholders acquire new powers when they take office without feeling or being changed:

> Just as when he takes office, the mayor of a city neither feels nor undergoes any sensible change, but nonetheless has a power for governing the whole city and preventing harm to it newly committed to him in virtue of his office by the king. (Somerset 129)

> Sicud maior ciuitatis nullam sensibilem mutacionem in se sentit nec habet, quam prius non habuit, et tamen potestatem habet nouiter sibi commissam uirtute officii a rege totam ciuitatem gubernandi et nociua illa ciuitati compescendi. (Dymmok, *Liber* 128)

Fiona Somerset cites this passage to discuss how analogy has functions other than explanation.[58] By comparing regal and papal dignities to the mayorality, Dymmok suggests that members of the urban oligarchy may be among his intended addressees. More specifically, he promotes the mayor's authority—even potentially against the king who invested him, if needed to defend the city. The affirmative version of Dymmok's "nociua . . . compescendi" would be in the vernacular "worschippe . . . kepte"[59]— widely recognized as the mayor's prime responsibility to the city. Meanwhile, while there may be no inward change in a new officeholder—king, pope, or mayor—there is indeed an outward change, marked by dress. All don furs as part of the outward display of high status. The mayor, in particular, accedes to a new exemption from sumptuary limits—including new access to some of the same furs as the king himself—such that his appearance can change quite noticeably.

---

57. *Chronicles of London*, 259 (edited from Cotton Vitellius A XVI, f. 205).

58. Somerset, *Clerical Discourse*, 129–30.

59. E.g., William Gregory, "Chronicle of London," *The Historical Collections of a Citizen of London*, ed. James Gairdner, 1876 (New York: Johnson Reprints, 1965), 55–239, see especially 223. (NB, Gregory continues to be cited as the author of this chronicle, British Library MS Egerton 1995, but this attribution is disputed: Mary-Rose McLaren, *The London Chronicles of the Fifteenth Century: A Revolution in English Writing* [Woodbridge: Boydell & Brewer, 2002], 29–33).

When the man who becomes a mayor is himself a purveyor of furs, the biopolitics of civic Law acquires yet another dimension. For Henry Barton, mayor from 1416–17 and 1428–29, sumptuary law and personal commerce converge to create a doorkeeper able to stand on all sides of the Law, by virtue of its specific exclusions. He enforces a law that holds him in exclusion; he is sovereign within the city: as Richard Maidstone has Richard II declare, "Maior electus qui regat urbem, / Regis" (An elected mayor reigns over the city as king).[60] The mayor frames the door that only he can walk through (there is only one mayor at a time). Whether the story changes if his collar has fleas, we can only hope.

---

60. Maidstone, 78.

5

Saintly Ecologies

TRACING COLLECTIVITIES IN THE LIFE OF KING OSWALD OF NORTHUMBRIA

᠅

Mary Kate Hurley

We should not limit in advance
the sort of beings populating the social world.[1]
—Bruno Latour

ACCORDING TO the field of Actor-Network Theory, an ultimately false system of binaries can be found at the very heart of so-called Enlightenment thinking. Actor-Network Theory—hereafter, ANT—proceeds from a relatively simple premise. Rather than artificially divide the world into false dichotomies—"nature" and "society," "subject and object," etc.—ANT asserts that in order to better describe the world around us, we must first recognize the multiplicity of different kinds of entities that compose the world and then trace the relationships that emerge between them. According to Latour, we must "compose the common world from disjointed pieces instead of taking for granted that the unity, continuity, agreement is already there, embedded in the idea that 'the same nature fits all.'"[2] Thus, the old terms of "society" and "culture" might best be replaced by the figure of "collectivity," a provisional collection of human and nonhuman entities that act upon one another in a network, rather

---

1. Bruno Latour, *Reassembling the Social: An Introduction to Actor-Network Theory* (Oxford: Oxford University Press, 2005), 16.

2. Bruno Latour, "An Attempt at a 'Compositionist Manifesto,'" *New Literary History* 41 (2010): 485.

than in a series of distinct two-party relationships. The primary focus of
ANT is not necessarily on the actors themselves within these networks;[3]
rather, ANT's focus is to consider "*at once* the actor *and* the network"[4]
by exploring the alliances that make such assemblages possible at all.
This understanding of the composition of the social world, which Latour
has described elsewhere under the rubric of political ecology, constitutes
one of the implications of ANT, traced most famously by Latour, John
Law, and Michel Callon.[5]

Put another way, ANT seeks to expose the connections that build
the world rather than the divisions. The classic example of the stakes
of ANT comes from Michel Callon's "Some Elements of a Sociology of
Translation." In his analysis of a fishing village and scallop production
in Northern France, Callon demonstrates the ramifications of thinking
through a collectivity rather than a society or a community. Where a
focus on human community examines only the actions and desires of
fisherman, researchers, and the scallop-enjoying public, a focus on collectivity takes into account the actions of scallops, water currents, tides, and
so on. Moreover, it acknowledges the ways in which scallops and water
currents can affect the behaviors of human beings, rather than examining only the human effects on the natural environment. ANT, in other
words, provides an alternative way of thinking about agency within particular systems. Callon's argument reveals not only the interdependence
of these groupings of actors but also the power relationships that emerge
when analyzing such systems, power relations that privilege—naturally,
but unfairly—human interests.

In the world outside of texts, ANT offers a powerful method for
demonstrating how a subtle shift in focus can lead from a primacy of
human interests to an understanding of the diversity of nonhuman relationships that shape the world. In this chapter, I argue that a literary

---

3. And herein lies one of the major charges leveled against Latour by Ian Bogost in
his *Alien Phenomenology*, where he positions Latour's credo—that "actors do not stand
still long enough to take a group photo"—as deemphasizing *things in themselves* (and for
themselves) in favor of their "couplings and decouplings. Alliances take center stage, and
things move to the wings" (*Alien Phenomenology or What It's Like to be a Thing* [Min-
neapolis: University of Minnesota Press, 2012], 7).

4. Latour, *Reassembling the Social*, 169.

5. Michel Callon, "Some Elements of a Sociology of Translation: Domestication of
the Scallops and the Fishermen of St. Brieuc Bay," in *Power, Action, and Belief: A New
Sociology of Knowledge?* ed. John Law (Boston: Routledge, 1986), 196–233; Bruno Latour,
We Have Never Been Modern, trans. Catherine Porter (Cambridge, MA: Harvard University Press, 1994); John Law, "Notes on the Theory of Actor-Network: Ordering, Strategy,
and Heterogeneity," *Systems Practice* 5 (1992): 379–93.

text can represent—and even function as part of—such a system, and I explore some of the implications of understanding a text in this manner. In traditional studies of the hagiography of King Oswald of Northumbria written by the Venerable Bede and Ælfric of Eynsham, the majority of critical attention has been paid to the methods by which Oswald's fragmented body functions as a sign of the holiness of England.[6] Oswald's actions and his martyrdom create holy sites that function as centers for cultic activities, such as pilgrimage, creating a Northumbrian and later English community around this holy king. One aspect of his story to which too little attention has been paid, however, is the soil and vegetation that his holy actions sanctify. This soil—sanctified by the king's holy life and death, and distributed around Northumbria—suggests that more than just a human community comes into being around the saint. Rather, I argue that it points to a distribution of agency that highlights the interconnection of human and nonhuman actors in the text.

My argument will unfold in two complementary parts. First, I will lay out the fundamental tenets of ANT and their intersection with literary study. In particular, I will focus on the concept of "collectivity" and its importance for describing the function and content of literary texts. Second, using the case study of the life of Saint Oswald as it is portrayed in Bede's *Historia Ecclesiastica* and Ælfric's *Lives of the Saints,* I will examine the ways in which the study of "collectivity"—when used to describe the group formations in the story of Oswald—allows for a more careful consideration of how agency is assigned within them. Moreover, I will argue that this approach reveals the processes by which texts literally and figuratively compose the social world through their narrative constructions.

---

6. See Kent G. Hare, "Heroes, Saints and Martyrs: Holy Kingship from Bede to Ælfric," *Heroic Age* 9 (Oct 2006): np. See also Daniel Donoghue, *Old English Literature: A Short Introduction* (Oxford: Wiley-Blackwell, 2004), 71; James Riggins Hurt, "*Ælfric and the English Saints*" (unpublished Ph.D. diss., Indiana University, 1965); Marianne Malo Chenard, "King Oswald's Holy Hands: Metonymy and the Making of a Saint in Bede's Ecclesiastical History," *Exemplaria* 17, no. 1 (Spring 2005): 33–56; Catherine Cubitt, "Sites and Sanctity: Revisiting the Cult of Murdered and Martyred Anglo-Saxon Royal Saints," *Early Medieval Europe* 9, no. 1 (2000): 53–83; Edward Christie, "Self Mastery and Submission: Holiness and Masculinity in the Lives of the Anglo-Saxon Martyr-Kings," in *Holiness and Masculinity in the Middle Ages,* ed. P. H. Cullum and Katherine J. Lewis (Toronto: University of Toronto Press, 2005), 143–57; John M. Hill, "The Sacrificial Synecdoche of Hands, Heads, and Arms in the Anglo-Saxon Heroic Story," in *Naked before God: Uncovering the Body in Anglo-Saxon England,* ed. Jonathan Wilcox and Benjamin C. Withers (Morgantown: West Virginia University Press, 2003): 116–37; and John E. Damon, "*Desecto Capite Perfido*: Reciprocal Violence in Anglo-Saxon England," *Exemplaria* 13, no. 2 (2001): 399–432.

I will demonstrate that while Oswald's holiness undoubtedly affects the soil, both the provenance and the consequences of that holiness are necessarily mediated by other (nonhuman) actors. As a result, an ecology of holiness reveals the interdependency of humans (even saintly ones) and their environments.

Composing the Common World

Medievalists have proven particularly sensitive to the utility of ANT and its implications for studying the medieval world.[7] Jeffrey Cohen—picking up on one of the many philosophical threads that inform the recent interest in ANT, as well as offshoots including "vibrant materialism" and "object-oriented ontology"—suggests that "things, especially things that appear to hold themselves in silence, must possess a power indifferent to language: something that comes from themselves, not via human allowance. Silent things must be able to speak, exert agency, propel narrative" in order to account for the remoteness of the nonhuman from the human in medieval texts, despite their very specific agency.[8] Cohen's recuperative project, which seeks to allow for an animation that, he argues, is both endemic in and specific to the animals, vegetables, and minerals that populate the essays in his collection, offers a fitting point of departure for the discussion of ANT, in part because the latter theory takes the diversity of the material world as its own *sine qua non*. Cohen's argument posits a kind of interiority to the mute animals or stones that inhabit medieval texts: this "vibrancy"[9] implies an utterly nonhuman desire.

As a result, the usual distinction between subject and object breaks down. As Karl Steel puts it, "Subjects are objects that are cared about. Each subject organizes its world, its polity, in its own way, unwilling and indeed unable to let everything into its borders and supremacy without sacrificing its own existence."[10] Object-oriented ontology posits the resis-

---

7. In part, this is likely a result of the variety of objects with agency that appear in medieval literature, ranging from saints' lives and the relics of the saint to texts such as the Anglo-Saxon Riddles, in which objects quite literally demand an answer from readers or listeners in response to their actions.

8. Jeffrey Jerome Cohen, "Introduction: All Things," in *Animal, Vegetable, Mineral: Ethics and Objects*, ed. Jeffrey Jerome Cohen (Washington, DC: Oliphaunt Books, 2012), 6.

9. See Jane Bennett, *Vibrant Matter: A Political Ecology of Things* (Durham: Duke University Press, 2010), vii–xix.

10. Karl Steel, "With the World, or Bound to Face the Sky: The Postures of the Wolf-Child of Hesse," in Cohen, *Animal, Vegetable, Mineral*, 33.

tance of "objects" to human ides of hierarchy, creating what Ian Bogost calls a "flat ontology," which "makes no distinction between the types of things that exist but treats all equally."[11] However, this "flatness" is not solely the province of scholars working in the realm of object-oriented ontology; rather, this "rhizomatic" approach to tracing relations between (for lack of better terminology) subjects and objects is also a fundamental tenet of ANT.[12] While object-oriented approaches offer a necessary corrective to anthropocentric readings of medieval and nonmedieval texts alike, the descriptive accuracy that ANT offers can function as an epistemologically grounded accounting of the collectivities in which lively nonhumans and humans participate and by which both humans and nonhumans are shaped. That is, while object-oriented ontology is useful for understanding things *in themselves,* ANT offers a useful model for understanding narratives of their interactions.

In *We Have Never Been Modern,* Latour suggests that taking into account the multiplicity of actors in the "social" world reveals that human beings are not the only—and sometimes not even the most important—entities within it. Originally aimed at practitioners of sociological studies and the emerging field of science studies, *We Have Never Been Modern* argues that in order to understand the composition of the world, we have to trace the networks that emerge between entities whose characteristics we do not limit in advance.[13] Thus, in her consideration of "medieval things," Kellie Robertson suggests that the relationship between the Merchant in Chaucer's *Canterbury Tales* and his beaver hat is not only about clothing and self-fashioning.[14] For Robertson, the beaver cap serves as an impetus "to ask about the particular kinds of thoughts that certain hats may instill in their wearers, the kinds of object networks that these things gather to themselves and maintain, and the myriad ways that objects shape human perception and knowledges rather than

---

11. Bogost, *Alien Phenomenology,* 17. Also see the discussion of a rhizome, which offers a more ANT-like understanding of "flat ontology" that they call the "plane of consistency," in Gilles Deleuze and Félix Guattari, *A Thousand Plateaus: Capitalism and Schizophrenia,* trans. Brian Massumi (Minneapolis: University of Minnesota Press, 1987).

12. The idea of the "rhizome" is, in part, a product of the thought of Deleuze and Guattari; see *Thousand Plateaus.* See also Graham Harman, *Prince of Networks: Bruno Latour and Metaphysics* (Melbourne: re.press, 2009); Bogost, *Alien Phenomenology,* 7, 14; and Bennett, *Vibrant Matter,* especially 52–61.

13. Bogost argues that Latour's focus here is too much on the human. See, for example, *Alien Phenomenology,* 7.

14. Kellie Robertson, "Medieval Things: Materiality, Historicism, and the Premodern Object," *Literature Compass* 5/6 (2008): 1060–1080.

being merely shaped by them."[15] That is, while the human uses of beaver hats remain an important realm of inquiry in studies of the Merchant, the hat might "use" humans as well, in some senses, to further causes and associations that are not only human in orientation. My study of Oswald suggests that humans are not only the creators of relics; rather, relics also "use" humans to distribute the holiness they possess through an expanded territory, drawing humans and nonhumans together in an assemblage provoked by—but not limited to—the saintly actions that set it in motion.

For Latour, the entities that compose the world can thus take a variety of forms: "They are collective because they attach us to one another, because they circulate in our hands and define our social bond by their very circulation."[16] Latour's tracing of these networks takes "into account, at the same time and in the same breath, the nature of things, technologies, sciences, fictional beings, religions large and small, politics, jurisdictions, economies, and unconsciousnesses."[17] Connected not only to humans but also to one another, these nonhuman actants can be understood to operate on the same plane of consistency[18] as humans: therefore, they are as capable of shaping the human (and nonhuman) environment as humans are. In order to understand the changing environment of humans, therefore, one must trace not only the creation and disintegration of alliances that shape interactions between humans but also those between human and nonhuman actants—and sometimes even solely between nonhuman actants.

To this end, Latour argues, it is necessary to do away with the ultimately fictive bifurcation of "nature" and "society"—a system composed of only these two mutually exclusive parts cannot be commensurate to the task of reassembling and describing the world *as it is,* because they are effects of the human exclusion of nonhuman actants from explanations of the world. Latour suggests that scholars should thus reconsider the terms by which they/we describe the world brought into a consistent plane:

> Whereas objects could only face out at the subjects—and vice versa— non-humans may be folded into humans through the key processes of translation, articulation, delegation, shifting out and down. What name can we give to the house in which they have taken up residence? Not nature, of course, since its existence is entirely polemical [. . .] Society

15. Ibid., 1075.
16. Latour, *We Have Never Been Modern*, 89.
17. Ibid., 129.
18. See above, footnote 11.

will not do either, since it has been turned, by the social scientists, into a fairy tale of social relations, from which all non-humans have been carefully enucleated [. . .]. In the newly emerging paradigm, we have substituted the notion of collective—defined as an exchange of human and non-human properties inside a corporate body—for the tainted word "society."[19]

That is, by replacing the idea of society with the idea of collectivity, Latour draws attention to the ways in which the human and the nonhuman are necessarily interconnected. It is important to note that Latour is not making a naïve argument that all things are connected, nor does he simply grant to nonhumans the subjectivity that has long been the sole province of humans. In fact, he deliberately eschews the simplistic argument that would "extend subjectivity to things, to treat humans like objects, to take machines for social actors."[20] By contrast, he suggests that when tracing collectivities, networks, and alliances, we ought to "*avoid using* the subject-object distinction *at all* in order to talk about the folding of humans and non-humans. What the new picture seeks to capture are the moves by which any given collective extends its social fabric to *other* entities."[21] That is, by breaking down the fictive distinction between subjects and objects, we are better able to see that multiple agents might simultaneously build and shape a given environment, even one nominally wrought by humans. Ultimately, describing the complexity of networks in this sense allows for a firmer understanding of humans' place within their environment without relegating that environment to the position of a static, singular entity.

It is from this distinction then—and the necessity of creating a method whereby we can "take into account" the multiplicity of actors in the social world—that ANT intersects with questions of political ecology and can profitably inform studies of medieval environments (as it can any other). Benjamin, in his "Theses on the Philosophy of History," remarks, "[A] chronicler who recites events without distinguishing between major and minor ones acts in accordance with the following truth: nothing that has ever happened should be regarded as lost to history."[22] In an analogous move for his considerations of political ecologies, Latour's version

---

19. Bruno Latour, *Pandora's Hope: Essays on the Reality of Science Studies* (Cambridge, MA: Harvard University Press, 1999), 193.
20. Ibid., 193–94.
21. Ibid.
22. Walter Benjamin, "Theses on the Philosophy of History," in *Illuminations*, ed. Hannah Arendt, trans. Harry Zohn (New York: Schocken Books, 1969), 254.

of this statement remade for ANT would argue that no *actant* should ever be regarded as lost to history (or sociology, or politics). I suggest that in the consideration of narratives, an ANT approach is not only possible but necessary because it allows for a better understanding of the ways that a given text imagines the composition of the communities that populate it. Therefore, rather than inquiring into which literary effects cause networks to form in the "real world,"[23] the ANT approach to literary criticism that I will demonstrate here examines the networks of actants represented in texts in order to understand what kinds of agency medieval texts grant to humans and nonhumans alike.

Oswald, Warrior and King

A brief recounting of the basic narratives in both versions of the Oswald story—Bede's *Historia Ecclesiastica* and Ælfric's *Life of Oswald, King and Martyr*—will help situate my reading of the collectivity that exists at its heart.[24] Centering on the life of the seventh-century King Oswald—Edwin of Northumbria's nephew and successor—the narrative traces three key moments in the life of the sainted warrior: his battle at Heavenfield, the works of Bishop Aidan in reconverting Northumbria, and Oswald's final battle at Maserfeld. Oswald spends much of his childhood in exile in Scotland, where he is converted to Christianity. Upon his return to England, Oswald wins a battle against the forces of the heathen king Cadwalla, who killed his uncle Edwin. The battle at Heavenfield occasions the erection of a cross that promotes healing among those who either visit it or receive the moss that grows upon it. As king, Oswald turns his energy to converting his people, inviting the bishop Aidan from Scotland to help in this matter. Oswald's reign—though not his power to protect his people—ends with his death in battle at Maserfeld in 642. He is defeated by the Mercian king Penda, who had been allied with Cadwalla

---

23. In this vein, I am partially distinguishing ANT analyses from the idea of "textual community," as defined in Brian Stock's *The Implications of Literacy: Written Language and Models of Interpretation in the Eleventh and Twelfth Centuries* (Princeton, NJ: Princeton University Press, 1983), and his *Listening for the Text: On the Uses of the Past* (Baltimore: Johns Hopkins University Press, 1990).

24. Old English text from *Ælfric's Lives of the Saints,* vols. 1 and 2, ed. W. W. Skeat, (EETS: London, 1881). All translations from the Old English are my own. Latin text and translation throughout is from *Bede's Ecclesiastical History of the English People* [hereafter referred to as *HE*], ed. Bertram Colgrave and R. A. B. Mynors (Oxford: Clarendon Press, 1969).

in the Heavenfield battle. Much of the narrative energy of Oswald's life (in both Bede and Ælfric) focuses on the miracles that surround both Oswald and the sites of his battles. In many instances, those miracles take place as a result of soil or vegetation that becomes holy through its proximity to the saint's material body, whether that body is alive or dead.

That the *Life of Oswald* is interested in the methods by which communities are formed is now a critical commonplace.[25] Ælfric wrote of Oswald at a key point during the second Viking invasion, when the question of community was particularly pressing to his Anglo-Saxon audience. In a territory beset by an invading force, it is not, as John E. Damon observes, "surprising to find that one major concern of a book written by a leader of the Church during this period would be the proper Christian attitude to warfare, the legitimate use of force against illegitimate violence."[26] Thus, Ælfric's *Lives of the Saints* might quite rightly be considered "a book about the relationship between warfare and sanctity."[27] Oswald's status as a king falls into this contested zone, in which legitimate kingship is inextricably tied to the violence necessary to protect and preserve human community. The broader context into which Oswald's kingdom is set, however, suggests a larger scale of understanding that, in its own right, exceeds Oswald's human interactions. His life, works, and finally his bodily death sanctify the soil he defends. That soil, made holy, extends Oswald's protective efficacy to other lives: it promotes healing, helps ward off death, and precipitates other events that take place around it and—crucially—on account of its presence. Meanwhile, the narrative draws attention to the ways in which land and vegetation precipitate action in the text when brought into association with humans, preserving a sense of the collectivity, even in a narrative that is ostensibly solely committed to outlining and understanding human communities.

Holy Soil, Holy Kingdom

The holy soil of Oswald's narrative in both Bede and Ælfric participates in a complex process by which the landscape as well as its people become Christian. Gillian Overing and Clare Lees argue in *A Place to Believe In*

---

25. See especially Hare, "Heroes, Saints and Martyrs," and Hurt, "Ælfric and the English Saints."

26. John E. Damon, *Soldier Saints and Holy Warriors: Warfare and Sanctity in the Literature of Early England* (Burlington, VT: Ashgate, 2003), 195.

27. Ibid.

for the interrelation of relics and their places of worship, noting that the "emplaced relic" can reveal much about the "identity and the *locus* of Northumbria and about literal as well as sacred topography."[28] When Oswald defeats Cadwalla at Heavenfield, he initiates a pattern of sanctification and subsequent transmission of that holiness that brings to light the importance of the soil of England—or its vegetation, or the pieces of the martyr's body itself—in creating and coalescing human communities. This movement of the soil highlights the participation of an otherwise static environment in the political structures that create human identities.[29] The distribution of that sanctifying power becomes a key component of Oswald's cult and extends through the battle at Maserfeld, where Oswald's death marks him as a martyr for the Christian faith in England. His ongoing concern for and involvement with his kingdom, however, persist in the physical objects that perpetuate his power. That is, the miracles that take place through Maserfeld originate in the soil but are not completely contained there. Rather, like the relics of the saint and the story of his life, Oswald's holy power is transferable from believer to believer because the dust in which it is embodied is portable—just as the story Ælfric tells about the dust is portable. Its movement from the battlefield to various locations around Northumbria and England emphasizes its role as a mediator between the supernatural power of the saint, his physical (and physically lifeless) body, and the other humans who this soil can physically act upon. In this way, it functions not solely as relic but also as an agent of healing change that provokes alteration in human behavior: humans behave differently around this sanctified soil, regardless of whether or not they recognize its power.

Oswald's actions at Heavenfield literalize the ways in which human action can modify the landscape, but they also provoke the reciprocal agency of the natural world to heal human bodies. Before his victory in battle over Cadwalla, Oswald raises a cross in order to honor God: "Oswold þa arærde ane rode sona / gode to wurðmynte ær þan þe he to ðam gewinne com" (26.17–18) [Oswald there quickly reared up a cross, to give worship to God before the coming battle]. We are then

---

28. Clare Lees and Gillian Overing, "Anglo-Saxon Horizons: Places of the Mind in the Northumbrian Landscape," in *A Place to Believe In: Locating Medieval Landscapes,* ed. Clare Lees and Gillian Overing (University Park: Pennsylvania State University Press, 2006), 21.

29. One might recall here Schmitt's analysis of sovereignty, in which "the exception in jurisprudence is analogous to the miracle in theology" *Political Theology: Four Chapters on the Concept of Sovereignty,* 1st rev. ed., 1934; trans. Charles Schwab (Chicago: University of Chicago Press, 2005), 36.

told of Oswald and his "geferum" (26.19) [companions], "Hi feollon þa ealle mid oswolde on gebedum" (26.24) [They all fell down with Oswald in prayer] and, thereafter, "gewunnon þær sige" (26.26) [they won the battle] against Cadwalla. Afterward, the cross becomes a site of healing:

And wurdon fela gehælde
untrumra manna and eac swilce nytena
þurh ða ylcan rode swa swa us rehte beda.
(26.31–33)

And many were healed, unwell men and also beasts through that same cross, as Bede has told us.

The cross, that is, functions in two modes. First, it testifies to Oswald's God-granted ability to defeat the invading Cadwalla and marks out his status as a chosen victor and monarch. Second, and perhaps more importantly, it creates a community—of humans *and* animals—that is brought together through its healing power.

Geography provides one way that we can assess the narrative's presentation of the connections between humans, objects, and the natural world. John Howe argues that sacred Christian geographies "converted" previously pagan sites by erecting crosses over them, among other strategies: Howe thus draws attention to the methods by which non-Christian places were literally and symbolically Christianized. Although his argument is meant to illuminate the ways in which geographical references in texts about saints can provide "a series of snapshots witnessing cultic developments over time,"[30] his methodology is instructive to the ANT interpretation of Oswald's story.

According to Howe, crosses such as the one at Heavenfield hold a particularly prominent place in the conversion of land. Such structures "[proclaimed] Christian territory," but they also proclaimed the identity of the people or rulers of such territories.[31] In Oswald's story, however, we can begin to see the dual motion of such a proclamation: the cross sanctifies the land both literally and metaphorically, but that sanctification changes both the ways that humans interact with the land and the ways that humans modify it. Most importantly, it creates holy space and

---

30. John M. Howe, "The Conversion of the Physical World: The Creation of a Christian Landscape," in *Varieties of Religious Conversion in the Middle Ages,* ed. James Muldoon (Gainesville: University Press of Florida, 1997), 68.

31. Ibid., 71.

holy soil that can allow us to decenter the hagiographic narrative from its focus on humans, bringing to the forefront the connections between humans and the world around them.

While the location of the cross becomes an important site for miracles, it is not the only entity claimed for a Christian shrine or symbol. Rather, the moss that grows on the cross also gains this power and becomes the vehicle for the saint's now-portable healing potential:

> Sum man feoll on ise þæt his earm tobærst
> an læg þa on bedde gebrocod forðearle
> oðþæt man him fette of ðære forsæden rode
> sumne dæl þæs messes þe heo mid beweaxen wæs
> and se adliga sona on slæpe wearð gehæled
> on ðære ylcan nihte þurh oswoldes geearnungum.
> (26.34–39)

> A man fell on ice so that his arm was broken, and he lay then in bed very
> much injured until a man fetched for him from that afore-mentioned cross
> a part of the moss that was growing on it, and the sick one soon became
> healed in his sleep on that very night through Oswald's worthiness.

The narrative of the Heavenfield miracle emphasizes the traversal of physical space between humans or human-made objects: the space between the cross and the injured man, for example. What is most interesting from an ANT perspective, however, is the destabilizing potential of the cross for anthropocentric hagiography. The cross, a wrought object itself, draws attention to its own composition through its role as mediator between the saint and his followers who are healed—but also between the saint and the natural world that his holiness infects.

The moss functions as an actor that can be "fetched"—*fette* (from the infinitive *fetian*).[32] Moved from its initial location to wherever it might be needed to work its healing power, the moss extends the holiness of Heavenfield and the efficacy of its healing cross to locales far from its initial environment. The interaction between the king, the cross, and the land creates an intermediary relic, but this relic is hardly static. The verb *fetian* calls attention to one way in which the moss functions as an actant in this holy network. The literal meanings of *fetian* contain valences of

---

32. See *Dictionary of the Old English Corpus*, s.v. "fetian, feccan," http://www.doe. utoronto.ca.

being sought out, fetched, brought, or moved—all of which prioritize the uses to which the moss is put.[33] *Fetian,* however, also falls into a semantic range that includes "to marry."[34] In a sense, this less common use of the word *fetian* functions as an apt metaphor for the way the moss helps create the collectivity that includes humans, elements of the natural environment, the supernatural (God), and the holy king. The moss performs a linking of humans with both the natural and supernatural worlds, and the result is a sense of the agency—nonhuman, but still present—that is attributed to it. Whether it is supernatural or not is ultimately not the fundamental issue. The moss has power.

In the miracles that take place at Oswald's second holy site—Maserfeld—the interaction of the martyr-king with the soil suggests a very specific network of relationships between the saint, the land, and those that inhabit it—all as actants in a single collectivity rather than as a community of humans that interacts with the "passive" land. Both Ælfric and Bede highlight that many men and animals are healed by the place of holiness at Maserfeld. In one sense, the diversity of actors that experience the place's healing potential draws animals and humans into relationship with the holy soil. Moreover, the soil changes how humans act: they become different entities in relation to it. In essence, humans are made to work toward achieving the ends of this holy soil—they are made into carriers of the dirt. Bede relates that because of the soil's healing properties, believers begin to take away so much earth from the holy place that in their piety they create a hole "as deep as a man's height" (Bede, *HE* III.9.242) [ad mensuram staturae uirilis altam]. The soil's holiness enables humans to reassert and connect to a cultic community around the long-dead Oswald. In the hole, "as deep as a man's height," however, there remains a latent possibility that these humans are not—or are not only—reaffirming a meaning that is solely for them. The soil, which heals the broken bodies of believers, draws these same believers into a relationship with their physical environment. In this relationship, humans become agents of the soil's distribution. They change land that also *changes them*—provokes them to remove it to other locales and English climes and make it a part of their ongoing Christian lives.

The material status of the soil as a specifically holy object, however, poses a problem for its interpretation. As Patrick Geary observes, the value of relics was not in their materiality but in "the relationships they

---

33. s.v. fetian[1, 3, 4–5]

34. s.v. fetian[6]

could create as subjects."[35] According to Robertson, "they are perhaps
the most written about of all medieval and early modern objects that
trouble the clarity of the division between human and non-human."[36]
Relics (like Oswald's soil) forge, Robertson observes, "networks among
donors, patrons, merchants, worshipers, and (occasionally) thieves."[37] In
the case of the holy dirt, however, it also connects humans to the world
at large and the other entities that inhabit it. Even the very first healing
that takes place there is not necessarily *for* a human, although a human is
present. A rider with a horse in "agonizing pain" touches the spot where
Oswald died at Maserfeld and the horse's pain is immediately cured.
Bede relates that the rider recognizes "that there must be some special
sanctity associated with the place in which the horse was cured" (Bede,
HE III.9.242) [aliquid mirae sanctitatis huic loco, quo equus est curatus],
although he does not directly link the site to Oswald. Nevertheless, he
still marks this place with a sign before he leaves it, presumably so that
others might return to it.

Two processes of interaction operate within this depiction of the
Maserfeld environment. A narrative distance suggests that Bede's—and
his readers'—knowledge of Oswald's story strengthens the specificity of
the interaction that takes place there. That is, the generic conventions
of hagiography are still present and condition the reader to expect such
miracles at a martyr's place of death. Beneath the surface of the narrative,
however, we can once again see the ways in which the story of Oswald
lays bare the connections between humans and what might otherwise be
thought of as background—the moss, soil, and even grass that can heal
through its sanctity. A horse in agonizing pain touches a single spot of
earth, which is simultaneously like and unlike every other spot of earth
in the narrative. That contact—the moment in which a network forms
between two actants, the horse and the earth—provokes the healing of
the horse. A human mind stands witness to it, another actant in this collectivity, creating yet another actant: the story of what has transpired
here, represented by the sign that the man erects at this place of miraculous healing. The formation of this collectivity changes the humans in
the narrative: they respond to these holy events by digging a hole that
marks this anonymous space as something different, special—a holy place.
Although this is one human outcome of collectivity, the soil's power is

35. Patrick J. Geary, *Living with the Dead in the Middle Ages* (Ithaca: Cornell University Press, 1994), 216.
36. Robertson, "Medieval Things," 1071.
37. Ibid.

not dependent on it. The horse would have presumably been healed even in the absence of a rider.

Ælfric's version of the story foregrounds the role of collectivity even more starkly, in part because it omits the rider's sense of the place being particularly holy. The horse is cured when "becom hit embe lang þær se cynincg oswold / on þam gefeohte feoll swa swa we ær forsædon" (26.208–9) [it came before long to the place where the king Oswald fell in the fight, as we said before]. The narrative does not imply that the rider actually knows that the place is holy or that Oswald died there. Moreover, the rider does not erect any sign to mark the place; rather, he simply "þa ferde forð on his weg" (26.212) [went forth afterwards on his way]. Although the reader is consistently reminded that the ground is holy (and that it is holy because of Oswald's death) the narrative itself only foregrounds the agency of the land to heal another nonhuman entity—the horse. The rider (if not the reader) is drawn into this holy collectivity completely unawares.

The miracles that Oswald's soil precipitates extend the text's vision of collectivity by lending protective power not only to animals and humans but even to vegetable and inanimate matter. The ground where Oswald dies at Maserfeld is marked out, in the *Historia Ecclesiastica,* as being qualitatively different from other ground of the same field. It is, the text observes, "greener and more beautiful":

> The story is told that about this time another man, a Briton, was travelling near that place where the battle had been fought, when he noticed that a certain patch of ground was greener and more beautiful than the rest of the field. He very wisely conjectured that the only cause for the unusual greenness of that part must be that some man holier than the rest of the army had perished there. (Bede, *HE* III.10.244)

> Eodem tempore uenit alius quidam de natione Brettonum, ut ferunt, iter faciens iuxta ipsum locum, in quo praefata erat pugna conpleta; et uidit unius loci spatium cetero campo uiridius ac uenustius, coepitque sagaci animo conicere, quod nulla esset alia causa insolitae illo in loco uiriditatis, nisi quia ibidem sanctior cetero exercitu uir aliquis fuisset interfectus.

Oswald's death alters the growth of the grass in this place of holiness: it grows more beautifully, more green. In some sense, it is more full of life—literally and metaphorically—through its association with the sainted king. The Briton man correctly interprets this altered growth pattern to

mean that a holy man had died at Maserfeld. Neither is it unimportant
that the man who recognizes this altered growth pattern is "de natione
Brittonum"—the power of the soil extends across geographical space and
also across the identities for which such spaces serve as a marker.

The power of the soil in this instance protects not just human or
animal life, but even the buildings that humans construct. Each of these
actants becomes part of ecological fabric of an anthropically neutral network. The Briton binds some of the dirt from that site in a cloth, and
takes it to a feast at a house in an unnamed village. During this feast, "it
happened that the sparks flew up to the roof which was made of wat-
tles and thatched with hay, so that it suddenly burst into flames" (Bede,
HE III.10.244–45) [contigit uolantibus in altum scintillis culmen domus,
quod erat uirgis contextum ac foeno tectum, subitaneis flammis impleri].
No human power could save the house, but the dust from Oswald's site
of death retains a protective function: "So the whole house was burnt
down with the single exception that the post on which the soil hung,
enclosed in its bag, remained whole and untouched by the fire" (Bede,
HE III.10.244–45) [Consumta ergo domu flammis, posta solummodo,
in qua puluis ille inclusus pendebat, tuta ab ignibus et intacta remansit].
Understandably curious, the witnesses seek more information and the
source of the miracle is revealed: "After careful inquiries they discovered that the soil had been taken from that very place where Oswald's
blood had been spilt" (Bede, *HE* III.10.244) [Qua uisa uirtute mirati sunt
ualde, et perquirentes subtilius inuernerunt, quia de illo loco adsumtus
erat puluis, ubi regis Osualdi sanguis fuerat effusus].

Bede's narration of this moment in the afterlife of Oswald shows the
ways in which the power of the soil distributes agency across the collectivities in which it participates. Oswald's holiness infects the land, as
before, but this time the interaction changes the way the very grass grows.
Greener, more beautiful, this particular portion of land attracts a human
to it by virtue of its increased vitality. The human in question—the man
"de natione Brittonum" (of the race of the Britons)—demonstrates the
ability of the soil to attract not only native Northumbrian believers but
even a member of the problematically anterior Briton race,[38] suggesting

---

38. Bede's emphasis here is remarkable in part because of his disdain for the ethnic
group, whom he blames for the Anglo-Saxon invasion on account of their lack of proscly-
tizing. Bede claims that "to other unspeakable crimes, which Gildas their own historian
describes in doleful words, was added this crime, that they never preached the faith to the
Saxons or Angles who inhabited Britain with them. Nevertheless God in His goodness did

that it can bring together all with eyes to see its difference from quotidian dirt. Within a scene of human community (the feast at which this man is ultimately received), the fire—which might otherwise be thought of as a human tool—escapes human control and exercises its unique power over the wrought objects of humans, consuming them as fuel for further energy. The dust, however, shows its power over both fire and the materials of the house, and in this sense demonstrates greater agency than the humans who created both. The holy dust stops the flames from consuming a single post and prompts the human minds who witness the "miracle" to wonder by what power it does so. The dust, then, signals the operation of collectivity: it changes grass, human action, fire, and even human cogitation. In this narrative, there is no way to separate the human actors from the nonhuman ones. Both are bound to the same network of associations that forms as a consequence of the saint's miraculous—divine—abilities.

Networks in/of Time

This kind of collective composition of humans, environment, plants, housing, God, and soil suggests another way in which considering collectivities differs from considering communities. In the connections that emerge between these actors, different scales of time are made evident: the time frame of the land outlives the time frame of the martyr-king, suggesting connections in a kind of *longue durée* scale that exceeds the saint's physical life and actions. The two stories of Heavenfield's name similarly telescope times and destinies, ultimately suggesting that time frames exceeding human ones operate within the text. The effect of these naming stories continues to create a narrative about human interaction with land. In Ælfric's *Life of Oswald*, the name of the place is part-and-parcel of its material presence and interaction with the saint:

Seo stow is gehaten heofon-feld on englisc

---

not reject the people whom He foreknew, but He had appointed much worthier heralds of the truth to bring this people to the faith" (*HE* I.22.68) [qui inter alia inenarrabilium selerum facta, quae historicus eorum Gildas flebili sermone describit, et hoc addebant, ut numquam genti Saxonum siue Anglorum, secum Brittaniam incolenti, uerbum fidei praedicando committerent. Sed non tamen diuina pietas plebem suam, quam praesciuit, deseruit; quin multo digniores genti memoratae praecones ueritatis, per quos crederet, destinauit].

wið þone langan weall þe þa romaniscan worhtan
þær þær oswold oferwann þon wælhreowan cynincg.
(26.40–43)

The place is called Heavenfield in English, against the long wall which the
Romans wrought, there where Oswald overcame the cruel king.

Heavenfield's name does not precede Oswald's encounter with it—in one
sense, at least, Oswald puts Heavenfield on the map.

Oswald's cross is one of two different kinds of material remains that
stand as a testament to the ways in which humans are connected to the
environment as more than static backdrop: the other, of course, is the
"long wall that the Romans wrought." In the *Historia Ecclesiastica*, by
contrast, Heavenfield's name, and its subsequent holiness, are inscribed
in relation to future action that has not yet taken place:

> This place is called in English Heavenfield, and in Latin *Caelestis campus*,
> a name which it certainly received in days of old as an omen of future
> happenings; it signified that a heavenly sign was to be erected there, a
> heavenly victory won, and that heavenly miracles were to take place there
> continuing to this day. The place, on its north side, is close to the wall
> with which the Romans once girded the whole of Britain from sea to sea,
> to keep off the attacks of the barbarians as already described. (Bede, *HE*
> III.2.216)

> Vocatus locus ille linuga Anglorum hefenfeld, quod dici potest latine Cae-
> lestis Campus, quod certo utique praesagio futuorum antiquitus nomen
> accepit; significans nimirum quod ibidem caeleste erigendum propeum,
> caelestis inchoanda uictoria, caelestia usque hodie forent miracula cele-
> branda. Est autemlocus iuxta murum illum ad aquilonem, quo Romani
> quondam ob arcendos barbarorum impetus totam a mari ad mare prae-
> cinxere Brittaniam, ut supra docimus.

Bede suggests that the name of Heavenfield is an inheritance—even a
prophecy—rather than an innovation. He avers that the very name—*Caelestis campus*—is a sign of the holiness of the place, bestowed in the time
of the Romans. They too sought to keep a barbarous people from attacking their lands; one need only substitute "pagan" for "barbarian" in the
passage above to make clear the similarity. Humans, that is, leave marks
on the landscape that connect them across time. Oswald's fight against

the pagan forces of Cadwalla is an instance of the use of human artifice to change the landscape to make it more defensible. Moreover, it points to the eventual coming of Christianity and remaking of the place through its Christianization—an occurrence, Bede suggests, that was long destined to be fulfilled.

This telescoping of times takes place between the past of a people and their future and is made manifest in the creation of networks of actors in the Oswald narratives. The same device is further evident in the movement of Oswald's body from his place of death to its resting place at Bardney Abbey. The physical displacement of Oswald's body also changes the ways in which time and identity cohere around the saint's relics, creating networks of the people, places, and groups that come into contact with them that are not limited to a single time. In both Bede's and Ælfric's versions of the narrative, the monks at Bardney initially refuse to give Oswald's body a resting place after his niece has brought him to them. The reasoning for this refusal highlights the changing efficacy of Oswald's corpse as it stands in for a spreading holiness.

In Ælfric's version of the story, the refusal of the monks is attributable to *menniscum gedwylde*—human error:

Ac þa mynstermenn noldon for menniscum gedwylde
þone sanct underfon. ac man sloh an geteld
ofer þam halgan ban binnan þære licreste.
(26.179–81)

But the monks would not receive the saint, because of human error. But they pitched a tent over the holy bones that were in the tomb.

Gedwyld itself has a variety of meanings, most often associated with error and heresy.[39] Importantly, however, the primary definition of the word has to do with "a wandering; a journeying."[40] In this sense, of course, it maintains the familiar definition of its Latin equivalent, *errare*. The sense of wandering, particularly as it intersects with a semantic range of error and heresy, here implies a movement that is simultaneously enabled by the presence of Oswald's body and attenuated by it. These monks refuse his body on account of wandering from the path they should follow.

In Bede's version of the narrative, the same sense of movement that

---

39. *DOE*, s.v. gedwyld²
40. s.v. gedwyld¹

allies the monks' refusal of the bones with heresy and error is assigned a cause that transcends mere error and, instead, focuses on a secular enmity that conditions their understanding of the sacred. Bede explains the relationship between the Queen of Mercia and the kingly martyr but pauses to remark on the monks' hesitation to accept the relics:

> The carriage on which the bones were borne reached the monastery toward evening. But the inmates did not receive them gladly. They knew that Oswald was a saint but, nevertheless, because he belonged to another kingdom and had once conquered them, they pursued him even when dead with their former hatred. (Bede, *HE* III.11.246)

> Cumque uenisset carrum, in quo eadem ossa ducebantur, incumbente uespera in monasterium praefatum, noluerunt ea, qui erant in monasterio, libenter excipere, quia, etsi sanctum eum nouerant, tamen quia de alia prouincia ortus fuerat et super eos regnum acceperat, ueteranis eum odiis etiam mortuum insequebantur.

Bede's story directly addresses the problematic relationship between Oswald's kingdom and the kingdom of the Mercians, suggesting the difficulty created for rival groups when a warrior king becomes a saint. In Bede's narrative, secular allegiance is not always trumped by Christian brotherhood.

Ælfric frames Oswald's death with the careful observation that Penda, king of the Mercians, was responsible for it. This information makes the omission of the reason behind the monks' resentment seem somewhat odd. He has already affirmed that Penda—nominally the "enemy"—was not Christian. This moment in Ælfric's text draws attention to a single phrase in the earlier segment that describes the relationship between the Mercians and Christianity. The Mercians are described in a single line: "and eall myrcena folc wæs ungefullod þa git" (26.152) [And all of the Mercian people had not yet been baptized]. The inclusion of the phrase *þa git* in this earlier description suggests a continuity granted by Christianity. Although the Mercians had not yet been baptized, they were still possible subjects of Christianity's healing grace. Thus when Ælfric reduces the monks' refusal of the bones to simple "human error," he changes the character of their response to the queen's request. Where Bede figures a Mercian response to a former adversary, Ælfric glosses over the monks' secular allegiance to a worldly kingdom in order to emphasize the more important allegiance all men owe to God, and by

extension, to his chosen saints.[41] The Mercians had not yet been baptized when Penda killed Oswald; when the monks refuse the holy relics,
Ælfric's smoothing over of their Mercian sentiments fulfills the promise of what was "yet" to come: just as the Mercians would eventually
come to believe in Christ after Penda kills Oswald, the monks would also
redeem their rejection of the bones by later accepting them. The telescoping of times in this instance emphasizes the power of the Mercians'
future over their past, the ability of their future conversion to redeem
their past failures and sins. Time itself seems to have some kind of agency
in human events, making the "human error" of a people's past legible in
terms of its ultimate salvific significance.

This movement across time leads back to our understanding of how
knowledge of the saint's life and death disseminate through the natural world. In the *Historia Ecclesiastica,* God's revelation to the Mercian
monks follows the manifestation of a miracle meant to garner the relics'
acceptance: "But a sign from heaven revealed to them how reverently
the relics should have been received by all the faithful. All through the
night, a column of light stretched from the carriage right up to heaven
and was visible in almost every part of the kingdom of Lindsey" (Bede,
HE III.11.246) [Sed miraculi caelestis ostensio, quam reuerenter eae sus-
cipiendae a cunctis fidelibus essent, patefecit. Nam tota ea nocte columna
lucis a carro illo ad caelum usque porrecta omnibus pene eiusdem Lindis-
sae prouinciae locis conspicua stabat]. The monks see the error of their
ways writ in the healing light and recant their position, taking the bones
to be housed in their monastery. The monks and people of Lindsey are
not the only ones who see the healing light, however. Several chapters
after its first appearance, the light returns: "Not only did the fame of this
renowned king spread through all parts of Britain, but the beams of his
healing light also spread across the ocean and reached the realms of Germany and Ireland" (Bede, *HE* III.13.252) [Nec solum inclyti fama uiri
Brittaniae fine lustrauit uniuersos, sed etiam trans oceanum longe radios
salutiferae lucis spargena Germaniae simul et Hiberniae partes attigit].
In itself, this mention of the light is probably metaphorical: "the beams
of his healing light" need not be the same beams that were present at
the Lindsey monastery when the monks saw the error of their ways.[42]

---

41. Although the implication is still clear in Ælfric that the Mercians would have not
welcomed a conquering king's bones, no matter how sacred, the important aspect is his
glossing of that smaller allegiance in favor of his dominant theme of a Christian, English,
kingdom.

42. Peter Clemoes, "The Cult of Saint Oswald on the Continent," *Bede and His*

However, if they are taken literally, then the light itself is implicated in a network that exceeds the physical borders of Britain, extending Oswald's efficacy and fame across the sea.

While the light literally acts on the surroundings at Bardney, it also signifies a metaphorical light, exemplified by the dissemination of Oswald's *hlisa*—his fame. The light thus bears some similarity to narratives concerning Oswald that are borne abroad by believers in his holiness. In Ælfric, it is both localized and localizable:

Hwæt þa god geswutelode þæt he halig sanct wæs
swa þæt heofonlic leoht ofer þæt geteld astreht
stod up to heofonum swilce healic sunnbeam
ofer ealle þa niht. and þa leoda beheoldon
geond ealle þa scire swiðe wundrigende.
(26.182–86)

Lo, then God made clear that [Oswald] was a holy saint, so that a heavenly light stood straight up over the tent like a high sun beam throughout all the night, and the people all around that shire beheld it with great wondering.

The specificity of Ælfric's description of the light leaves little to no room for misinterpretation. The beams are a "heavenly light" sent by God in order to make Oswald's sanctity clear. The light is confined to the province in which Bardney is located. When the spread of Oswald's cult becomes a focus some seventy lines later, the light is conspicuously absent: "Þa asprang his hlisa geond þa land wide / and eac swilce to irland and eac suþ to franclande" (26.239–40) [Then his fame sprang throughout this land widely, and also even to Ireland, and also south to Frankland].[43] People from Ireland and Frankland are healed through his holy relics—but moreover, they are brought to a firmer faith as a result of his *hlisa,* which carries the semantic range of "sound, rumour, report, reputation, renown, fame, glory."[44] Joining the collectivity that

---

World: The Jarrow Lectures 1958–1993, ed. Michael Lapidge (London: Aldershot, 1994): 587–610.

43. See T. Northcote Toller and Joseph Bosworth, *An Anglo-Saxon Dictionary Based on the Manuscript Collections of the Late Joseph Bosworth* (Oxord: Clarendon Press, 1898), 544. The definition given in Toller and Bosworth for *hlisa* is "sound, rumor, report, reputation, renown, fame, glory"—clearly not the beams of healing light referenced in Bede.

44. Toller and Bosworth, s.v. hlisa.

binds together men with soil—changing both in the process—and that yokes the past of a people to their eventual Christian futures, we can begin to see the implications of the circulation not only of relics but also of narratives as agential objects. The very story of Oswald—his *blisa*—is as capable of bringing together people in divergent geographical areas as the initial distribution of his soil was of bringing together the English people through the working of miracles.

Conclusion

Oswald's story is, of course, entirely comprehensible without the use of Actor-Network Theory. In fact, it follows a traditional hagiographic model that requires proximity to the saint to create miracles: everything and everyone that is saved or cured is ultimately saved or cured by Oswald himself, or more precisely, through God's grace inherent in him. What that narrative ignores, and what Actor-Network Theory allows to come to the forefront, are all of the nonhuman actors that connect the saint to the person or thing being saved. Without the horse, the soil, the moss, the grass, and the fire, the full extent of the saint's involvement in the world cannot be recognized. An essential element of his sanctity is its manifestation of that sanctity on the world as a whole, not just on humans who inhabit it.

Ælfric's version of the story of the house fire throws Bede's narrative into high relief by adding one final element to the saintly ecology I have sought to trace through these narratives. As with the *blisa* that brings together a transnational group of believers after Oswald's bones are housed at Bardney, Ælfric adds, in essence, the *story* of Oswald to the collectivity implicit in the text:

> Eft siððan ferde eac sum ærendefæst ridde
> be þære ylcan stowe and geband on anum claþe
> of þan halgan duste þære deorwurðan stowe
> (26.221–23)

> And again, a horseman [was] bound on an errand by that same place, and [he] bound up some of the holy dust from that precious place in a cloth.

The implication in this description—with the absence of any physical marker to show the holy difference of the place in question—is that the "errand-bound man" *already knows* that this place is special. The dust, hung on a post, saves that post from a fire in this account of the miracle as well, but human minds are not prompted to seek the truth of the holy soil in this instance. Rather, "se post ana ætstod ansund mid þam duste, / and hi swyðe wundrodon þæs halgan werce gearnunga / þæt þæt fyr ne mihte þa moldan forbærnan" (26.234–36) [The post alone stood whole, with the dust, and they greatly wondered at the merit of the holy man that the fire could not burn the earth]. The implication in this scene is that the narrative of Oswald's holiness—perhaps even the hagiography itself—has already disseminated widely enough for these people to know without asking that the martyr-king's holiness is what makes these miracles possible. A last component of the collectivity is thus illuminated: joining the house, the fire, the dust, Oswald's blood, and the human minds who perceive it all, we have Oswald's *story*: the accumulation of miraculous evidence that occurs both during and after his holy life. This scene, finally, points to a narrative's participation in the holy ecology these versions of Oswald's life trace. The story is revealed to be as portable as the holy dust itself, collecting believers into a network of actors that implicates them in worlds natural, human, and divine.[45]

45. The author would like to thank her colleagues in the English Department at Ohio University as well as John Burden (Yale University) for their excellent suggestions and comments on drafts of this essay.

6

Undeadness
and the Tree of Life

THE ECOLOGICAL THOUGHT
OF SOVEREIGNTY

꧁꧂

Kathleen Biddick

perpetual spirals of power and pleasure
—Michel Foucault[1]

DEAD TREES started to drift into gallery spaces in the late 1960s, and they continue to wash up with tidal frequency on gallery shores today. Take for example the arresting tree-works of artists Robert Smithson (1969) (figure 6.1) and Anselm Kiefer (2006) (figure 6.2).[2]

---

1. Michel Foucault, *The History of Sexuality: Volume 1, An Introduction,* trans. Robert Hurley (New York: Random House, 1978), 45.

2. My reading of these dead trees is shaped by and also questions the boundary between creaturely life and animal life posed by the work of Eric L. Santner, *On Creaturely Life: Rilke, Benjamin, Sebald* (Chicago: University of Chicago Press, 2006), and his *The Royal Remains.: The People's Two Bodies and the Endgame of Sovereignty* (Chicago: University of Chicago Press, 2011). I am grateful to Randy Schiff for reminding me of Agamben's beautiful meditation on vegetative life (Aristotle believed in a plant soul): Giorgio Agamben, *The Open: Man and Animal,* trans. Kevin Attell (Palo Alto: Stanford University Press, 2004), 12–19. For the work of Robert Smithson, see *Robert Smithson: The Collected Writings* (Berkeley: University of California Press, 1996); on Anselm Kiefer, see review of Palmsonntag in *Whitehot Magazine,* April 2007, http://whitehotmagazine. com/articles/white-cube-mason-s-yard/396. Smithson's installation was lovingly reenacted in 1998 at the Pierogi Gallery, Brooklyn: Frances Richard, "A Tree Dies in Brooklyn," *Art Forum,* 36 (Feb. 1998): 19–20. See also the tree-work of artist Zoe Leonard, es-

As I contemplated writing this chapter on the medieval "tree of life," I could not help but notice that dead trees began falling into galleries synchronous with the so-called revolution in molecular biology, one which redefined the molar (arboreal root and branch) to the molecular (the spiral and rhizome).[3] In the temporal shadow of sculptures of the DNA spiral, widely exhibited in the late 1950s, Michel Foucault, a seer of this shift from molar to molecular, from classical sovereign power (to make die and let live) to modern biopower (to make live and let die), construed molecular biopolitics as a question of both power and pleasure: "perpetual spirals of power and pleasure."[4] Such spirals have become the instigation for important recent studies on vibrant matter and ecological thought by scholars such as Jane Bennett and Timothy Morton.[5] As Foucault envisioned his suggestive image, biotechnology laboratories had just figured out how to cut up spirals of DNA into molecular scraps to be spliced into engineered matter—think of such matter as souvenirs of the "undeadness" of the classical sovereign who can make die and let live. It is such undeadness, or what Eric Santner has called, an archive of creaturely life, that animates, I argue, the undeadness of dead trees in gallery spaces, and, as I shall unfold, the undeadness of historical trees of life.[6]

---

pecially *Tree* (1997), http://www.artseensoho.com/Art/COOPER/leonard97/leonard1. html. Leonard's tree became part of the "After Nature" exhibition, New Museum of Contemporary Art, New York, 2008. Mark Godfrey examines Zoe Leonard's work in "Mirror of Displacements: On the Art of Zoe Leonard," *Artforum International* 46 (March 2008): 292–302.

3. Bruno J. Strasser reflects on fifty years of DNA in "Who Cares about the Double Helix?" *Nature* 422 (24 April 2003): 803–4; see also Suzanne Anker and Dorothy Nelkin, *The Molecular Gaze: Art in the Genetic Age* (New York: Cold Spring Harbor Laboratory Press, 2003).

4. This essay addresses some of the intriguing points made in Tim Dean's "The Biopolitics of Pleasure," *South Atlantic Quarterly* 111 (2012): 477–95, although what is a vanishing act for Dean is a toggle for me. Dean draws attention to Foucault's molecular imagination of power and pleasure seen in *History of Sexuality*, 45. Relevant to the subject of trees, molar and molecular, see Mark A. Ragan, "Trees and Networks before and after Darwin," *Biology Direct* 4, no. 43 (2009), http://www.biology-direct.com/content/4/1/43.

5. Bennett, *Vibrant Matter: A Political Ecology of Things* (Durham: Duke University Press, 2010), 110–22. Bennett's imagination of the "pluriverse" is one of vortices, spirals, and eddies. In Timothy Morton, *Ecological Thought* (Cambridge, MA: Harvard University Press, 2010), ecological thought is a spiral (3). See also Bennett's recent exchange with Morton in "Systems and Things: A Response to Graham Harman and Timothy Morton," *New Literary History* 43 (2012): 225–33.

6. Santner, On *Creaturely Life*, xiii.

FIGURE 6.1. Robert Smithson, *Dead Tree* (1969).
Art © Estate of Robert Smithson / Licensed by VAGA, New York, NY.

FIGURE 6.2. Anselm Kiefer, *Palmsonntag* [Palm Sunday] (2007).
Art © The Gagosian Gallery. Photo by Douglas M. Parker Studio.

This chapter, therefore, returns to the fearful symmetry proposed by Foucault—the classical sovereign who could make die and let live and the modern biopolitical sovereign who can make live and let die—in an effort to think the cut and the spiral together in an untimely way. This chapter further asks whether it might be necessary to supplement the notion of *homo sacer* (the one who can be killed but who may not be sacrificed) with a diffracting notion of *res sacra* (the thing that can be cut but may not be sacrificed).[7] Imagined another way, how to think of the undeadness of the tree of life.

> *Although you cannot be my bride . . . you will assuredly be my own tree.*
> —*Ovid*, Metamorphoses *1:669–70*[8]

Scholars of sovereignty have written much on the becoming beast of the sovereign but have little to say about the becoming tree of sovereignty's objects, even though dendranthropy has fascinated fabulists such as Ovid.[9] Such muteness may be partially explained by the uncanny misfortunes of scholarly preservation: take, for example, the loss of Aristotle's book on plants, the companion piece to his well-studied book on animals; a forgetting of Wolfgang Goethe's amazing treatise on the *Metamorphosis of Plants* (with his exuberant imagining of the leaf as the evolutionary toggle). Likewise, Charles Darwin's prescient pedagogical immersion in botany is mostly marginalized in favor of his zoological dramas.[10] The animal turn of contemporary critique leaves plants on the verge. Yet, contemporary media rustles with trees, or what students of vibrant matter might call arboreal actants.[11] Take for example, the great

---

7. For *homo sacer* as the sign of sovereignty, the one who decides on bare life, see Giorgio Agamben, *Homo Sacer: Sovereign Power and Bare Life*, trans. Daniel Heller-Roazen (Stanford: Stanford University Press, 1998).

8. I am using Charles Martin's stunning translation of *Ovid, Metamorphoses* (New York: W. W. Norton and Company, 2004), 38. I use his line numbers in the citations that follow in this chapter.

9. Jacques Derrida, *The Beast and the Sovereign*, vol. 1, ed. Michel Lisse, Marie-Louise Mallet, and Ginette Michaud, trans. Geoffrey Bennington (Chicago: University of Chicago Press, 2009); Diego Rosselo, "Hobbes and the Wolf-Man: Melancholy and Animality in Modern Sovereignty," *New Literary History* 43 (2012): 255–79.

10. For an overview, see Ragan, "Trees and Networks."

11. Graham Harman has now become the arbiter of Bruno Latour's Actor-Network Theory (ANT). In his *Prince of Networks: Bruno Latour and Metaphysics* (Melbourne: RE.Press, 2009), *actants* are also called objects (not things): "No actor, however trivial, will be dismissed as mere noise in comparison with its essence, its context, its physical

FIGURE 6.3. Jessica Chastaine in *Tree of Life* (Fox Searchlight Pictures, 2011).

tree-rhizome, the Home Tree, the star of James Cameron's recent film, *Avatar* (2009), whose proposed fate was to be bombed to splinters for the sake of colonial mineral extraction ("It is only a god-damn tree" as the sovereigns of the film exclaim). Or consider the mystical dance of Mrs. O'Brien (Jessica Chastaine), who levitates around the trunk of a tree of life in Terrence Malick's film of that name (see figure 6.3). Malick had uprooted this majestic oak (all thirty tons of it), whose trunk, branches, and leaves punctuate *The Tree of Life,* and transported it eight miles to the movie set for transplanting. As I watched the You Tube video of the transport,[12] I understood that Malick wished to assure his viewer that this is *the* tree of life, not *a* life of a tree. These stories of arboreal uprooting trace the anxiety around the differences between life and "a life" closely observed by Jane Bennett: "A life thus names a restless activeness, a destructive-creative-force presence that does not

---

body, or its conditions of possibility. Everything will be absolutely concrete, all objects and all modes of dealing with objects will now be on the same footing" (13). A strong incarnational theology informs this vision. For a differing vision see Bennett's "Systems and Things: A Response to Graham Harman and Timothy Morton," *New Literary History* 43 (2012): 225–33.

   12. "Tree of Life." www.youtube.com/watch?v=sOfKsg7SH8c.

coincide fully with any specific body."[13] It is this noncoincidence of "a life"—its spectral materialism—that I track in the following four splices.

Splice One: 2049 BCE Bronze Cuts Wood: Seahenge, Northwest Coast of Norfolk

It is springtime 2049 BCE on the northwestern coast of Norfolk, England. The climate is warming; the sea is rising. The oak groves growing behind the shoreline are failing in the wet conditions. Enter a troupe of fifty or so bronze-axe-wielding men and women with the aim to fell over twenty oak trees to build a palisade and, also, to uproot a massive trunk (measuring five feet across by five-and-a-half feet wide with an estimated weight of two tons) of a huge century-old oak tree.[14] These armed folk were an elite entourage. Smiths were only just introducing into Britain the sovereignty of metallurgy (smelting the metals of copper and tin).[15] Just like an I-Phone today, bronze axes then were signs of the most avant-garde tool/weapon. In 2049 BCE, local Norfolk inhabitants would have been content with their elegantly knapped flint tools, which were capable of murder, butchery, deforestation, and harvesting. Industrial-scale mining in nearby Grimes Grave, Norfolk, furnished an ample supply of flint. So why inaugurate an exceptional metal ritual of tree cutting, one that excluded the use of contemporary flint tools (archaeologists have detected not one flint mark on the well-preserved timber from this excavation)?

Let us carry this question forward. Once "tree fellers" finished their task, they used honeysuckle ropes to drag the several tons of harvested wood along a timber pathway toward the coastal barrier, a distance of eighty feet or so. They had meticulously stripped the uprooted stump

---

13. Bennett, *Vibrant Matter*, 54.

14. My reconstruction is based on the survey by Charlie Watson, *Seahenge: An Archaeological Conundrum* (Swindon, UK: English Heritage, 2005). Della Hooke draws our attention to Seahenge in the opening of her *Trees in Anglo-Saxon England: Literature, Lore, and Landscape* (Rochester, NY: Boydell & Brewer, 2010), 3.

15. Here I am thinking of the reflections of Gilles Deleuze and Félix Guattari on early metallurgy: how early smiths linked the metal with the weapon to constitute a new assemblage, a constellation that deducts from the flow, thus the notion of a sovereign cut. In their words, "Metal is neither a thing nor an organism, but a *body* without organs" (*A Thousand Plateaus: Capitalism and Schizophrenia*, trans. Brian Massumi [Minneapolis: University of Minnesota Press, 1987], 411).

of its bark and relocated it *upside down* (roots facing upward) within the center of the henge-like palisade. They arranged the unstripped palisade posts with their bark-side facing outward. An eight-inch gap in the palisade, barely an entrance way, was blocked by a post set before the gap. This metal-wielding community had thus created an uprooted stripped trunk set within a fabricated trunk-like structure of the palisade that for all intents and purposes sealed it off.

Fast forward to 1998, when the tidal currents exposed the submerged henge and it became the subject of archaeological investigation. The palisade measured 24.5 feet in diameter and reached an estimated height of 13 feet. Druids and environmentalists protested what they saw as a desecrating excavation, but they lost their legal case. Eventually, earth-moving equipment lifted the central, inverted stump (see figure 6.4). No evidence of Bronze Age burial or habitation was found within the woodhenge. What to think, then, about this palisade designed by its makers to appear like a tree trunk enclosing the monumental inverted and stripped tree stump? The new life of metal ("we are walking, talking metals"[16]) collides in this woodhenge with the vegetal world. I like to think of this Seahenge as a chapter in the history of phenomenology before phenomenology. The bronze axe wielders used the new life of metals to investigate the properties of arboreal life. With their understanding of the "polycrystalline structure of nonorganic matter,"[17] these "metal people" uprooted and stripped a life of a tree in order to produce an enclosure that captured it as the tree of life and punctuated the undead seam of the organic and the nonorganic—a seam of violence and intensive labor. It is at such a seam that Franz Kafka wrote his short story of trees to remind us not to mistake a life of *a* tree for *the* tree of life: "For we are like tree trunks in the snow. In appearance they lie sleekly and a little push should be enough to set them rolling. No, it can't be done, for they are firmly wedded to the ground. But see, even that is only an appearance."[18]

---

16. My interpretation is inspired by Jane Bennett's discussion of a life of metal. Bennett quotes, here, Lynn Margulis and Dorion Sagan from their discussion of Soviet mineralogist Vladimir Ivanovich Vernadsky's notion of a "biosphere" in their book *What Is Life?* (Berkeley: University of California Press, 2000), 49; see Bennett, *Vibrant Matter*, 52–61.

17. Bennett, *Vibrant Matter*, 60.

18. The brief meditations on trees first appeared in 1912 in a collection called *Betrachtung*, see *Franz Kafka: The Complete Stories*, trans. by Nahum Norbert Glatzer (New York: Schocken Books, 1995), 474.

FIGURE 6.4. Excavation at Seahenge, Norfolk, UK (1998).
Photo copyright © Eastern Daily Press.

Splice Two: 8 CE Tomis on the Black Sea: "All that remains of her is a warm glow"

My second splice (8 CE) comes from Rome via the flat marshlands that border the Black Sea. There stood the Roman outpost of Tomis (now Constanta, Romania) to which the alleged traitor, Ovid (43 BCE–17/18 CE) was exiled by the personal command of his patron and sovereign, Augustus Caesar. Tomis might seem far afield from Britain, the topic of this edited collection, nevertheless, Ovid stood behind the writing desk of the leading British botanist of the late-eighteenth century, Erasmus Darwin, just as he also guided Carl Linnaeus on his ethnobotanical tour of Lapland.[19] In the medieval school curriculum, the *Metamorphoses* powerfully shape-shifted medieval pedagogy and commentary and spawned a world of medieval talking trees.[20]

One strand of the contested publication history of the *Metamorphoses* claims that Ovid put his finishing touches on his *Metamorphoses* as he went into exile. Ovid, the rhetorical master of becoming animal and

---

19. See, for example, Erasmus Darwin's proem to his "Loves of Plants" taken here from the second edition of his *Botanic Garden. A Poem in Two Parts* (New York: T&J. Swords, 1798):

> Whereas P. OVIDIUS NASO, a great Necromancer in the famous Court of AUGUSTUS CAESAR, did by art poetic transmute Men, Women, and even Gods and Goddesses, into Trees and Flowers; I have undertaken by similar art to restore some of them to their original animality, after having remained prisoners so long in their respective vegetable mansions; and have here exhibited them before thee. Which thou may'st contemplate as diverse little pictures suspended over the chimney of a Lady's dressing-room, connected only by a slight festoon of ribbons. And which, though thou may'st not be acquainted with the originals, may amuse thee by the beauty of their persons, their graceful attitudes, or the brilliancy of their dress. (265)

Likewise Linnaeus self-consciously fashioned his ethnobotany of Lapland after Ovid; see Lisbet Koerner, *Linnaeus: Nature and Nation* (Cambridge, MA: Harvard University Press, 1999).

20. See the important reference essay by Alison Keith and Stephen Rupp, "After Ovid: Classical, Medieval, and Early Modern Reception of the Metamorphoses," in *Metamorphosis: Changing Face of Ovid in the Medieval, and Early Modern Periods, ed.* Alison Keith and Stephen Rupp (Toronto: University of Toronto Press, 2007), 15–32, and also James G. Blark, Frank T. Coulson, and Kathryn L. McKinley, eds., *Ovid in the Middle Ages* (New York: Cambridge University Press, 2011). Dante relied on Ovid in his encounter with the thornbush in the Wood of the Suicides (into which Pier delle Vigne has metamorphosed); see Janis Vanacker, "'Why Do You Break Me'? Talking to a Human Tree in Dante's *Inferno*," *Neophilologus* 95 (2011): 431–45.

becoming vegetal (whom, surprisingly, Deleuze and Guattari silently pass over) knew all too well the state of exception, since the sovereign had decided on his treason and exiled him. Bodies, understandably, are up for grabs in the *Metamorphoses*. For the purposes of this splice, I focus on the story of Apollo and Daphne, because it addresses a famous example of the "becoming tree" of a human and because her metamorphosis raised for Ovid the pain of the radical exposure to sovereign jouissance. The poet rhetorically stages the sovereign cut in his description of Apollo's pursuit of Daphne. The closer the god gets to the fleeing nymph the more wolf-like he becomes. Ovid compares Apollo to a Gallic hound, a large canine breed valued by Romans and used for animal games in their amphitheaters: "He clings to her, is just about to spring, with his long muzzle straining at her heels, while she, not knowing whether she's been caught, in one swift burst, eludes those snapping jaws" (I.738–42). At the moment the sovereign Apollo rhetorically metamorphoses into a wolf-like creature, Daphne implores her father, Peneus, the river god, to transform her. She is arrested in midair as a thin layer of bark girdles her trunk and her feet turn into the roots of a laurel tree. Her head becomes a canopy: "Remanet nitor unus in illa" (I:761) [All that remains of her is a warm glow]. Ovid suggests with "nitor" that her glow, her luster, her brightness is the elusive phenomenal intensity and excess of Daphne's shape-shifting. Walter Benjamin, a student of aura, helps us to understand how Daphne's aura belongs to a disjunctive temporality and medium of perception that glows with the biopolitical excitation of her radical exposure to Apollo's sovereign jouissance.[21] The becoming tree of Daphne flickers with her sovereign abandonment. For Benjamin there was something human about objects, something vibrant, something noncoincidental with human labor. In a letter (7 May 1940) written just a few months before his death, Benjamin attempted to explain aura to the skeptical Theodor W. Adorno: "The tree and the shrub vouchsafed to people are not made by them. Thus there must be something human about objects that is *not* bestowed by the work done."[22]

---

21. The discussion that follows is inspired by the meticulous and radiant essay of Miriam Bratu Hansen on "Benjamin's Aura," *Critical Inquiry* 34 (2008): 336–75. Hansen questions the narrow understanding of Benjamin's notion of aura as articulated in many contemporary studies of technology of aura. Her excavation of Benjamin's work offers a rich analysis of the disjunctive temporalities of aura that cannot be contained by any essentialist notion of technological media. See also Beatrice Hanssen, *Walter Benjamin's Other History: Of Stones, Animals, Human Beings, and Angels* (Berkeley: University of California Press, 1998).

22. Gershom Scholem and Theodor W. Adorno, eds., *The Correspondence of Walter Benjamin* 1910–1940, trans. Manfred R. Jacobson and Evelyn M. Jacobson (Chicago:

Splice Three: "āstyred of stefne mīnum. . . . syđđan [stefn] ūp gewāt" [ripped up by my roots. . . . after the voice departed up][23]

The *Dream of the Rood,* an Anglo-Saxon poem about a talking tree, opens with an interjection: "Attention!" The careful listener can hear the brutal sound of its uprooting at the hands of the enemy. It is then transported to Calvary, where it will serve as the cross. Alone, it stands vigil by the corpse of Christ after the echoes of mourners have ceased to ring in the air. The poet plays with the Anglo-Saxon word *stefn,* a homograph which can mean "root" and also "voice." This splice pays attention to that homographic toggle as it tacks back and forth between the poem and its neighbor, a famous sculptured talking tree known as the Ruthwell Cross. These stone trees speak of radical exposure to the state of exception, the scene of the crucifixion. They were reexcavated by exiled German-Jewish art historians in the 1930s and '40s (and beyond to Ernst Kantorowicz's belated intervention in the debate in 1960s).[24] This splice asks how these talking trees serve as aerials that transmit a trauma, such that it becomes transitive, or what Bracha Ettinger has called "transtraumatic."[25] Their transmission of the sovereign cut paradoxically opens up a linked border space between talking trees and scholars in exile.

The Ruthwell Cross (see figure 6.5) brings us again to the beach, this time to the west coast of Scotland. When, in the early 1790s, the

---

University of Chicago Press, 1994), Letter 238, p. 629.

23. Anglo-Saxon text from *The Dream of the Rood,* ed. Michael Swanton (Exeter, UK: University of Exeter Press, 1996), ll. 30, 71; English translation from Elaine Traherne, ed., *Old and Middle English: An Anthology c.* 890–*c.* 1400 (Oxford: Blackwell, 2004), 108–15.

24. Ernst Kitzinger, "Anglo-Saxon Vine Scroll Ornament," *Antiquity* 10, no. 37 (1936): 61–71; Fritz Saxl, "The Ruthwell Cross," *Journal of the Warburg and Courtauld Institutes* 6 (1943): 1–19; Meyer Schapiro, "The Religious Meaning of the Ruthwell Cross," *Art Bulletin* 26 (1944): 232–45; Ernst H. Kantorowicz, "The Archer in the Ruthwell Cross," *Art Bulletin* 42 (1960): 57–59. On the emigration of German-Jewish art historians in 1933, when the Nazis instituted their race laws in the university see, Karen Michels, "Art History, German-Jewish Immigrants, and the Emigration of Iconology," in *Jewish Identity in Modern Art History,* ed. Catherine M. Soussloff (Berkeley: University of California Press, 1999), 167–79; Christopher S. Wood, "Art History's Normative Renaissance," in *The Italian Renaissance in the Twentieth Century,* ed. Allen Gerieco, Michael Rocke, and Superbi Fiorella (Florence, IT: Olschki, 2002): 65–92.

25. Bracha L. Ettinger, *The Matrixial Borderspace* (with a foreward by Judith Butler) (Minneapolis: University of Minnesota Press, 2006), 167. Her revisions of Lacanian psychoanalysis can productively open the boundaries between creaturely life and animal life defended by Eric L. Santner (see footnote 2).

FIGURE 6.5. Ruthwell Cross, South Face. © Trustees of the British Museum.
Photograph Project Woruldhord, University of Oxford
(http://projects.oucs.ox.ac.uk/woruldhord).

pastor of the parish of Ruthwell, Dumfriesshire (map ref. 54.9933163,-3.4084523), prepared his returns for the *Statistical Account of Scotland,* he reported that once upon a time an ancient, broken "obelisk" carved with runes and holy images, stood at the seashore. According to "tradition," the obelisk was eventually uprooted and drawn by oxcart to the churchyard, where it stood until the Reformation.[26] The locals remembered well; archaeologists tell us that the sea lapped the shores of early medieval Ruthwell, which lay much closer to the shoreline than the early modern parish.[27] In 1642, the cross was uprooted again and broken into pieces, one more casualty of sovereignty during the civil war between King Charles I and his Parliament. In protest of the king's tyrannical and popish activities, the General Assembly of the Church of Scotland had ordered in July 1640 that idols, such as the Ruthwell Cross, be destroyed, and so it was (July 1642). In 1887, the cross was reconstituted by sovereignty and dubbed an official "national monument."[28] Antiquarians and scholars have been putting together the fragments of these sovereign cuts ever since. It is a "Humpty Dumpty" network of pieces. Seeta Chaganti has already linked up part of the inscriptional network in her wonderful study of stone, metal, and parchment at stake in the Ruthwell Cross and the *Dream of the Rood.*[29] My splice seeks to link her network with other nodes: deathly fleas, ships, Jerusalem, Saracens, and the Warburg Institute in exile, in order to get at the performance of enemy, speech, and root that intertwine the tree of life incised on the Ruthwell Cross and the letters of the text of the *Dream of the Rood.*

---

26. John Sinclair, *Statistical Account of Scotland*, vol. 10 (Edinburgh, UK: William Creech, 1794): 226–27.

27. Christopher Crowe, "Early Medieval Parish Formation in Dumfries and Galloway," in *The Cross Goes North: Processes of Conversion in Northern Europe* CE 300–1300, ed. Martin Carver (Rochester, NY: Boydell & Brewer, 2003), 195–206.

28. Brendan Cassidy, "The Later Life of the Ruthwell Cross: From the Seventeenth Century to the Present," in *The Ruthwell Cross,* ed. Brendan Cassidy (Princeton, NJ: Princeton University Press, 1992), 3–34.

29. Seeta Chaganti, "Vestigial Signs: Inscription, Performance and *The Dream of The Rood,*" *PMLA* 125 (2010): 48–72. I would add to her exemplary bibliography: Andy Orchard, "The Dream of the Rood: Cross-References," in *New Readings in the Vercelli Book,* ed. Samantha Zacher and Andy Orchard (Toronto: University of Toronto Press, 2009), 225–53; Christopher Crowe, "Early Medieval Parish Formation in Dumfries and Galloway," in *The Cross Goes North: Processes of Conversion in Northern Europe AD 300–1300,* ed. Martin Carver (Rochester NY: Boydell & Brewer, 2003): 195–206. I have found Kate Thomas's use of queer theory to imagine networks especially helpful: "Post Sex: On Being Too Slow, Too Stupid, Too Soon," in *After Sex? : On Writing Since Queer Theory,* ed. Janet Halley and Andrew Parker (Raleigh, NC: Duke University Press, 2011), 66–75.

The tale of death begins thirty miles to the east of Ruthwell, at Bewcastle (map ref. 35.118979,-90.7234437), where, to this day, a sculptured
stone cross's shaft stands in the south churchyard. Scholars attribute the
Bewcastle and Ruthwell sculptures to the same workshop and reckon
them to be close in date. Éamonn Ó Carragáin emphatically dates the
Ruthwell Cross between 730–760 BCE.[30] The Bewcastle sculptors inge-
niously transformed a liturgical aid (known as a *liber vitae*) used for keeping track of the names of the dead to be commemorated during the
celebration of the Mass into a solid stone cube, sentiently bound on the
eastern face of the shaft by the incised branches of the tree of life. The
main western face of the Bewcastle shaft features a commemorative panel
etched in runes, now faded, on which, it is conjectured, were inscribed
the names of the dead. Blank panels on the southern and northern sides
of the Bewcastle Cross await the inscription of more names. These blanks
punctuate a network of actants that linked fleas and sputum, the bacil-
lus, *yersina pestis,* brought by ship from the Mediterranean and over-
land routes to northern Britain. The infected flea bite and the cough
devastated the densely occupied Northumbrian monasteries in the late
seventh century.[31] This epidemiological catastrophe, archaeologists tell
us, resulted in the reorganization of rural and monastic settlement, and
I argue, the commemoration of the dead. Fleas and sputum proliferated
into a "forest" of sculptured stone crosses in the north of Britain.[32]

The devastation of plague and the urgency of commemoration
inspired a major project of sacred topography undertaken by Adomnán,
abbot of Iona, who completed his treatise *De locis sanctis* in the plague-
ridden years of the 680s. He presented his study to Aldfrith, King of Northumbria (685–704), so that it could be more widely disseminated. The

30. Éamonn Ó Carragáin, *Ritual and the Rood: Liturgical Images and the Old English Poems of the Dream of the Rood Tradition* (Toronto: University of Toronto Press, 2005), 213.

31. J. R. Maddicott, "Plague in Seventh-Century England," *Past and Present* 156 (1997), 7–54.

32. In 1927, W. G. Collingwood estimated that there were about fifty such carved crosses and the work of the Corpus of Anglo Saxon Sculpture continues to add to the total: *Northumbrian Crosses of the Pre-Norman Age* (London: Faber, 1927). The University of Durham has now posted online volume 1 (Durham and Northumberland) of the *Corpus of Anglo-Saxon Stone Sculpture* by Rosemary Cramp (British Academy, 1977): http://www/ascorpus.ac.uk. See also, Richard N. Bailey, *England's Earliest Sculptors* (Toronto: Pontifical Institute for Medieval Studies, 1996), and the important study by Fred Orton and Ian Wood, eds., *Fragments of History: Rethinking the Ruthwell and Bewcastle Monuments,* with the assistance of Clare A. Lees (Manchester: Manchester University Press, 2007).

tales of one shipwrecked Arculf, who had visited the Holy Land (679–82) and washed up at Iona, served as the alibi for Adomnán's project on Christology and memorial topography.[33] Scholars have noted with great interest that Adomnán had access to recent information about Jerusalem as he wrote *De locis sanctis.* He mentions the Umayyad Caliph Mu'āwiya and reports that the "Saracens" had built in Jerusalem "a rectangular house of prayer (perhaps, the early stages of construction of the Dome of the Rock, which was completed c. 692 CE). Likewise in Damascus, he observed that there existed a church of the nonbelieving Saracens ["quaedam etiam Sarracinorum ecclesia incredulorum"].[34] Bede abridged Adomnán's treatise which further ensured its popularity. Thus we know that Bede was also apprised that Saracens were occupying the Holy Land. In his other writings, he tracked contemporary Muslim expansion in Europe. When he wrote his *Historia ecclesiastica,* he described Saracens as a "gravissima lues Sarracennorum" [a terrible plague of Saracens].[35] Put in sovereign terms, when Bede explained the etymology of *Agareni,* another term for Saracens, he used the Hebrew word for "enemy" [*ger*].[36] In sum, an exegetical interest in sacred topography, especially tombs, refined by Adomnán during the plague years of the 680s collided with new information about the expansions of a people of the desert [deserta Sarracenorum], whose inroad into Christian European space is likened to plague—the biological and the metaphorical intertwine.

If we listen attentively to the Ruthwell talking tree, does it speak of the "terrible plague of Saracens" voiced by monastic contemporaries? To explore this question, I turn to the projects and publications of German-Jewish scholars and associates of the Warburg Institute, who, in 1933 had fled from Nazi Germany to London, where the Warburg was relocated.[37]

---

33. See Thomas O'Loughlin, *Adomnán and the Holy Places: The Perceptions of an Insular Monk on the Locations of the Biblical Drama* (London: T&T Clark, 2007); and Katharine Scarfe Beckett, *Anglo-Saxon Perceptions of the Islamic World* (Cambridge: Cambridge University Press, 2008).

34. Quotations of Adamnan taken from Denis Meehan, ed. and trans., *Adamnan's De Locis Sanctis* (Scriptores Latini Hiberniae, 3) (Dublin: Dublin Institute for Advanced Studies, 1958), 98.

35. See Beckett, *Anglo-Saxon Perceptions,* for a summary of Bede's contemporary references, 123–24.

36. John V. Tolan, *Saracens: Islam in the Medieval European Imagination* (New York: Columbia University Press, 2002), 73.

37. See *Common Knowledge* 18 (Winter 2012), a special issue on the Warburg Library, edited by Anthony Grafton and Jeffrey Hamburger; Katia Mazzucco, "1941: English Arts and the Mediterranean: A Photographic Exhibition by the Warburg Institute in London," *Journal of Art Historiography* 5 (2011): 1–28. Georges Didi-Huberman's argu-

Under pressure to earn its way in its new British home and to prove its
Englishness, in 1941 the Warburg organized an exhibit of over five hundred photos devoted to the topic "English Art and the Mediterranean."
The exhibit proved to be a huge success and travelled to eighteen British cities over a two-year period. Pride of place belonged to freshly taken
photographs of the Ruthwell Cross. The reexposure of the cross launched
a spate of publications. But to backtrack for a moment: as early as 1936,
Ernst Kitzinger, a German-Jewish art historian in exile at the British
Museum, broached the scholarship of the Ruthwell Cross. He treated
only the vine scrolls of the tree of life carved into two narrow sides of its
shaft. His focus on ornament was exclusive. He ignored the incised runes
that ran in a band around the vine scrolls; nor was he interested in the
figural sculpture on the main faces of the cross. "Ornament" was what
mattered. Through the careful comparison of photographs of different
late-Antique and Coptic sculptures, Kitzinger crafted an intricate argument that brought the Ruthwell tree of life into direct contact with the
mosaic vine scrolls of the Dome of the Rock and the carved vine scrolls
on the wooden trusses that supported the roof of the Mosque at Damascus. What networks might have brought together Muslim Jerusalem and
Ruthwell were not his concern.

Warburg scholars and associates radically shifted the study of the
Ruthwell Cross in the 1940s.[38] Two scholars, Fritz Saxl and Meyer Schapiro, devoted themselves to interpreting the ensemble of figural images
carved on the two main faces of the cross. Schapiro mentioned the runic
inscriptions only in passing and ignored the tree of life vines on the narrow shafts of the cross. Saxl also maintained his distance from the tree
of life. In his introduction, Saxl claimed to "fight shy" of former discussions of Ruthwell ornament (the tree of life carvings) in order to study
the neglected figural program. The exclusions of the Ruthwell tree of life
from both these essays had the effect of silencing the nodes of Ruthwell
carvers and Muslim Jerusalem. But, as I shall show, things Islamic would
come to haunt Ruthwell scholarship again in 1960.

In debate with the war-time articles of Saxl and Schapiro, Ernst Kantorowicz, the noted author of *The King's Two Bodies: A Study in Medieval
Political Theology* (1957) and himself a German-Jewish exile at Berke-

---

ments about the domestication of iconology and his concept of the "rend" of an image
have influenced my reading: *Confronting Images: Questioning the Ends of a Certain History of Art,* trans. John Goodman (University Park: The Pennsylvania State University
Press, 2005).

38. Cassidy, "Later Life of the Ruthwell Cross," 29–34.

ley and Princeton, belatedly (in 1960) turned his attention to the figure known as the Ruthwell archer. Saxl had admitted that he was stumped by the figure: "but who is the archer beneath the cross-beam?"[39] Schapiro, in contrast, staunchly maintained that the archer was "the oldest medieval example of secular imagery."[40] Kantorowicz offered an uncharacteristic (for him) midrashic reading of the archer. Relying on the scholarship of Louis Ginzberg, a noted Talmudist and author of the multivolume series, *Legends of the Jews,* Kantorowicz argued that the archer represented Ishmael, son of Hagar, around whom a midrash of a wilderness archer and wild man grew up. Kantorowicz, the scholar of premodern sovereignty extraordinaire, suggests here that the archer marks a sovereign cut on the Ruthwell Cross, meaning that the archer raises the question of the enemy. His suggestion makes sense when we recall that Bede called the sons of Hagar the "enemy." The talking tree of life, too, murmurs of this sovereign cut when it tells us through its runic inscription that it was "wounded with arrows." Scholars have wondered why the runes give the plural form of arrow, since in the singular, it might be possible to translate "spear" which would fit perfectly into the Gospel narrative. Here, I think, the rend of the Ruthwell image comes into play. The shipwrecked Arculf, who had recited to Adomnán the stories of his stay in contemporary Muslim Jerusalem, might have related how the skilled archers of the seventh-century Rashidun army, with their superior composite bows, struck fear into the hearts of their enemy. The archer of the Ruthwell Cross, I argue, may be read as a toggle between a militant and sovereign Christian church and the sons of Hagar, noted and feared archers, whom Christians would name as the "enemy."[41] In his article on the Ruthwell archer, published three years before his death, Kantorowicz finally revisited his own (suppressed) ancestral roots—first his genealogy as a grandson of a noted Poznan rabbi, since he uncharacteristically cited Jewish midrash, and second, his work as a young military attaché in

---

39. Saxl, "Ruthwell Cross," 6.

40. Schapiro, "Religious Meaning," 238.

41. See Kathleen Biddick, "Dead Neighbor Archives: Jews, Muslims, and the Enemy's Two Bodies," in *Political Theology and Early Modernity,* ed. Graham Hammill and Julia Reinhard Lupton (Chicago: University of Chicago Press, 2012), 124–42; on early Muslim archers and their large composite bows see, John W. Jandora, "Archers of Islam: A Search for 'Lost' History" *Medieval History Journal* 13 (2010): 97–114; and David Nicolle, *The Great Islamic Conquests ad 637–750* (Oxford: Osprey Publishing, 2009). My interpretation pays attention to what Didi-Huberman calls the "rend" of an image and offers an alternative account of the archer to the one given by Carragáin in *Ritual and the Rood,* 141–43.

Istanbul during World War I, where he supervised the German work on the Orient Express and went on to write a dissertation on Muslim craft guilds. Kantorowicz was interested in the tree of life. It appears as an entry in the index of *The King's Two Bodies,* "Tree, Inverted."[42]

Splice Four: "The men of your confederacy have deceived you" (Cloisters Cross)

The story of my fourth tree of life, the Cloisters Cross, a ceremonial cross carved out of walrus ivory, brings us on the beach one more time, now to the ice floes of the White Sea, where the Sami harvested walrus and paid tribute in ivory tusks to Norsemen—or so Ohthere, Norwegian chieftain and merchant, told King Alfred of Wessex (c. 890 CE) as he presented him with samples of this highly valued raw material.[43] Sometime during the plague years of the late seventh century, Sami hunters felled a walrus (radio carbon date 676 x 694 CE). Its tusks, an heirloom, would be carved, several centuries later, into a ceremonial cross standing 23 inches high with an arm span of 14.25 inches.[44]

Shady post-World War II art markets put the Cloisters Cross into modern circulation. Since its "unprovenanced" purchase in 1963 by the

---

42. Ernst H. Kantorowicz, *The King's Two Bodies: A Study in Medieval Political Theology,* rev. ed. (Princeton: Princeton University Press, 1957; rev. 1997), 565.

43. Else Roesdahl, "Walrus Ivory," in *Ohthere's Voyages: A Late 9th-Century Account of Voyages along the Coasts of Norway and Denmark in Its Cultural Context,* ed. Janet Bately and Anton Englert (Roskilde, DK: Viking Ship Museum, 2007), 92–93; Danielle Gaborit-Chopin, "Walrus Ivory in Western Europe," in *From Viking to Crusader: The Scandinavians and Europe* 800–1200, ed. Else Roesdahl and David M. Wilson (New York: Rizzoli, 1992), 204–05. Once Greenland was colonized, c. 985 CE, its icy waters also served as a source of walrus tusks, which were traded along European trade routes: Marek E. Jasinski and Fredrik Søreide, "Norse Settlements in Greenland from a Maritime Perspective," in *Vinland Revisited: The Norse World at the Turn of the First Millennium,* ed. Shannon Lewis-Simpson (St. John's, NL: Historic Sites Association of Newfoundland and Labrador, 2000), 123–32.

44. For a comprehensive overview and bibliography of scholarly debates about the cross to 2006, see Elizabeth C. Parker, "Editing the *Cloisters Cross," Gesta* 45 (2006), 147–60. For excellent photographs and details of the inscriptions, see Elizabeth C. Parker and Charles T. Little, *The Cloisters Cross: Its Art and Meaning* (New York: Metropolitan Museum of Art, 1994). The tree of life of the Cloisters Cross is not of the vine type of the Ruthwell Cross; instead, it is a carved "rough-hewn" cross whose trunk sections are rendered as the knots and stumps of the palm tree: Jennifer O'Reilly, "The Rough-Hewn Cross in Anglo-Saxon Art," in *Ireland and Insular Art ad 500–1200,* ed. Michael Ryan (Dublin: Royal Irish Academy, 1987), 153–58.

FIGURES 6.6 AND 6.7. The Cloisters Cross. 12th century.
Front and Back. English, attributed to Bury Saint Edmunds,
Suffolk. The Cloisters Collection, 1963 (63.12). The Metropolitan Museum of Art. Image copyright © The Metropolitan Museum of Art. Image source: Art Resource, NY.

Metropolitan Museum of Art, debate over the cross has released an excess of historicism, as if historicism could resolve its enigma. Scholars disagree over the basics of geographical provenance—it could have been carved anywhere from England to Sicily, and it could date anytime between 1050 CE and 1180 CE. I propose in this splice an untimely reading of this tree of life, not to decide its provenance or date but instead to investigate how the Cloisters tree of life, another talking tree, transmits through its inscriptions a crisis of sovereignty and treason.[45]

The tree of life is carved on the front shaft of the Cloisters Cross. On its back, carvers have arranged in niches the busts of thirteen Old Testament prophets and the evangelist Matthew. Each bears a scroll engraved with scriptural verses. The Cloisters Cross prophets draw upon and compound a performative tradition of liturgical drama of the *Ordo Prophetarum,* with its roots in the anti-Semitic Advent liturgy, as well as popular "quasi-liturgical" drama (*Jeu D'Adam*), sculptural programs embellishing the portals of Romanesque cathedrals, and the genealogies represented in stained-glass windows representing the Tree of Jesse.[46] The performative impulse of the Cloisters Cross with its noisy riot of inscription plunges the cross into complex dramaturgy with an anti-Semitic genealogy.

The tree of life is carved on the front panel of the Cloisters Cross. Running down both of its sides are two engraved Latin couplets that read, "The earth trembles, Death defeated groans with the buried one rising. / Life has been called, Synagogue has collapsed with great foolish effort" [Terra tremit mors victa gemit surgente sepulto / Vita cluit synagoga ruit molimine stult(o)]. On the narrow sides of the shaft two other couplets are inscribed: "Cham laughs when he sees the uncovered limbs of his parents. / The Jews laughed at the pain of God suffering" [Cham ridet dum nuda videt pudibunda parentis. / Iudeis risere dei

---

45. My reading engages the work on the transtraumatic developed by Ettinger, *Matrixial Borderspace.*

46. The selection of prophets on the Cloisters Cross is most akin to the Rouen *Ordo Prophetarum,* which Edith Armstrong Wright thinks that Archbishop Hugues (1130–64) introduced to Rouen from the earlier Laon version: *Dissemination of the Liturgical Drama in France* (Geneva: Slatkine Reprints, 1980), 56. For background on the anti-Semitic aspects of this liturgical drama and the Laon tradition, see Regula Meyer Evitt, "Eschatology, Millenarian Apocalypticism, and the Liturgical Anti-Judaism of the Medieval Prophet Plays," in *The Apocalyptic Year* 1000: *Religious Expectation and Social Change,* 950–1050, ed. Richard Landes, Andrew Gow, and David C. Van Meter (Oxford: Oxford University Press, 2003), 205–29; Robert C. Lagueux, "Sermons, Exegesis, and Performance: The Laon *Ordo Prophetarum* and the Meaning of Advent," *Comparative Drama* 43 (2009): 197–220.

penam mor(ientis)]. At the foot of the trunk of the tree of life, Adam and Eve hold on for dear life. The tree of life climbs up the shaft of the cross and passes through a medallion depicting Moses and the Brazen Serpent. Its trunk culminates in an innovative iconographical scene that features the gospel story of Pilate and Caiaphas, arguing over the wording of the titulus to be affixed to Christ's Cross. Pilate holds a scroll inscribed with the text of John 19:22: "What I have written, I have written" [Quod scripsi scripsi]. The scroll held by Caiaphas features the text of John 19:21: "Write not the King of the Jews, but that he said: I am the King of the Jews."[47] The carved titulus above this scene features inscriptions in Latin, Greek, and Hebrew; it is important to note that the titulus contradicts the theology of the scene below between Pilate and Caiaphas. Its mistranscriptions offer clues to the sovereign cut of this tree of life. The Hebrew inscription of the titulus is undecipherable (the carver invented an unreadable Hebrew script); however, the Latin and Greek inscriptions are correct and contradict the Scripsi Scripsi scene; they inscribe their titulus to read: "Jesus of Nazareth, King of the Confessors [Rex Confessorum]"—not King of the Jews [Rex Iudeorum].

The *Rex Confessorum* titulus concerns me here because its deliberate unscripting of *scripsi scripsi* links the cross to the fierce rhetorical and iconographic debate fomented by the sovereign crisis between King Henry II and his Archbishop of Canterbury, Thomas Becket. Their battle culminated in Becket's assassination in December 1170 (during the liturgical season, when the anti-Semitic treatises on the *Ordo Prophetarum* were read). By the time of his assassination, Becket was regarded by the royal party as a traitor; his own followers immediately dubbed him with the martyrial title of "the Confessor." Henry II had precipitated the crisis when he tried to bully his lay and ecclesiastical barons into accepting the Constitutions of Clarendon in 1164. Infamous among the articles was Article 15, which ruled as follows: "Pleas concerning debts, which are due through the giving of a bond shall be in the jurisdiction of the king" [Placita de debitis quae fide interposita debentur, vel absque interpositione fidei, sint in justicia regis].[48] With Article 15,

---

47. Sabrina Langland, "Pilate Answered: What I Have Written I Have Written," *Metropolitan Museum of Art Bulletin* 26, no. 19 (1968): 410–29. See also Colum Hourihane, *Pontius Pilate, Anti-Semitism, and the Passion in Medieval Art* (Princeton, NJ: Princeton University Press, 2009), 201–3.

48. Latin text taken from William Stubbs, ed., *Selected Charters and Other illustrations of English Constitutional History from Early Times to the Reign of Edward the First* (Oxford: Clarendon Press, 1870), 167. Translation taken from Ernest F. Henderson, trans. and ed., *Select Historical Documents of the Middle Ages* (London: George Bell & Sons,

the king was arrogating to himself sovereign control over oral promises (*nuda pacta*) in cases of debt. Article 15 announced that an oral faith-plight (otherwise known as bare promise) in a debt transaction could *not* be the grounds for sending such disputes over money-lending to the church courts, where such disputes were traditionally heard. Thus when it came to debt, both faith-plight and documentary writing (an early notion of binding contract, *pactum vestitum* [a clothed or veiled pact] depended on written instruments) became the domain of the king's justice. Henry II thus declared oral promises to be the state of exception when it came to debt litigation and, in so doing, radically repositioned the sovereignty of written records and the nature of the archive itself.[49] Becket's resistance to the Constitutions famously resulted in his trial for treason in October 1164. During this trial and after his precipitous flight from London court to exile in France, the Becket circle, a transnational clerical group with broadly based theological, juridical, and artistic connections, polemicized against Henry II by staging him as a Jew.[50] The

---

1892), 15. For a discussion of the long-term sovereign and liturgical conflict over debt claims, see Richard H. Helmholz, *Roman Canon Law in Reformation England* (Cambridge: Cambridge University Press, 1990), 25–33.

49. This argument about the overriding (overwriting) of what came to be known as the "nudum pactum"—the naked pact made on faith between two legal persons—challenges us to rethink arguments about "memory to written record" as a crisis of sovereignty and faith rather than some accretion of governmentality. See Michael Clanchy, *From Memory to Written Record: England* 1066–1307, 2nd ed. (Oxford: Blackwell, 1993).

50. I think that Becket's Sicilian connections might be key for the provenance of the Cloisters Cross. On the transnational complexities of Becket's circle, see Anne Duggan, "Thomas Becket's Italian Network," in *Pope, Church and City: Essays in Honour of Brenda M. Bolton*, ed. Frances Andrews, Christoph Egger, and Constance M. Rousseau (Leiden, NL: Brill, 2004), 177–204; and more generally her essay, "*De consultationibus:* The Role of Episcopal Consultation in the Shaping of Canon Law in the Twelfth Century," in *Bishops, Texts and the Use of Canon Law around* 1100, ed. Bruce C. Brasington and Kathleen G. Cushing (Burlington, VT: Ashgate, 2008), 191–214. Becket as a patron of the arts is discussed by Ursula Nilgen, "Intellectuality and Splendour: Thomas Becket as a Patron of the Arts," in *Art and Patronage in the English Romanesque*, ed. Sarah Macready and F. H. Thompson (London: Society of Antiquaries, 1986), 145–58. For the relations of the Becket circle and the new professional illuminators of Paris, see Christopher F. R. de Hamel, *Glossed Books of the Bible and the Origins of the Paris Booktrade* (Woodbridge, UK: Boydell & Brewer, 1984), especially 55–63. More recently Patricia Stirnemann has linked these manuscripts associated with Becket to the Cistercian abbeys around Sens: "Indeed, the impetus behind the sudden emergence of this professional booktrade is not the schools, but the presence in Sens of three episcopal courts in the 1160s and 1170s: the papal court of Alexander III, the exiled archiepiscopal court of Thomas Becket, and the archiepiscopal court of Guillaume aux Blanches Mains, the brother of Henry the Liberal, who was elected to the bishopric of Charters in 1164 and then simultaneously held the archbishopric of Sens as of 1168" ("The

Becket circle mapped Henry II and his supporters as members of *syna-goga*. They accused Henry and his baronial henchmen (lay and clerical) of behaving worse than the Jewish High Priests, Annas and Caiaphas, and the Roman procurator, Pontius Pilate, at the trial of Jesus. They compared the scandal of suffragan bishops judging Becket in the king's court (October 1164) to Ham (also called Cham), the son of Noah, who laughed at the "the uncovered parts of a father." By celebrating, on the morning of his trial, the mass of St. Stephen Protomartyr, whose story of martyrdom had him dying at the hands of Jews, Becket further dramatized this Judaization of sovereignty.[51] Although Becket did not directly compare himself to Moses and the Brazen Serpent (the central medallion on the front of the Cloisters Cross), there are iconographic overtones of that image when he, vested in his liturgical apparel, carried his own archiepiscopal cross (usually a cross-bearer carried the cross in front of the archbishop) into the royal trial-chamber. A fellow juror and royal partisan, the outraged Bishop of London (dubbed the *archisyna-goga* in correspondence among Becket's friends), reprimanded Becket for coming to court "armed with a cross." After Becket's assassination, his circle intensified their Judaization of the sovereignty that was at stake in his trial (see Appendix to this essay for a selection of examples). They dubbed those clerics who allegedly betrayed Becket as Caiaphas and Pontius Pilate: "Indeed it is believed that his murder was arranged by the disciples who betrayed him, and planned by the chief priests: they outbid Annas and Caiaphas, Pilate and Herod in wickedness, in proportion as they took more pains to see that he was not brought before a judgment seat, was not summoned by accusers, did not appear before the face of a judge" [Et quidem, ut creditur, necem ipsius traditores procuravere discipuli, sacerdotum principes formaverunt, tanto in malitia, Annam et Caipham, Pilatum et Herodem amplius praecedentes, quanto diligentius praecauerunt ne in iudicium traheretur, ne conveniretur ab accusatoribus, ne appareret ante faciem presidis].[52]

Study of French Twelfth-Century Manuscripts," in *Romanesque Art and Thought in the Twelfth Century,* ed. Colum Hourihane [Princeton, NJ: Princeton University Press, 2008], 82–94; citation, 87).

51. For a discussion of thirteenth-century depictions of Stephen, see Karen Ann Morrow, "Disputation in Stone: Jews Imagined on the St. Stephen Portal of Paris Cathedral," in *Beyond the Yellow Badge: Anti-Judaism and Antisemitism in Medieval and Early Modern Visual Culture,* ed. Michael B. Merback (Leiden, NL: Brill, 2007), 64–86.

52. Excerpted from the *Letters of John of Salisbury,* vol. 2, (1163–80), ed. W. J. Millor and C. N. L. Brooke (Oxford, UK: Clarendon Press, 1979), Letter # 305 (1171), John of Salisbury to John of Canterbury, Bishop of Poitiers.

The Pilate and Caiaphas scene of the Cloisters Cross captures, I argue, the state of exception decided on by Henry II with the Constitutions of Clarendon and then uses iconography to Judaize the sovereign decision. Sovereignty could be Judaized, and Jews could be biopoliticized in this traumatic juncture of sovereignty and biopolitics in the twelfth century.[53] All this could be pinned by sovereignty on the tree of life. The Cloisters Cross speaks of making die and making live, the cut and the spiral.

The undeadness of the Cloisters Cross lives on most notably in the recent and widely travelled installation by Anselm Kiefer entitled *Palmsonntag* (2006). Kiefer installed a massive palm tree, preserved in resin and fiberglass, on the gallery floor. Large glass covered panels, displayed like the pages of an ancient text, show dead palm fronds, seedpods, and dried roses, beautifully arranged upon parched, cracked earth. The words *aperiatur terra et germinet salvatorem* are scrawled in handwriting against the dusty backdrop. They are the words of Isaiah 45:8: "You heavens above, rain down righteousness; let the clouds shower it down. Let the earth open wide, let salvation spring up, let righteousness grow with it; I, the Lord, have created it." When the artists of the Cloisters Cross invoked Isaiah on the central medallion depicting Moses and the Brazen Serpent, they chose a darker verse: "Why is thy apparel red and thy garments like theirs that tread in the winepress?" (Isaiah 63:2). These moments of arboreal undeadness link the untimeliness of the sovereign cut and biopolitical spiral in the border spaces of spectral materialism.

Appendix

Judaizing the Becket Conflict (some selected references)

Accounts of the Trial, October 1164
From *English Law Suits from William I to Richard I,* edited by R. C. Van Caenegem, vol. 2 (Selden Society, 1991), 446–57.

    a. Cham ridet reference

        However, he [Thomas Becket] complained much more about his suffragan bishops than about the judgment and the judging magnates,

---

53. Kathleen Biddick, "Arthur's Two Bodies and the Bare Life of the Archive," in *Cultural Diversity in the British Middle Ages: Archipelago, Island, England,* ed. Jeffrey Jerome Cohen (New York: Palgrave, 2008), 117–34.

declaring that it was an innovation and a new procedure to let an archbishop be judged by his suffragans and a father by his sons, adding that it would be a *lesser evil to laugh at the uncovered parts of a father than to judge the person of the father himself.*

Veruntamen multo magis quam de judicio seu proceribus judicantibus, de confratribus suis suffraganeis coepiscopis querebatur, novam dicens formam hanc et ordinem judiciorum novum, ut archipraesul a suis suffraganeis, pater a filius, judicetur. Minus fore malum, adjiciens, verenda patris detecta derider, quam patris ipsius personam judicare. (447)

b. Thomas and St. Stephen, protomartyr (who dies at the hands of Jews)

And on the advice of wise men, he in the morning before going to court, celebrated with great devotion the mass of St. Stephen the protomartyr, in whose office one reads the phrase: "Princes did sit and speak against me (Ps. Cxix, 23), and he commended his cause to the highest judge of matters, who is God.

Per consilium cujusdam sapientis, in crastino antequam ipse ad curiam pergeret, cum summa devotione celebravit missam de Sancto Stephano protomartyre, cujus officium tale est: "Etenim sederunt principles et adversum me loquebantur etc."; et causam suam summo judici, qui Deus ist, commendavit. (434)

c. Becket's archiepiscopal cross at the trial (overtones Moses and Brazen Serpent on the cross)

As he entered the hall, the archbishop soon took the cross from the cross-bearer who walked in front of him and openly in the sight of everybody carried his cross himself, as the standard-bearer of the Lord carries the standard of the Lord in the battle of the Lord.

Archipraesul vero, ut aulam ingreditur, a crucis bajulo ante ipsam crucem mox accipit et palam in omnium conspectu crucem ipsemet bajulavit, tanquam in praelio Domini signifer Domini vexilium Domini erigens. (452)

From Letters of John of Salisbury (some selected examples)

Excerpted from the *Letters of John of Salisbury,* vol. 2 (1163–80), edited by W. J. Millor and C. N. L. Brooke (Oxford, UK: Clarendon Press, 1979).

a. Synagoga

About Gilbert Foliot (Bishop of London and opponent of Becket):

Archisynagago

(Letter # 174 [July 1166] to Bartholomew, Bishop of Exeter); (Letter # 187 [Late 1166] to Baldwin, Archdeacon of Totnes)

PART III

Politics, Affect, and Life

7

Sovereign Ecologies

MANAGING THE KING'S BODIES IN
ANGLO-NORMAN HISTORIOGRAPHY

Joseph Taylor

ON 2 AUGUST 1100, King William II, known hereafter as Rufus (the red-haired), went late hunting in New Forest. A hunting companion, the Frenchmen Walter Tirel, fired an arrow at a stag that glanced off the deer and struck the king, who fell forward, further driving the missile into his body and killing him. In his *Gesta Regum Anglorum,* William of Malmesbury reports:

> Walter ran up, but finding him unconscious and speechless, he leapt hast-
> ily on his horse, and with good help from his spurs got clean away. Nor
> indeed was there any pursuit, one party conniving at his flight, others
> pitying him; but all had other things to think about, some fortifying their
> own places of refuge, some in secret carrying off what spoils they could,
> some looking about them every moment for a new king. A handful of
> country folk, with a horse and cart, picked up the king's body and carried
> it to the cathedral at Winchester, with blood dripping freely the whole
> way. There it was laid in the ground, within the tower, many nobles being
> present, but few to mourn him.[1]

---

1. All quotations of William of Malmesbury's *Gesta Regum Anglorum* come from William of Malmesbury's *Gesta Regum Anglorum,* vol. 1, ed. and trans. R. A. B. Mynors, R. M. Thomson, and Michael Winterbottom (Oxford: Clarendon Press, 1998), 574–75.

Accurit Walterius; sed quia nec sensum nec uocem hausit, perniciter cor-
nipedem insiliens benefitio calcarium probe euasit. Nec uero fuit qui per-
sequeretur, illis coniuentibus, istis miserantibus, omnibus postremo alia
molientibus; pars receptacula sua munire, pars furtiuas predas agere, pars
regem nouum iamiamque circumspicere. Pauci rusticanorum cadauer, in
reda caballaria compositum, Wintoniam in episcopatum deuexere, cruore
undatim per totam uiam stillante.

Here, William does not merely recount Rufus's death but also how his
corpse signals a sudden absence of law and rule and, at the same time, a
strange enmeshment of all England's subjects—both nobles and "country folk," both the deer with whom he shares his fateful arrow and the
ground that his blood now fertilizes, in a singular forest space. Dying in
the wilderness his father "constructed" as a royal hunting ground, Rufus
comes to illustrate the symbolic and fleshly remnants of the king's two
bodies. The playful paradox of Rufus as both hunter and game is complicated by accounts of his death, or more precisely, by the troublingly
bloody scenes of Rufus's body on parade through the greenwood. Early
Norman England frequently bore witness to the untimely deaths of kings
and heirs that complicated an already fragile state still reconciling the
1066 invasion and the melding of two disparate cultures: English and
Norman. In this chapter, I will suggest that this close attention to the
sovereign's fleshly body connotes an anxiety in this period over royal succession and the dissemination of sovereign power, signified in the biological fact of the flesh and blood king.

Rufus's blood-on-the-ground and the confused dispersion of royal
retainers and noblemen in the wake of his death point to the trouble with
a king's two bodies. Twelfth-century historians had not taken up exactly
Ernst Kantorowicz's problematic political theology of the sovereign—
that the sovereign's sempiternal body politic and his natural, biological body form a singular entity in kingship—but certainly their accounts
bear witness to a sense of the natural body of men passing through the
office of the king.[2] Reading the famous mirror scene from Shakespeare's
Richard II, Kantorowicz notes how the divested Richard is reduced to

---

2. Kantorowicz examines the obscure and singular work of the so-called Norman
Anonymous (c. 1100) before moving forward to what he sees as a late-twelfth- and thirteenth-century shift to a juridical model of kingship (Ernst Kantorowicz, *The King's Two
Bodies: A Study in Medieval Political Theology*, rev. ed. [Princeton: Princeton University
Press, 1957; rev. 1997]). For a recent revisitation of Kantorowicz's project, see the essays
in the forum "Fifty Years of *The King's Two Bodies*," collected in *Representations* 106, no.
1 (2009): 63–142.

"the banal face and insignificant *physis* of a miserable man, a *physis* now void of any metaphysis whatsoever."[3] But this explanation is problematized by the need of his usurper to murder Richard in Act V. If his body is simply "physis," then it would fail to threaten Bolingbroke. Only because a little touch of the king remains in Richard must he receive the death blow from Exton. Similarly, the graphic display of sovereign remnants in William of Malmesbury's account of Rufus's death suggests that something of the king's power lingers in his dead body, a corpse that the historiographer's audience must have desired to witness as dead and bleeding. Revisiting Kantorowicz's study more recently, Eric Santner contemplates the possibility of a sovereign residue left when the natural body is effectively spit back out. Rereading Richard II's divestiture, Santner contends, "What remains at the end of Richard's degradation is the appearance of the semio-somatic surplus that comes to amplify the human body when it is invested with a symbolic office, as minimally conceived as that office might be."[4] In other words, a little bit of the sublime body of the king still clings to Richard's natural body-now-divested, denying his altogether escape from the specter of sovereignty. Santner continues, "Richard's deposition is correlated with his exposure to a dimension of lawlessness immanent to the dynamics of sovereignty and sovereign succession that we have come to know as the state of exception or emergency, one that becomes manifest in the (state of) emergence of the new sovereign, Bolingbroke."[5] Much like Rufus's bizarre death and unceremonious funeral parade, Richard II's divestiture becomes an example of the uncanny proximity of the sovereign and the creature or, in Giorgio Agamben's formulation, *homo sacer* (the sacred man).

There is more in Rufus's hunted corpse, however, than the natural body of the king stained with a remnant of the so-called *corpus mysticum*. In explaining the king's two bodies as the natural body and the mystical body, composed of all English subjects "past, future, and present, actual and potential,"[6] Kantorowicz notes what he calls the "classic example" of the commonweal's description as organism, as human body, in John of Salisbury's *Policraticus* (c. 1150s). John of Salisbury's example proves influential to the explosion of legal thought witnessed in the thirteenth century in the theories of Thomas Aquinas, among others. John explains:

---

3. Kantorowicz, *The King's Two Bodies,* 40.

4. Eric L. Santner, *The Royal Remains: The People's Two Bodies and the Endgames of Sovereignty* (Chicago: University of Chicago Press, 2011), 47.

5. Ibid., 48.

6. Kantorowicz, *King's Two Bodies,* 195.

In this, nature, that best guide to living, is to be followed, since it is
nature which has lodged all of the senses in the head as microcosm, that
is, a little world, of man, and has subjected to it the totality of the members in order that all of them may move correctly provided that the will
of a sound head is followed.[7]

While John's formulation explains a body politic with the monarch as
its head, it also finds the entirety of this corpus to emerge from nature.
Like many twelfth-century political thinkers and historians, John derived
from Cicero, Virgil, and Seneca, the view that nature, including natural law, sanctioned human society and, consequently—as Cicero gleaned
from Plato's *Timaeus*—nature initiates the state.[8] In his adherence to the
theory that man is by nature a social and political animal, John "fore-
shadowed the Aristotelian-Christian naturalism of the later thirteenth
century."[9] It is John's clarity on the matter that I want to emphasize
here. In his figuration, the king himself is comprised of, and constructed
by, nature. Although John views nature as, ultimately, a product of God,
he sees it acting largely in autonomy. In so doing, as Cary Nederman and
Catherine Campbell contend, John "demonstrated a considerable degree
of respect for the value of temporal life apart from supernatural experience or revelation," and this was part of his "fashion[ing] a far more
flexible approach to the relationship between secular and ecclesiastical
realms."[10] Sovereign power, in this light, is not otherworldly but metanatural. When the king "goes bad" in these twelfth-century chronicles he
gives over to his natural proclivities, to his wildness. That Henry I was
wanton of women, that Rufus's court was rumored to be a place of sexual
depravity and excess, and that the Conqueror cruelly slaughtered thousands of his subjects were common criticisms of these Anglo-Norman
kings. But their sins do not so much result from a supernatural power
held in the divinity of the king—in his sublime body—as they result from
the sovereign's taking his nature too far. The realization is that, rather
than uniquely divine, the king becomes the one who gets to fully sub-
scribe to what is most natural about him. As I read these premodern

7. John of Salisbury, *Policraticus: Of the Frivolities of Courtiers and the Footprints of
Philosophers,* ed. and trans. Cary J. Nederman (Cambridge: Cambridge University Press,
1990), 28.

8. Gaines Post, *Studies in Medieval Legal Thought: Public Law and the State, 110–
1322* (Princeton, NJ: Princeton University Press, 1964), 503.

9. Ibid., 515.

10. Cary Nederman and Catherine Campbell, "Priests, Kings, and Tyrants: Spiritual
and Temporal Power in John of Salisbury's *Policraticus,*" *Speculum* 66 (1991): 588.

accounts of sovereignty, I want to press beyond the king's two bodies to get at the nature that drives *both* of them.

In Rufus's death, we see that the king's body maintains a complex relationship with the land he rules, not merely symbolic but real. Indeed it is difficult to ignore other forms of life that inhabit and perform in accounts of sovereign excess and death read in chroniclers such as William of Malmesbury, Orderic Vitalis, and Henry of Huntingdon. The active presence of nature in these crises-laden narratives calls us to revisit them with an ecological awareness. The immediate crisis of belief at the dawning of the twelfth century concerned the royal succession of an already problematic and still relatively new line of Norman kings, and it manifests a disturbing attention to the somatic substance of the sovereign in the early histories of Anglo-Norman England. Santner's arguments on the royal remains prove relevant for medieval kingship and the problems brought on by a detached symbolic flesh of the sovereign.[11] Yet, I want to push his metaphor of the flesh—the sacral soma or sublime flesh—to a more literal grounding. What we witness in these twelfth-century provocations of a king's two bodies is not the "physis now void of any metaphysis"; rather, we see the metanatural body of the sovereign reduced to ecological material. Rather than a remnant of the king's metaphoric sempiternity, what emerges is an ecology of the sovereign. Here, the king's body is sublime in its *super-natural* state, that is, its excessively natural presence. And its dying necessitates an equally excessive attention to that body's naturalness. I will, thus, examine these texts for how nature both informs and troubles historical representations of the new Norman state and how, as Lawrence Buell frames it, "artistic representation envisages human and nonhuman webs of interrelation."[12] The biologic facticity of Rufus's dead body that leaches into the ground

---

11. Though my reading aims to focus on the literal ooze of the king's natural body, rather than his symbolic form, it is important to note here a study that captures the extenuating presence of dead kings: Paul Strohm's insightful examination of Richard II's death, burial, and reburial in *England's Empty Throne: Usurpation and the Language of Legitimation, 1399–1422* (New Haven, CT: Yale University Press, 1998), especially 101–27. As Strohm notes, alluding to the disturbing interments and disinterments undergone by the corpses of Richard II, Henry IV, and Hotspur Percy, among others, "The problem with these unburied bodies . . . is that their troubled spirits continue to walk the earth" (102). Strohm probes this paradox via the Lacanian concept of the two deaths within which these figures are caught, physically dead yet symbolically undead. As was the case with the specter Richard II for the early Lancastrians, such restless spirits receive "an uncanny and disturbing power" (ibid.).

12. Lawrence Buell, *The Future of Environmental Criticism* (West Sussex, UK: Blackwell, 2005), 138.

of the New Forest—his "blood dripping freely the whole way" to Winchester—illustrates the manner by which the king's bodies are bound up
in what Timothy Morton calls an ecological mesh. As Morton claims,
"All life forms are in the mesh, and so are all dead ones, as are their habitats, which are also made up of living and nonliving beings."[13] Where we
find sovereign dead bodies, we also find these writers manifesting their
anxiety in accounts that weigh the biological fact of the king's body—
and perhaps its insignificance—within a larger life system, of which he is
a part.

Corpses Rot in Caen

That we should find early Anglo-Norman historians fleshing out the death
of kings is not surprising. As Paul Binski notes, "In the Anglo-French
sphere, division of the body reflected political and territorial divisions."[14]
Thus, proper burial and dispersing of the corpse and its parts—which
became more commonplace in the late-twelfth and thirteenth centuries[15]—demonstrated prestige and power in ways that mirrored the living king. But we must also consider these early historiographic accounts
in light of the events during and after the reign of Henry I within which
many of them wrote. Only a few years prior to these writers taking up
their historical projects, in 1120, Henry I's son, William Adelin, perished
when his boat sank leaving the Norman port of Barfleur. The White
Ship Disaster, as it is known, proved a stunning blow to Henry's rule.
As Warren Hollister clarifies, "Much of Henry I's political activity during the years that follow was shaped, directly or indirectly, by that single
disaster."[16] Both Adelin and his mother, Queen Matilda II, who had died
in 1118, served as Henry I's regents when the king was outside of England.[17] But the monarch's loss was more than convenient administration;

---

13. Timothy Morton, *The Ecological Thought,* (Cambridge, MA: Harvard University
Press, 2010), 29.

14. Paul Binski, *Medieval Death: Ritual and Representation* (Ithaca, NY: Cornell
University Press, 1996). Richard I, for example, kindly granted his guts to the abbey of
Chaluz in 1199, which caused "positive offence" (63).

15. Daniel Westerhof, *Death and the Noble Body in Medieval England* (Cambridge,
UK: Boydell & Brewer Press, 2008), 65.

16. C. Warren Hollister, *Henry I,* ed. and completed by Amanda Clark Frost (New
Haven, CT: Yale University Press, 2001), 280.

17. C. Warren Hollister and John W. Baldwin, "The Rise of Administrative Kingship:
Henry I and Philip Augustus," *American Historical Review* 83, no. 4 (1978): 267–305.

Adelin was his sole legitimate male heir, a fact not lost on the aging monarch. Awareness of his own mortality is witnessed in Henry I's founding the Cluniac Abbey at Reading—which he determined to be his own burial place—within a year of the disaster.[18] Foremost, however, we see the king "looking impatiently for fresh heirs from a new wife" (William of Malmesbury, *Gesta Regum Anglorum*, 762–63) [Parens enim celibatui renuntiauit, cui post Mathildis studuerat, futuros heredes ex nova coniuge iamiamque operiens], which he found in 1121 with his marriage to Adeliza of Louvain. Yet by the mid-1120s, the lack of a male to be produced between the couple lead the monarch to consider his sole remaining legitimate child, Matilda (also called Maud). Henry's eldest bastard son, Robert earl of Gloucester, might have made an excellent king himself, but illegitimacy dampened enthusiasm for his succession. Henry also had to fight off the claims of his nephew, William of Clito, the son of his troublesome older brother Robert Curthose. In 1126, Henry met with King David of Scotland; Conan, duke of Brittany; Rotrou, count of Perche; Geoffrey, archbishop of Rouen; and John, bishop of Lisieux, the head of his Norman administration, along with other important family and administrative subjects, in order to discuss succession.[19] This resulted in a regal ceremony of oath-taking performed by powerful men within and without the realm to support Matilda's claim. Her marriage to Geoffrey of Anjou in 1128 aimed to provide Matilda further support for her claim rather than to enlist a male ruler.[20]

If a crisis of succession was briefly averted, it returned with greater force following King Henry's death on 1 December 1135, while in a feud with his daughter and her husband. Chroniclers demonstrate suspicion at Matilda's capabilities. Noting the anonymous *Gesta Stephani*, among other historical works of the period, Carolyn Anderson explains how these texts "reveal anxiety over the prospect of Matilda as a powerful queen by elaborating and maximizing the roles of male relations and rivals, thus reducing the representation of Matilda as a powerful figure."[21]

---

18. Elizabeth Hallam, "Royal Burial and the Cult of Kingship in France and England, 1060–1330." *Journal of Medieval History* 8 no. 4 (1982): 369.

19. See Hollister, *Henry I*, 313.

20. Orderic Vitalis notes that Geoffrey of Anjou was to be a "stipendiary commander in his wife's behalf" (*The Ecclesiastical History of Orderic Vitalis*, 5 vols., ed. and trans. Marjorie Chibnall [Oxford: Oxford University Press, 1981], 6:482). For a fuller discussion, see Hollister, *Henry I*, 323–25.

21. Carolyn Anderson, "'Lady of the English,' in the *History Novella*, the *Gesta Stephani*, and Wace's *Roman de Rou*: The Desire for Land and Other," *CLIO* 29, no. 1 (1999): 47.

The dispute with her father and Matilda's already-problematic claim created opportunity for the king's nephew Stephen of Blois to assert his own legitimacy. What follows in England is the so-called Anarchy, a period of civil war and royal contestation between Matilda's and Stephen's factions that only ceases when, in 1154, Matilda's son Henry (the future Henry II) is crowned king following peaceful negotiations with Stephen. As Hollister notes, "Contemporary chroniclers were obviously confounded by the unanticipated turn of events after Henry's death and found it difficult to deal with the ethical and political implications."[22] Explaining what she calls the "analogical substantiation" of society's working through its encounters with pain, Elaine Scarry contends,

> At particular moments when there is within society a crisis of belief—that is, when some central idea of ideology or cultural construct has ceased to elicit a population's belief either because it is manifestly fictitious or because it has for some reason been divested of ordinary forms of substantiation—the sheer material factualness of the human body will be borrowed to lend that cultural construct the aura of "realness" and "certainty."[23]

It is this crisis of succession, spun from the White Ship disaster and leading up to the Anarchy, that, I argue in part, inflects historiographers' troubled accounts of the sovereign flesh. When the sovereign's political body is threatened in its very sempiternity through succession or civil war, his natural body's presence grows unwieldy.

The confusion of these writers, however, does not rest solely on such singular moments of crisis. Historians of the so-called Twelfth Century Renaissance of Latin historiography confronted in the Norman Conquest a historical rupture that already problematized sovereign legitimacy and simple identity. Refuting the critical commonplace that contemporary and near-contemporary histories of Anglo-Norman England fail to acknowledge the Conquest, Monika Otter explains how "their conspicuous avoidance" actually draws attention to the transition. Otter shows how, for example, the late-eleventh-century *Vita Aedwardi* projects Edward the Confessor's divinity and simultaneously "[launches] itself and its readers into the future . . . sidestepping the offensive moment of

---

22. Hollister, *Henry I,* 478.

23. Elaine Scarry, *The Body in Pain: The Making and Unmaking of the World* (Oxford: Oxford University Press, 1985), 14.

crisis itself."[24] Robert Stein further finds this historiographic enterprise aimed at smoothing over the sutures of Anglo-Saxon and Norman England. In reaction to the dilemma of the Conquest, then, Anglo-Norman historiographers fashioned English histories that contributed to what Stein has called "the master narrative . . . of the rise of the Norman state in England" wherein "an English state is . . . both the precursor and preordained outcome of the story."[25] Such a narrative depended on a seamless, if fictitious, genealogy of the English people surviving through centuries of invasion and cultural incursion and culminating in the new Anglo-Norman England. On the necessity of the English to maintain continuity, Stein notes, "The patriotic sense that 'there will always be an England' easily elides into the historical assurance that there always was an England."[26] Of course, such a formulation parallels the very notion of the medieval political theology of the king's sublime body, which assures that there will always be a king when the natural body dies: "The king is dead. Long live the king!"

As these authors imagine connectivity between the stories of Anglo-Saxon and Norman England, however, their accounts of union become troubled by omens that frequently and uncomfortably meld the natural and what we are apt to call the supernatural in implicit political commentary. Again, we can look to William of Malmesbury for such theatrics. Inching closer to an account of the Norman invasion in the *Gesta Regum Anglorum*, William tells of a witch at Berkeley who, foreseeing a terrible death at the hands of devils, begged her children to lessen her eternal torment with prayer and by letting her body lay in peace for three days. They place her, wrapped in a deerskin, within a stone coffin chained shut to protect her from the devils' onslaughts. She hoped that they might bury her after three days, but she worries, "The earth itself, which I have so often burdened by my wickedness, may refuse to take and hold me in its bosom" (William of Malmesbury, *Gesta Regum Anglorum*, 379) [Ita, si tribus noctibus secure iacuero, quarto die infodite matrem uestram humo; quamquam uerear ne fugiat terra sinibus me recipere et fouere suis, quae totiens grauata est malitiis meis]. For three nights, devils attack the church where her coffin sits, finally wresting her body away and dragging her to hell. We might write such tales off as fantasy or

---

24. See Monika Otter, "1066: The Moment of Transition in Two Narratives of the Norman Conquest," *Speculum* 74, no. 3 (1999): 585.

25. Robert Stein, *Reality Fictions: Romance, History, and Governmental Authority, 1025–1180* (Notre Dame: University of Notre Dame Press, 2006), 89–90.

26. Ibid.

divine intervention, but these stories also point to a concern that the
unnatural bonds forced between the Normans and the English, as well
as a spectral lingering of English identity—which I suggest, the witch's
body signifies—cannot be buried. In other words, nature will not cover
over these fissures, and the historian must labor anxiously to hide the
historical seams in the parchment of his books.

While we may locate the king's body in these texts, no attempt is made
to hide its wounds or its dying; rather, this body is made to bleed and
the corpse is frequently and explicitly on display in a larger panorama of
nature. Supplementing the initial books of his *Historia Anglorum* in the
late 1130s, Henry of Huntingdon begins Book X (c. 1138)[27] by alluding
to competing views of Henry I's reign that emerged just after his death.
Critical of those who viewed the King as excessively greedy and "at all
times subject to the power of women," Henry contends that these criticisms pale in comparison to the "dreadful time that followed" his death.
Specifically, Henry refers to Stephen's seizure of the English throne and
the "mad treacheries of the Normans" encapsulated in Stephen's and his
followers' failure to adhere to oaths of loyalty sworn to Matilda as the
rightful claimant. Henry follows this lament with an account of the grim
afterlife of the King's corpse:

> His body was brought to Rouen, and there his entrails, brain, and eyes
> were buried together. The remainder of the corpse was cut all over with
> knives, sprinkled with a great deal of salt, and wrapped in ox hides, to
> stop the strong, pervasive stench, which was already causing the deaths
> of those who watched over it. It even killed the man who had been hired
> for a great fee to cut off the head with an axe and extract the stinking
> brain, although he had wrapped himself in linen cloths around his head:
> so he was badly rewarded by his fee. (Henry of Huntingdon, *Historia
> Anglorum*, 702–3)

> Cuius corpus allatum est Rotomagum. Et ibi uiscera eius et cerebrum
> et oculi consepulta sunt. Reliquum autem corpus cultellis circumqua-
> que dissecatum, et multo sale aspersum coriis taurinis reconditum est,

---

27. Accounting for surviving manuscripts of the many iterations of the *Historia An-
glorum*, Greenway notes that version 3², in which Books VIII, IX, and X first appear, can
be dated around 1138, the year to which Book X runs in this instantiation. See Henry,
Archdeacon of Huntingdon, *Historia Anglorum*, ed. and trans. Diana Greenway (Oxford:
Clarendon Press, 1996), ivii–ixi. All subsequent quotations and translations of Henry of
Huntingdon are taken from this edition.

causa fetoris euitandi, qui multus et infintius iam circumstantes inficiebat. Vnde et ipse qui magno precio conductus securi caput eius diffiderat, ut fetidissimum cerebrum extraheret, quamuis lintheaminibus caput suum obuoluisset, mortuus tamen ea causa precio male gauisus est.

The king had died in the Forest of Lyons, far from Reading Abbey, and, with unfavorable winds keeping his body in France, his attendants were forced to eviscerate the remains to keep them from rotting.[28] While we see in Henry's illustration, the dead sovereign flesh penetrated repeatedly with blades, and though its salting makes it seem a piece of bovine or cervine meat to be eaten as much as buried, the king still "makes die" those who would dare to treat his remains as merely dead flesh. Henry notes in sovereign terms the gravity of the decapitator's dying: "He was the last of many whom King Henry put to death."[29] We might want to hear, in this account, the ghostly performance of murder by a revenant king. Yet even Henry of Huntingdon seems clear that death derives from the corpse's natural decomposition—the "pervasive stench" derived from organisms compelling putrefaction of the body kills unwitting servants. In its being consumed back into the ecological fabric, the body deals— or rather emits—one last sovereign blow. Indeed, the royal remains of Henry I refuse to go quietly even as the body is wrapped in animal skins and taken to St. Stephen's in Caen, where the Conqueror's body was also brought in 1087. Here it leaks a black liquid through the animal hides until it is finally taken to England, to Reading Abbey, where it is buried. Certainly, a latent morality inheres in Henry's of Huntingdon's account. As Danielle Westerhof suggests, "The misbehaving royal corpse in its unstoppable process of decomposition thus became a testament to [the king's] sinful behavior in life and a mirror of his wretched soul in death."[30] Henry's explicit textual autopsy of the king's body, how- ever, does more than merely didactic work. Melding the "physis" of the king's remains to a still-lethal power, Henry recounts an ecological body reabsorbed—seemingly desperate to be reabsorbed—into a larger system of life. What is more, his jarring narrative testifies to a broader historio- graphic attention to the sovereign flesh of Anglo-Norman kings emerg- ing in the wake of the crises brought on by the White Ship disaster and subsequent questions of succession.

---

28. See Hallam, "Royal Burial," 364.

29. "Hic est ultimus e multis quem rex Henricus occidit" (Henry of Huntingdon, *Historia Anglorum*, 703).

30. Westerhof, *Death and the Noble Body*, 31.

Henry wrote his *Historia Anglorum* at the request of Bishop Alexander of Lincoln. Diana Greenway explains that this commission probably came quickly after Henry's appointment to Lincoln in 1123 and before 1130, when he composed the epilogue to the work.[31] He wrote the final three books of the *Historia* in the years of Anarchy following Henry I's death. William of Malmesbury, according to Rodney Thomson, began his *Gesta Regum Anglorum* around 1124 and likely completed his first draft by 1126, revising the work until at least 1135.[32] Though the manuscript history of the *Gesta Regum* is winding and difficult, we can presume William began this text not long after the White Ship Disaster.[33] If the scope of Orderic Vitalis's *Historia Ecclesiastica* is significantly larger than Henry's or William's projects, Marjorie Chibnall has asserted that Orderic wrote much of the work—eleven-and-a-half of the thirteen books—between 1123 and 1137. Books VII–IX, which include much of the history of the Normans in his own time, date between 1130 and 1135.[34] Of Orderic, Chibnall long ago explained, "[He] wrote in the midst of the most momentous changes in Anglo-Norman government and society, when the rights of succession in kingdom and duchy were still imperfectly defined, when customs of succession were merely in the process of hardening into law."[35] But we might apply her comments, as well, to Orderic's contemporaries writing in the same period. As I hope to show, this vertiginous political period deeply affected these historians' attention to sovereignty within their texts.

Disparity in the chronicles pertaining to the deaths of the Anglo-Norman kings reveals the shift that occurs in historical attention to the king's body as the crises of the 1120s begin to pervade political culture in both England and Normandy. *The Anglo-Saxon Chronicle,* for

---

31. Henry of Huntingdon, *Historia Anglorum,* lvii. Alexander notably asked Geoffrey of Monmouth to translate the prophecies of Merlin from Welsh to Latin, as well (lviii).

32. R. M. Thomson, *William of Malmesbury,* rev. ed. (Cambridge: Boydell & Brewer, 2003), 6.

33. The edition of William of Malmesbury's *Gesta Regum Anglorum* from which I cite was prepared by its editors, R. A. B. Mynors, Rodney Thomson, and Michael Winterbottom, not as a representation of a single manuscript but as a collation of several manuscript witnesses that testify, perhaps, to William's own vision for the work. As they note, their edition is "one that we hope he [William] would have regarded as by and large close to his wishes" (xxiii).

34. Orderic Vitalis, *The Ecclesiastical History of Orderic Vitalis,* 5 vols., ed. and trans. Marjorie Chibnall (Oxford: Oxford University Press, 1981), 1.46–47. All quotations and translations of Orderic are taken from this edition.

35. Marjorie Chibnall, *The World of Orderic Vitalis: Norman Monks and Norman Kings* (Woodbridge, UK: Boydell & Brewer, 1984), 182.

example, soberly dispatches the Conqueror, claiming that "fierce death, which spares neither powerful men nor mean, seized him. He died in Normandy on the day after the Nativity of St. Mary, and he was buried at Caen in St. Stephen's monastery: he had built it, and afterwards endowed it richly."[36] The chronicler notes the irony: "He who had been a powerful king and lord of many a land, had then of all the land only a seven-foot measure; and he who was once clad in gold and gems, lay then covered with earth."[37] Here, history testifies to the problematic position of an invader-turned-king, one whose ferocity in conquest and rule is laid low by death. *The Anglo-Saxon Chronicle*'s account is bloodless, yet, even this testimony hints at the exploded sovereign ecologies to come in later accounts. Framing the king's death as a return to the earth, the *Chronicle* signifies the sovereign's power in land made small by the final plot wherein his body will reduce itself to waste.

The Conqueror's demise was long and excruciating. Like the dying of his son thirteen years later, William I's death began a melee in the legal absence of the king wherein retainers and servants scrambled to safe lodging and, worse, looted the body of any wealth. Orderic relates how the Conqueror's corpse was "despoiled by his own followers in another's house, and abandoned on the bare ground from the hour of prime to that of terce" (Orderic Vitalis, *Ecclesiastical History*, 4.102–3) [Ecce potentissimus heros cui nuper plus quam centum milia militum auide seruiebant, et quem multae gentes cum tremore metuebant, nunc a suis turpiter in domo non sua spoliatus est, et a prima usque ad terciam supra nudam humum derelictus est]. Yet the saga of the king's remains was only beginning. As David Douglas pointedly claims, "The blend of the earthly and the sublime in . . . descriptions of the Conqueror's death were even more blatantly displayed in the circumstances of his burial."[38] When William, archbishop of Rouen, ordered the body conveyed to Caen, no attendants were left to complete the task. In fact, Orderic explains, "At last a certain country knight called Herluin was moved by his natural goodness, and actively took charge of the funeral preparations" (Orderic Vitalis, *Ecclesiastical History*, 4.104–5) [Tunc Herluinus quidam pagensis eques naturali bonitate compunctus est, et curam

---

36. All quotations of the *Anglo Saxon Chronicle* are taken from the E-text as translated in *The Anglo-Saxon Chronicle*, ed. and trans. Dorothy Whitlock, David C. Douglas, and Susie Tucker (New Brunswick, NJ: Rutgers University Press, 1961), 163.

37. Ibid.

38. David. C. Douglas, *William the Conqueror: The Norman Impact upon England.* Berkeley: University of California Press, 1964), 362.

exequiarum pro amore Dei et honore gentis suae uiriliter amplexatus est]. The irony of this simple knight paying for the readying of the sovereign's cast-off corpse is a striking example of the way the biologic fact of the king's natural body (its decomposition) suddenly engulfs the remainders—both his subjects and his sublime self—in a perplexing web of relations, a state of exception in which law, ritual, and rule prove absent. Like his former subjects now living in the lawless zone of the exception, the king's body itself becomes *sacer*. Santner warns of Richard II's divested frame, "This is not the natural human body left over once all of one's social vestments have been stripped away, but something more like the rotting flesh of the sublime body, what remains when its sublimity has wasted away."[39] The Conqueror's body, physically stripped of any wealth by greedy servants and left in the hands of a country knight, seems to maintain an excess piece of the sovereign's sublime substance. Though the body rots, Orderic notes that this Herluin pays for and transports the corpse out of his "love of God and the honour of his race."[40] Though the stinking, naked body of the monarch might be literally dead, it still evokes a sublime love for the king and his realm.

The Conqueror's mortal remains problematize his metaphysical power as they continue their journey in Orderic's account. Following their arrival in Caen, the remains were readied for internment at the abbey of St. George, founded by the king himself. When the corpse was lowered into the stone coffin, already in the ground, Orderic verifies:

> Next, when the corpse was placed in the sarcophagus, and was forcibly doubled up because the masons had carelessly made the coffin too short and narrow, the swollen bowels burst, and an intolerable stench assailed the nostrils of the bystanders and the whole crowd. A thick smoke arose from the frankincense and other spices in the censers, but it was not strong enough to conceal the foul ignominy. So the priests made haste to conclude the funeral rites, and immediately returned, trembling, to their own houses. (Orderic Vitalis, *Ecclesiastical History*, 4.106–7)

> Porro dum corpus in sarcofagum mitteretur, et violenter quia vas per imprudentiam cementariorum breve intolerabilis faetor circum astantes personas et reliquum vulgus implevit, fumus thuris aliorumque aromatum

---

39. Santner, *Royal Remains*, 44.

40. See full quote n. 42. "Pro amore Dei et honore gentis suae" (Orderic Vitalis, *Ecclesiastical History*, 4.104–5).

de thuribulis copiose ascendebat, sed teterrimum pudorem excludere non
prevalebat. Sacerdotes itaque festinabant exequias perficere, et actutum
sua cum pavore mappalia repetere.

Much like the ground in Berkeley that was thought to refuse a local
witch in William of Malmesbury's symbolic story, the coffin and ground
at Caen seem to refuse the Conqueror's body. So too did the claimant
to the church's property, a certain man name Ascelin, refuse at the very
burial ceremony to allow the Conqueror's interment without compensation for what he claimed, and which were later proven to be, his lands. In
his *Historia Novorum in Anglia,* written largely by 1109, Eadmer offers
a more detailed account of the king's death than does the *Anglo-Saxon
Chronicle.* The Canterbury monk briefly explains "with what railing of a
slave he was buried," that the Conqueror was abandoned by all but "one
serving man," and that the ground in which he was to be buried was "disputed by a yokel."[41] Yet like the *Chronicle,* Eadmer ignores the body's
grotesque interment. Orderic, however, does not miss the implications of
the Conqueror's fleshly end. He concludes his account, "All who beheld
the corruption of that foul corpse learnt to strive earnestly through the
salutary discipline of abstinence to earn better rewards than the delights
of the flesh, which is earth, and will return to dust."[42] Not only does the

---

41.

> Qui autem region funere interfuerint, quave pompa corpus eius Cadomum
> delatum sit, quamque libere, immo quam servili calumnia, in ecclesia Beati
> Stephani sepultum sit, et dictum lugendum et auditu fatemur esse miseren-
> dum. Quem enim condition sortis humane non moveat ad pietatem, cum
> auditum fuerit regem istum qui tantae potentiae in vita sua extitit, ut in tota
> Anglia, in tota Normannia, in tota Cinomanensi patria, nemo contra impe-
> rium ejus manum movere auderet, mox ut in terram spiritum exhalaturus
> positus est, ab omni homine, sicut accepimus, uno solo duntaxat serviente
> except, derelictum, cadaver ejus sine omni pompa per Sequanam in naucella
> delatum, et cum sepeliri deberet ipsam terram sepulturae illius a quodam rus-
> tico calumniatam, qui eam haereditario jure reclamans conquestus est illam
> sibi jam olim ab eodem injuria fuisse ablatam?

All Latin quotations of Eadmer are taken from *Eadmeri Historia Novorum in Anglia,
et Opuscula Duo De Vita Sancti Anselmi et Quibusdam Miraculis Ejus,* ed. Martin Rule
(Wiesbaden, DE: Kraus, 1965), 24–25. All translations of Edmer are taken from *Edmer's
History of Recent Events in England,* trans. Geoffrey Bosanquet (London: Cresset Press,
1964).

42. "Inspecta siquidem corruptione cenosi cadaueris quisque monetur, ut meliora
quam delectamenta sunt carnis quae terra est et in puluerem reuertetur, labore salutaris
continentiae mercari feruenter conetur" (Orderic Vitalis, *Ecclesiastical History,* 4.108–9).

Conqueror's body prove didactic in its somatic misery, its struggle to be interred clarifies the fallacy of the sovereign's fleshly pursuits, his excessive nature, and the ways that these entangle the human body, the body politic, and the land itself. A "country knight" doing duty to king and realm pays for the body while a second, audacious knight, seeking justice from a now-dead tyrant who had stolen his lands, is paid for the ground into which the body will decompose, become dust. Meanwhile, the landscape itself—and the stone drawn from it for the king's coffin—force the body into pieces, accelerating the corpse's decomposition and its return to the soil. Nature has revealed the frailty of the sovereign's body in its spectacular return to the earth.

If his own physical death and interment were drawn out by a series of political and ecological complications, such an end proves fitting for a sovereign who had perpetrated so much death and destruction over the course of his life. Arguably the Conqueror's greatest sin against life— though many are noted in accounts of his death—is his Harrying of the North of England in the winter of 1069–70; however, accounts of this destruction—notably the horrible ends of his subjects—offer uncannily similar images to the Conqueror's own interment. Attacking the North for its part in a misguided alliance with the Danes, the Conqueror brought forces to Nottingham before moving to York, after which he, in the words of John Le Patourel, "split up his army into detachments which he set deliberately and cold-bloodedly to devastate all the country within reach."[43] *Domesday Book* testifies to numerous "wastes" in the North nearly twenty years after that bloody winter. Though many historians question the long-term destruction wrought by the campaign, early-twelfth-century accounts are pointed in their description of widespread devastation. William of Malmesbury concludes: "Thus a province once fertile and a nurse of tyrants was hamstrung by fire, rapine, and bloodshed; the ground for sixty miles and more left entirely uncultivated, the soil quite bare even down to this day" (William of Malmesbury, *Gesta Regum Anglorum*, 464–65) [Itaque provintiae quondam fertilis et tirannorum nutriculae incendio, preda, sanguine nervi succisi; humus per sexaginta et eo ampilius miliaria omnifariam inculta; nudum omnium solum usque ad hoc etiam tempus]. Further, William encapsulates the web of political and ecological devastation: "As for the cities once so famous, the towers whose tops threatened the sky, the fields rich in pasture and

43. John Le Patourel, "The Norman Conquest of Yorkshire." *Northern History* 6 (1971): 7.

watered by rivers, if anyone sees them now, he sighs if he is a stranger, and if he is a native surviving from the past, he does not recognize them" (William of Malmesbury, *Gesta Regum Anglorum*, 464–65) [Vrbes olim preclaras, turres proceritate sua in caelum minantes, agros laetos pascuis irriguos fluuiis, si quis modo uidet peregrinus, ingemit; si quis superest uetus incola, non agnoscit]. For William, not only has this Harrying rendered the life-bearing soil "bare," it has further demolished the threatening towers of the northern ecclesiastic and political center of York and its surrounding domains. The account of the *Historia Regum*, attributed to Symeon of Durham, which runs to 1129, illustrates specifically the famine spurred by the Conqueror's ecological torments. Symeon calls to mind how,

> so great a famine prevailed that men compelled to hunger, devoured human flesh, that of horses, dogs, and cats, and whatever custom abhors; others sold themselves to perpetual slavery . . . others, while about to go into exile from the country, fell down in the middle of their journey and gave up the ghost. It was horrific to behold human corpses decaying in the houses, the streets, and the roads, swarming with worms.[44]

> adeo fames prævaluit, ut homines humanas, quines, caninas, et catinas carnes . . . alii vero in servitutem perpetuam sese venderent . . . alii extra patriam profecturi in exilium, medio itinere deficientes animas emiserunt. Erat horror ad intuendum per domos, plateas, et itinera cadavera humana dissolvi, et tabescentia putredine cum fœtore horrendo scaturire vermibus.

Symeon's physiological testimony of bodily hunger, exhaustion, and human decay portrays the North as a topography both ghostly—with its unburied bodies starved, burned, bashed, stabbed, and trampled to death—and biologically overturned, humans on their way down the food chain in either cannibalized consumption by fellow survivors or as food for bacteria and worms. Orderic also notes the sheer number of lives "young and old" starved to death, and John of Worcester reiterates

---

44. Latin quotations of Symeon of Durham is taken from *Symeonis Monachi Opera Omnia*, vol. 2, *Historia Regum*. ed. Thomas Arnold (London: Longmans & Co., 1885), 188; translation is taken from *A History of the Kings of England*, trans. J. Stephenson (Dyfed, UK: Llanerch Enterprises, 1987), 137.

the horror of human flesh-made-food.[45] These chroniclers attend to the human costs of the Harrying with a pointed attention to the flesh.

Most astounding of these many retellings of the Harrying, however, is the Conqueror's own. Orderic's account of the Conqueror's deathbed confession intimates a relationship very different from the ones we've already witnessed between the king and his subjects.[46] As Orderic relates, the Conqueror lamented:

> In mad fury I descended on the English of the north like a raging lion, and ordered that their homes and crops with all their equipment and furnishings should be burnt at once and their great flocks and herds of sheep and cattle slaughtered everywhere. So I chastised a great multitude of men and women with the lash of starvation and, alas! was the cruel murderer of many thousands, both young and old. (Orderic Vitalis, *Ecclesiastical History*, 4.94–95)

> Vnde immoderato furore commotus in boreales Anglos ut uesanus leo properaui, domos eorum iussi segetesque et omnem apparatum atque supellectilem confestim incendi, et copiosos armentorum pecudumque greges passim mactari. Multitudinem itaque utriusque sexus tam dirae famis mucrone multaui, et sic multa milia pulcherrimae gentis senum iuuenumque proh dolor funestus trucidaui.

The King's own memory of the Harrying, as alleged here, offers a significant meditation on sovereignty and bare life in early Norman England. The sheer brutality of the campaign suggests that it was an effort not merely to subdue a sense of northern autonomy but to destroy it completely. What is remarkable here, however, is the Conqueror's self-admitted animal-rage in murdering both human and nonhuman subjects. This

---

45. Normannis Angliam uastantibus, in Northymbria et quibusdam aliis prouinciis anno precedenti, sed presenti et subsequenti fere per totam Angliam, maxime per Northymbriam et per contiguas illi prouincias, adeo fames preualuit, ut homines equinam, caninam, cattinam, et carnem comederent humanam" (10–11). All quotations of John of Worcester are taken from *The Chronicle of John of Worcester*, vol. 3, *The Annals from 1067–1140 with The Gloucester Interpolations and the Continuation to 1141*, ed. and trans. P. McGurk (Oxford: Clarendon Press, 1998), 10–11.

46. Orderic's diocesan, Gilbert Maminot, bishop of Lisieux, was present at the Conqueror's deathbed, so it is at least possible that Orderic had firsthand knowledge or close secondhand knowledge of the confession. At the same time, as Chibnall suggests, "some ideas very characteristic of Orderic's outlook are developed at length" in the Conqueror's long speech recounted in Orderic's work (*World of Orderic Vitalis*, 186).

"raging lion" preys upon the livestock of the region, slaughtering "flocks and herds," but only after he has made war on the land, burning all arable ground in northeast England in order to afflict the bodies of remaining inhabitants with starvation, their flesh turned to consuming itself. The Conqueror's account of his killing with impunity human, animal, and vegetable admits not simply to the punishment of perceived traitors to the realm but to the destruction of a whole web of human and nonhuman life in the North of England. More telling, Orderic's juxtaposition in his history of this confession with the account of the Conqueror's own troubled interment, his rotting flesh, intimates the uncannily similar ends of the king and his corporate body. If he is called to answer for many sins in his dying moments, the King remembers most significantly his beastliness, which is to say his excessive naturalness.

Eats, Shoots, and Leaves

That the Conqueror demonstrates such ecological self-awareness at the end of his life testifies to the significant attention he gave to England's landscapes during his reign, particularly to his own creation: New Forest. The Conqueror set aside seventy-five thousand acres initially for the forest, followed by another twenty thousand acres for which some two thousand people were displaced, and then a final afforestation of thirty thousand acres.[47] As Douglas notes, the Conqueror "came from a province which . . . was plentifully filled with forests," the ducal rights of which are evident in eleventh-century charters."[48] Although, as Douglas concedes, "in England, before the Conquest, the royal forest artificially created and fiercely protected was a familiar institution," Norman forest laws were more systematic and far-reaching, and "no doubt the forest law which came to be characteristic of medieval England was in its essentials an importation from Normandy."[49] Forest law protected not only animals but also trees and the soil itself. In the New Forest, a careful rotation of animal presence contributed to a profitable landscape for all. Cattle, for example, were removed from the Forest in midsummer when hinds and does birthed their calves and during winter heyning when forage

---

47. Charles Petit-Dutaillis and George Lefebvre, *Studies and Notes Supplementary to Stubb's Constitutional History* (Manchester, UK: Manchester University Press, 1930), 169–70.

48. Douglas, *William the Conqueror*, 372.

49. Ibid.

was scarce. Pigs were turned out for two months in fall to eat up green acorns that proved dangerous, when consumed in large quantities, for the forest's primary animals, deer and cattle.[50] New Forest was built at the expense of the human life that once inhabited its spaces and later managed by the new Norman state to promote the life of its species, but both the means to create this space and its ends served a sovereign function. As Orderic bluntly states, the Conqueror "forced the peasants to move to other places, and replaced the men with beasts of the forest so that he might hunt to his heart's content."[51] Making a similar complaint, that the Conqueror placed the hunting grounds and its game above his own people, the *Anglo-Saxon Chronicle* marks his death in the so-called *Rime of King William,* part of the 1087 entry:

> He made great protection for the game
> And imposed laws for the same,
> That who so slew a hart or a hind
> Should be made blind.
> He preserved the harts and boars
> And he loved the stags as much
> As if he were their father.
> Moreover for the hares did he decree that they should go free.
> Powerful men complained of it and poor men lamented it,
> But so fierce was he that he cared not for the rancor of them all,
> But they had to follow out the king's will entirely
> If they wished to live or hold their land property or estate or his favour
>   great.[52]

Here the Conqueror offers the animals of his forest the protection of the sovereign's law, and yet this sanctuary comes at the expense of a whole population of subjects, both secular and religious, divested of their livelihoods at the whims of the king. William Perry Marvin observes of the punishments for poaching noted in the *Rime,* "Such severe penalties that ranked poaching with homicide, rape, and treason, obviously shot far off

---

50. Coin R. Tubbs, *The New Forest: An Ecological History* (Newton Abbot, UK: David and Charles, 1968), 65–88.

51. "Guillelmus autem primus postquam regnum Albionis optinuit, amator nemorum plusquam lx parrochias ultro deuastauit, ruricolas ad alia loca transmigrare compulit, et siluestres feras pro hominibus ut uoluptatem uenandi haberet ibidem constituit" (Orderic Vitalis, *Ecclesiastical History,* 5.284–85).

52. *Anglo Saxon Chronicle,* 164–65.

any economic scale for compensating the value of a deer carcass."[53] The juxtaposition of the animal and human here is striking. Not only do the beasts inhabit a royal forest from which most men are prohibited to hunt, they are like children to the King, his own flesh and blood.

The manufactured nature of the New Forest perfectly illustrates the complex network of life—the way all life intertwined—in Norman England. The forest, however, was not kind to its Norman benefactors, often seeming to act violently upon the royal family in its midst. Recounting the death of the Conqueror's second son, Richard, William of Malmesbury recalls that "he was an elegant boy and for a child of that age, had high ambitions." But, William continues, "while shooting stags in the New Forest he caught some sickness from breathing the foggy and corrupted air."[54] William juxtaposes Richard's unexpected death with a reminder of how the Conqueror "with villages abandoned, had reduced for thirty miles and more to woodland glades and lairs for the wild beasts."[55] The passage continues with a description of how the Conqueror's grandson, the bastard child of Duke Robert, also named Richard, met his end while hunting in the forest when he was either pierced accidentally by an arrow or "hanged by the throat on the branch of a tree."[56] Already a space wherein the peasant or freeholder might be thrown off his land for the sake of the King's deer—or maimed or killed for his own illicit hunting—and already a space wherein the wildlife are as children to the sovereign yet exposed to his and his nobles' bows, the forest becomes the site of the royal family's own sacralization as well, their being made into bare life, killed with impunity in this state of exception by the human and nonhuman entities with whom they are enmeshed: the corrupt air, a sharp tree branch, or as with Rufus, by the human companion with whom he hunts the deer. In a space so notoriously governed by *silva lex,* a state of exception appears within which even the sovereign is immune from the protection of law.

---

53. Marvin, *Hunting Law,* 52.

54. "Ricardus magnanimo parenti spem laudis alebat, puer delicates et, ut id aetatulae pusio, altum quid spirans. Sed tantum primeui floris indolem mors acerba cito depasta corrupit; tradunt ceruos in Noua Foresta terebrantem tabidi aeris nebula morbum incurrissee" (William of Malmesbury, *Gesta Regum Anglorum,* 502–5).

55. "Locus est quem Willelmus pater desertis uillis per triginta et eo amplius miliaria in saltus et lustra ferarum redegerat" (Ibid., 504–5).

56. "Vunde pro uero disseritur quod in eadem silua Willelmus, filius eius, et nepos Ricardus, filius Rotberti comitis Normanniae, mortem offenderint seuero Dei iuditio, ille sagitta pectus, iste collum traiectus, uel, ut quidam dicunt, arboris ramusculo equo pertranseunte fauces appensus" (Ibid.).

Thinking ecologically about sovereign power allows us to witness the king's interrelation with all living bodies within and beyond his realm. In the New Forest, the sovereign aims to make die and let live the displaced or poaching humans he encounters. At the same time, he aims to make live and let die the flora and fauna—even the deer's forage, which he protects from grazing cattle in the winter months. We become unclear as to what exactly is in-lawed and what is outlawed, and unsure of where the human, the animal (other nonhumans), and the vegetable begin and end. One cannot say whether New Forest is an exceptional and legally empty center to an otherwise legal state or whether it perfectly encapsulates the space of a retributive natural law—one in which even the king must pay for his encroachments on nature. As he began to clarify explanations on biopolitics at the end of his 1975–76 lectures, Foucault offers the illuminating example of Spanish dictator Francisco Franco, who had recently passed away after a rather substantial and elaborate wielding of medical technology to prolong his life. In Franco's final moments, Foucault claims, "The man who had exercised the absolute power of life and death over hundreds of thousands of people fell under the influence of a power that managed life so well, that took so little heed of death, and he didn't even realize that he was dead and was being kept alive after his death."[57] For Foucault, this image illustrates the clash between "sovereignty over death" and the "regularization of life."[58] Nearly a decade later, a grim hospital photograph of Franco in those final days, wherein he lies on life support under a bevy of tubes and wires, was published on the cover of the Spanish magazine *La Revista del Mundo*.[59] If Foucault claims Franco's 1975 death as a moment of biopolitical shift, then the magazine's 1984 reflection on this sovereign's dying intimates (and illustrates) both the biological fact of his flesh and blood—which all the technology engulfing him in the picture is meant to sustain—and the awe (and disdain) for Franco's sovereign power still evident nearly a decade after his actual death, which makes the photo so compelling. One's response might be, as Foucault seems to imply: "Is this the man who

---

57. Michel Foucault, *Society Must Be Defended: Lectures at the College de France 1975– 1976*, ed. Mauro Bertani and Alessandro Fontana, trans. David Macy (New York: Picador, 2003), 249.

58. Ibid.

59. *La Revista del Mundo,* 29 October 1984; for a fuller discussion of the photographs in the context of Franco's legacy, see Francisco Larubia-Prado, "Franco as Cyborg: 'The Body Re-formed by Politics: Part Flesh, Part Machine,'" *Journal of Spanish Cultural Studies* 1, no. 2 (2000): 135–52. As Larubia-Prado notes, the photographer remains unknown (n. 2).

slaughtered hundreds of thousands of his subjects?" But our gaze more specifically ponders what, if any, sublime remnants are still evident in the dying sovereign's body. Though they are not the transition point witnessed by Foucault in Franco's plight, twelfth-century historiographers' representations of Anglo-Norman *kings-in-death* nevertheless anticipate the disciplinary and regulatory sovereignty to come. Presenting an eerily similar scene to Franco's "agonia y muerte," these texts witness the "sovereign over death" subjected to the biological fact of his flesh, a reenveloping of the sovereign body by nature. The crises of sovereignty born of the Conquest—and of Henry I's failure to produce a male heir and the resulting Anarchy—provoke a realization of a much larger field of relations, calling into question the sublimity of the king's body and, in the process, pulling his flesh into an ecological fray that can never be ruled but only, at best, managed.

Perhaps no single event illustrates this better than William Rufus's death at the unwitting hands of a hunting companion in New Forest in 1100. Rufus and his companions had ignored the usual custom of an early morning hunt, choosing to eat later in the day and depart afterwards. By all accounts, Rufus was struck by the arrow of a fellow huntsmen and died almost immediately. William of Malmesbury first names Walter Tirel, the Count of Poix and a benefactor to Anselm's Abbey of Bec, as the unfortunate archer. Intrigue has surrounded Rufus's death, including suggestions of a plot between his brother Henry I and the Clare family—to whom Tirel was related—and, more profound, inklings of witchcraft that would have Rufus voluntarily dying as "a secret initiate in the 'Old Religion,'"[60] but contemporary chroniclers writing much closer to the event seem to view it as an accident or, perhaps more emphatically, "the avenging hand of God."[61] The unexpected—if welcome—manner by which Rufus dies is recounted in each chronicle, yet, much like his father's, Rufus's death becomes a site of excess flesh in the most literal sense. In an early account of Rufus's killing, the *Anglo-Saxon Chronicle* illustrates: "In the morning after Lammas, King William when hunting was shot with an arrow by one of his own men, and then brought to Winchester and buried in that bishopric—that was in the thirteenth year after his succession to the throne."[62] These plain lines in no way match the

---

60. See C. Warren Hollister, "The Strange Death of William Rufus," *Speculum* 48, no. 4 (1973): 638.

61. Frank Barlow, *William Rufus* (Berkeley: University of California Press, 1983), 425.

62. *Anglo Saxon Chronicle*, 176.

vitriol the chronicler directs at Rufus later in the very same entry, when
he complains that Rufus kept "counsels of wicked men" and "because of
his avarice, he was always harassing this nation with military service and
excessive taxes."[63] Further, as Michael Bruno has argued, Rufus "had
little regard or respect for the Church except when it came to supplementing his own revenue."[64] As the chronicler complains, Rufus "keeps
down God's Church, and all the bishoprics and abbacies . . . he sold for
money."[65] Given this frustration with such a tyrant, we might expect the
chronicler to revel somewhat in the physicality of Rufus's demise, yet the
scene of regicide proves uninteresting when compared to the larger case
against Rufus as, simply, a bad king.

Later historiographers, however, literally and graphically return Rufus
to his natural roots. William of Malmesbury recounts how, just before the
fateful day of his death, Rufus was accosted by a certain monk, Robert
Fitz Hamon, who reveals to Rufus a startling dream. Fitz Hamon claims
that he saw Rufus burst into a church, seize a crucifix, and begin gnaw-
ing the arms and legs of the figure. With its leg about to be bit off, the
image strikes Rufus with its foot, knocking him backwards. Dragon-like,
the king then emitted a burst of flame from his mouth, such that "the
rolling billows of smoke even reached the stars."[66] Rufus merely laughs at
the vision and pays the monk for his entertainment, but the image of his
nearly cannibalizing Christ's own figurative flesh is translated—beyond
allusions to his wars with the church—into Rufus's becoming animal in
further accounts of his death.[67] Offering a narrative not heard in other
versions of Rufus's unfortunate end, Orderic writes:

> The moment the king was dead many nobles made off from the wood to
> their estates, and prepared to resist the disorders they anticipated. Some
> of the humbler attendants covered the king's bloody body as best they
> might with wretched cloths and carried him like a wild boar stuck with

---

63. Ibid.

64. Michael J. S. Bruno, "The Investiture Contest in Norman England: A Struggle
between St. Anselm of Canterbury and the Norman Kings: Part II," *American Benedictine
Review* 61, no. 3 (2010): 308.

65. *Anglo Saxon Chronicle*, 176.

66. "Ex ore iacentis tam effusam flammam exisse ut fumeorum uolumninum orbes
etiam sidera lamberent" (William of Malmesbury, *Gesta Regum Anglorum*, 572–73).

67. Walter Map's *De nugie curialium* will play on this story with another in which
Rufus relates to Gandulph, Bishop of Rochester, a vision he had in which he visits a chapel
where he finds a naked man lying on the altar. Rufus eats the man's finger before depart-
ing. See Gualteri Mapes, *De Nugis Curialium, Distinctiones Quinque*, ed. Thomas Wright
(New York: AMS Press, 1968), 222–23.

spears from the wood to the town of Winchester. (Orderic Vitalis, *Ecclesiastical History*, 5.292–93)

Mortuo rege plures optimatum ad lares suos de saltu manicauerunt, et contra futuras motiones quas timebant res suas ordinauerunt. Clientuli quidam cruentatum regem uilibus utcumque pannis operuerunt, et ueluti ferocem aprum uenabulis confossum de saltu ad urbem Guentanam detulerunt.

Like his contemporary, William of Malmesbury, Orderic frequently embellishes historical phenomena with rigorous detail or supernatural flourish. What is interesting here, however, is the telling manner by which he melds nature and man. Rufus becomes game of the forest—a "wild boar"—like the cervine progeny associated with his father in earlier accounts of New Forest's creation. If we meld Orderic's account with that of William of Malmesbury, quoted at the beginning of this chapter, then, Rufus's boar-like body ambulates through New Forest "with blood dripping freely the whole way." He is born not by nobles but by "country folk" or "some of the servants" with a "horse and cart." In the swirling mesh of this state of exception, peasants are made into "the hunters"—in, other words, the nobles or kings licensed to hunt there—while the sovereign's corpse equates to a stuck pig.

The relation of the sovereign to his subjects and to the earth becomes most evident here. Like the victims of the Conqueror's Harrying of the North, whose corpses provide sustenance to their still-living human and animal acquaintances or fertilize the earth itself, these kings' bodies saturate the ground of the forest in blood or spill out into the depths of a cemetery ground. The Norman kings, with their afforestation and forest laws acted out a brutal, exhibitionist sovereignty based on the notion of "right" that so often characterizes medieval politics. At the same time, however, in their management of the forest itself, we witness regulatory aims played out across a population of forest life—its forage, its timber, its venison—considered essential to the sovereign state. We are also frequently reminded that the New Forest was fashioned. As John of Worcester recalls, at the command of the Conqueror, "Men were expelled, homes were cast down, the land was made habitable only for wild beasts,"[68] and Henry of Huntingdon similarly notes, "He had vil-

68. "Antiquis enim temporibus, Eduuardi scilicet regis, et aliorum Anglie regum predecessorum eius, eadem regio incolis Dei cultoribus et ecclesiis nitebat uberrime, sed, iussu regis Wilelmi senioris, hominibus fugatis, domibus semirutis, ecclesiis destructis,

lages rooted out and people removed, and made it a habitation for wild beasts."[69] The king creates a space of security for the beasts of his realm, defending them in law from the whims of the unsanctioned hunter. But the king only protects these creatures from being killed by anyone else; they are, in fact, sacralized by the very law that aims to sustain their lives. Finally, this ironic killing chain comes full circle, as the king himself is sacked in the act of hunting. As Morton claims of the ecological thought, "Thinking interdependence involves dissolving the barrier between 'over here' and 'over there,' and more fundamentally, the metaphysical illusion of rigid, narrow boundaries between inside and outside."[70] In the wake of a crisis of succession, as his historical body proves more and more vulnerable, the sovereign's fleshly mortality emerges to overwhelm his sublimity, if there ever was any to begin with. Indeed, the king becomes the hunted.

In his *Policraticus,* John of Salisbury had called huntsmen "half-beast[s]" and even "monsters," by which he asserts it is even more shameful for kings to lower themselves to the activity. Alluding to Rufus's death among others, John concludes, "For they who lived while they could like beasts have often died like dogs."[71] Rufus's dead body is more than a sign of his frivolity or divine retribution for a life of sin. John explains his rant against hunting: "The activity . . . is laudable when moderation is shown and hunting is pursued with judgment and, when possible, with profit. . . . The wise is called a mad man, the just unjust, if we pursue e'en virtue beyond the realm of sense."[72] John's humanist bent, his desire to see humanity as imitating nature's inherent rationality and balance, however, undoes his larger systematic explanation of the political organism that he claims emerges from nature. Robert Pogue Harrison explains, in his seminal study of the greenwood,

---

terra ferarum tantum colebatur habitatione, et inde, ut creditur, causa erat infortunii" (John of Worcester, *Chronicle,* 92–93).

69. "Vnde in siluis uenationum, que uocantur Noueforest, uillas eradicari, gentem exstirpari, et a feris fecit inhabitari" (Henry of Huntingdon, *Historia Anglorum,* 404–5). The public frustration with afforestation culminates in four clauses (44, 47, 48, 53) seen in the Magna Carta that demanded disafforestation of forests created in John's reign (47), as well a call for restoration of, or investigation into, lands dispossessed in afforestation (53).

70. Morton, *Ecological Thought,* 39.

71. John of Salisbury, *Frivolities of Courtiers and Footprints of Philosophers, Being a Translation of the First, Second, and Third Books and Selections of the Seventh and Eighth Books of the Policraticus,* trans. Joseph B. Pike (New York: Octagon Books, 1938). 26–27.

72. John of Salisbury, *Frivolties,* 34. John refers to Horace's *Epistles* I.vi; see *Horace: Satires, Epistles, and Ars Poetica,* trans. H. Rushton Fairclough (Cambridge, MA: Harvard University Press, 1929), 286.

The king embodies and represents in his person the civilizing force of history, but by the same token he harbors in his sovereignty a savagery that is greater and more powerful than the wilderness itself. Had he not this more primordial nature he could be neither the protector nor the ruler of his kingdom. As sovereign of the land, the king overcomes the wilderness because he is the wildest of all by nature.[73]

Harrison broaches the very ecological underpinnings of a political theology of kingship. The king is, first, *of* nature. If, as political theology contends, the sovereign finds his identity and power in the exception, then the forest space as the quintessential state of exception grounds sovereign identity in Anglo-Norman England.

The many political and ecological layers at work in the New Forest are made further evident by the context of Rufus's death. John of Worcester notes, "In the place where the king fell, in former times, a church had been built, but in his father's time, as we have said, it was destroyed."[74] These chronicle narratives try to reason with events, denoting, as Eadmer claims of Rufus's death, "justo judicio Dei" [the righteous judgment of God].[75] In the case of the Conqueror, for example, his untimely end and the rather difficult time had interring his body result, it is claimed, from his attacks on the church. Yet the very ground that refuses his body and against which his fat belly bursts is indeed that of the church he endowed. Rufus's own wars with the church are duly noted in most accounts of his reign. If he quarreled with his barons repeatedly, he also notoriously fought with his bishops, particularly Anselm, the archbishop of Canterbury.[76] These wars with the church lead Henry of Huntingdon, for example, to fume, "On the day of his death he had in his own hands the archbishopric of Canterbury, the bishoprics of Winchester and Salisbury, and eleven abbeys let out at farm."[77] As

---

73. Robert Pogue Harrison, *Forests: The Shadow of Civilization* (Chicago: University of Chicago Press, 1993), 74.

74. "In loco quo rex occubuit, priscis temporibus ecclesia fuerat constructa, sed patris sui tempore, ut prediximus, erat diruta" (John of Worcester, *Chronicle,* 92–93).

75. Eadmer*i Historia Novorum in Anglia,* 116.

76. Barlow, *William Rufus,* 339. The Conqueror had ceased, in the final years of his reign, to recognize a pope. Norman England thought it politically expedient to ignore the papal schism that followed Gregory VII's death in 1385, particularly since England had no immediate cause to appeal to Rome at that time. But Anselm pressed the issue, asking repeatedly to go to Rome to claim the pallium for his office, an act Rufus repeatedly refused because it signified his own recognition of Pope Urban II.

77. "Siquidem in die qua obiit, in proprio habebat archiepiscopatum Cantuarie, et

Frank Barlow notes, however, Rufus made several grants to monasteries and particularly to Battle Abbey in the memory of his father. And he was remembered reverently at Rochester, Westminster, and Bermondsey, for which he contributed to the foundations.[78] What is more, the particular synchrony of his falling dead on the very spot where his father tore down a church illuminates the manufactured nature of the greenwood itself. If monastic chroniclers such as Henry or William of Malmesbury, who have been known to give biased accounts of those historical figures who had failed to treat their own religious houses with respect, imply that Rufus's death ends a tit-for-tat, we must also marvel at how the church itself encroached on nature with its own manufactured spaces. What, in fact, was cleared to make space for this church? These environmentally conscious monastics are not angry because Norman kings encroached on their serene woodlands; rather, they are mad because sovereign woodlands encroached on their ecclesiastical edifice. In this sense, the wild spaces that made up New Forest become what Alfred Siewers has referred to as "palimpsest[s] of cultural interactions with the land in rich layers of time and noontime."[79] But as we peel back the legal and ecclesiastical layers of this space, we are only likely to find more *natural* remnants of the body politic, of the Romans, the Britons, or even the giants.

Rufus's reign seems most disturbing to William of Malmesbury, amidst the many crises both he and his contemporaries wrestled with in the early twelfth century. *The Anglo-Saxon Chronicle* claims, in the months prior to Rufus's death, that "there was seen blood bubbling out of the earth" at a village in Berkshire,[80] and the vision is repeated in several other histories. But William posits Rufus's reign as shot through with nature's unnatural omens in a startling string of images, one after the next, offered just before he illustrates Rufus's death in Book IV of the *Gesta Regum Anglorum*. An earthquake rocks England, according to William, in the second year of Rufus's reign, while a violent lightning storm in his fourth year seems to have brought the devil himself to a church in Winchcombe. In that same year, more than six hundred houses were destroyed by "a great confusion of quarrelsome winds,"

---

episcopatum Winceastrie, et Salesbirie, et undecim abbatias ad firmam datas" (Henry of Huntingdon, *Historia Anglorum*, 449).

78. Barlow, *William Rufus*, 113.

79. Alfred K. Siewers, "Ecopoetics and the Origins of English Literature," in *Environmental Criticism for the Twenty-First Century*, ed. Stephanie LeMenager, Teresa Shewry, and Ken Hiltner (New York: Routledge, 2011), 110.

80. *Anglo Saxon Chronicle*, 176.

and a year later, lightning blasted off the roof of the Salisbury Cathedral tower. Floods, famine, a comet, the aforementioned spring of blood and—repeating John of Worcester's text—the appearance of the Devil in the woods mark subsequent years leading up to Rufus's death. Barlow notes, "The several portents described helped to transfer the blame to the king himself, for he irreverently refused to take heed of repeated warnings."[81] This encapsulation of nature's speaking to the king is more compelling than Barlow lets on. The political, the divine, and the natural come together in Rufus's end in a disturbing, agential web. The Norman kings cut down churches and villages in order to better preserve the flora and fauna that are their pleasure—even their progeny—yet the land kills the royals for their overarching claims to ownership over the very nature that sanctions them; all the while, the divine speaks through nature, (or does nature prophesy?) portending the death of kings at the hands of both nature and God. In all of these histories, when the king lies dead, we see not merely John of Salisbury's natural body politic seeking a new head; rather, we witness the sovereign and the subject, the man and the animal, the human and nonhuman living and dying into an ecological mesh—nature, rather than *Nature,* enfolded over every singular identity or office.

Fish Food

Henry of Huntingdon's rendering of the White Ship disaster in his *De Contemptu Mundi,* first written around 1135, omits any direct illustration of the shipwreck itself. He had accounted for this briefly in Book VII of his *Historia.* Instead he explains England's loss via a juxtaposition of the accoutrements of kingship and the royal corpse. England's heir apparent, William Adelin, is enveloped by ecological forces beyond his control. As Henry laments, "Accordingly it came about that instead of wearing a crown of gold, his head was broken open by the rocks of the sea; instead of being dressed in gilded apparel, he was tossed about naked in the waves; instead of gaining the loftiness of kingly rule, he was buried in the bellies of fishes at the bottom of the ocean" (Henry of Huntingdon, *Historia Anglorum,* 594–95) [Contigit igitur ei quod pro corona auri rupibus marinis capite scinderetur, pro uestibus deauratis nudus in mari uolutaretur, pro celsitudine regni maris in fundo piscium uentribus

81. Barlow, *William Rufus,* 426.

sepeliretur].[82] Henry's sad narrative comprises two sovereign bodies, but these are not the distinct natural and political corpora of which Kantorowicz famously speaks.[83] Henry envisions the signifiers of sovereignty's sublime essence—his crown, his robes, and his power—yet, these signs are overwhelmed by Adelin's body in its biologic facticity: its smashed head, its nudity, and its becoming part of oceanic ecosystem. If the sublime attributes are not lost, given the dispersion of biological material, then they remain attached to the dead flesh with nowhere else to go.

What do these historians witness in their lingering attention to the sovereign's dead body? Writing of Eduardo Kac's genetically altered fluorescent green rabbit and the strange combinatory efforts of other genetic research, Morton asks: "Is the horror of this art simply shock value derived from the clichéd Frankenstein interpretation—that science has overstepped its bounds of human propriety? Or is it the revelation that if you can do that to a rabbit, then there wasn't that much of a rabbit in the first place?"[84] Morton's point is that such definitive identities as "rabbit" do not account for the essence of each creature: "We are faced with the extraordinary fact of increasing detail and vanishing fullness."[85] This attention to modern science equally informs medieval crises of sovereignty. In the details of royal death repeatedly offered by twelfth-century chroniclers, much like the disturbing photos of a fascist dictator on life-support, we sense a desire to find the sublime body of the king, a sempiternal flesh that supersedes the historical ruptures of conquest or the untimely death of the monarch's natural body. The terrible reality— the seeming Real—of these sovereign ecologies, however, is that there really wasn't much sovereignty there in the first place. Like the bizarre events that often seem to spell the coming of the Norman Conquest,

---

82. Henry likely knew King Henry I's bastard son Robert, who also perished in the White Ship Disaster. Henry spent part of his youth with Robert in the household of Bishop Robert Bloet, former Chancellor to Rufus. And, as Diana Greenway notes, Henry might have seen and known several other victims of the shipwreck and might have travelled with the royal court shortly after in 1122 and 1123. Greenway concludes, "In the course of a public career of over forty years, Henry was often at the centre of political life, in attendance at the royal court and present at ecclesiastical gatherings. He possessed the kind of knowledge of royal government that would have been familiar only to those close to the seat of power" (Henry of Huntingdon, *Historia Anglorum*, liii–lvi).

83. Interestingly, Strohm points to a similar illustration of the sovereign's body as food, noting an adherent of Richard II who reimagines the events of 1399, seeing Henry IV as shipwrecked in his channel-crossing and becoming food for birds and fish (*England's Empty Throne*, 102).

84. Morton, *Ecological Thought*, 36–37.

85. Ibid.

the portents surrounding the deaths of these kings aim to put more substance on their bodies than is really there. God must punish a king for
his misdeeds and, consequently, nature announces this punishment prior
to its delivery. Yet these very same accounts of divine (or is it natural?)
omens betray the very reality of the king's flesh. He is of nature, and his
death merely remakes his flesh and blood into a different component of
the ecological fabric. A more frightening realization, perhaps, for William
of Malmesbury, Orderic Vitalis, and Henry of Huntingdon, one which
they seem intent on covering over, is that there wasn't much Englishness
there in the first place either.

8

Radical Conservation and the Eco-logy of Late Medieval Political Complaint

Stephanie L. Batkie

THE OPENING LINES of Alan de Lille's oft-cited *Omnis mundi creatura* describe the relationship between man and the natural world as representational and reflective:

> Omnis mundi creatura
> quasi liber et pictura
> nobis est in speculum;
> nostrae vitae, nostrae sortis,
> nostri status, nostrae mortis
> fidele signaculum.
> (1–6)

> Every thing in the created universe is like a book for us, a picture, a mirror, a truthful sign of our life, our destiny, our condition, our death.[1]

The range of relationships here encompasses the interpretational, the visual, the providential, and the existential—all the ways Alan envisions

---

1. Texts and translations are taken from Alan de Lille's "Omnis Mundi Creatura," in *Literary Works,* ed. and trans. Winthrop Wetherbee, *Dumbarton Oaks Medieval Library* (Cambridge, MA: Harvard University Press, 2013), 601.

the natural world signifying meaning for the human. Moreover, he claims that it is the bounty of the created world which makes visible the conditions of individual existence inasmuch as it displays the varied permutations a life might take: given enough time and enough examples, nature will provide a comprehensive analogy to each individual's destiny and condition.[2] The variety contained within the possibilities of nature thus cast it as the "fidele signaculum" Alan uses as the framework for his short text. In the same way, the variety of nature is balanced by the variety found in the human condition. The poem insists on the plurality and collectivity of its readers; it is not the individual who sees himself in the natural world, but "nobis est in speculum, nobis vitae, nostrae sortis"— not "he" and "his," but "we" and "ours." Just as nature works as a mirror for the individual, it also reflects back the social collective as a whole, displaying both the *form* of human society and the *matter* of each man within it.[3]

In this, Alan, like others before him, finds nature/*Natura* to be a regulating force that exerts order upon a human world with clear ties to its earthly and cosmic surroundings.[4] In *De planctu Naturae,* he is careful to limit her powers—she is subject to the power of God and is less

---

2. Alan describes Nature's works as "*operatio multiplex*" [manifold] as opposed to God's "operatio simplex" [simplicity] in classic Neoplatonic language. See "*De planctu Naturae,*" in *Literary Works,* 21–217).

3. Nature herself makes this analogy in Alan's *De planctu Naturae.* She notes the similarities between the cosmos and political governmental structures: "Observe how, in this universe as in a noble city, a kind of majestic civil order is ensured by well-considered governance" (6.9) [Attende qualiter in hoc mundo velut in nobili civitate quaedam reipublicae maiestas moderamine rato sancitur], and the analogies between the universe and the individual, physical body: "In other respects, too, the form of the human body bears a hidden resemblance to the universe" (6.11) [In aliis etiam humani corporis forma mundi furatur effigiem]. Elsewhere she is called the "mirror for fallen creatures" [speculum caducis] and the "law of the world" [regula mundi] (7.1). There is a similar balance described in Nature's explanation of the generation of animals and people: God decrees that individual beings are born, live, and die, but life itself is an eternal force that is always present (8.29).

4. For the complexities of personified Nature as a force of ethical judgment in human affairs, see Joan Cadden, "Trouble in the Earthly Paradise: The Regime of Nature in Late Medieval Christian Culture," in *The Moral Authority of Nature,* ed. Lorraine Daston and Fernando Vidal (Chicago: University of Chicago Press, 2004), 207–31. See also the encyclopedic first chapter of Hugh White's explanation of the shifting attitudes towards depictions of *Natura* and her relation to reason and governance in *Nature, Sex, and Goodness in a Medieval Literary Tradition* (Oxford: Oxford University Press, 2000). For an account of the complex interactions between natural law and positive law, see the beginning of Helen Barr, "The Treatment of Natural Law in *Richard the Redeless* and *Mum and the Sothsegger,*" *Leeds Studies in English* 23 (1992), 49–80.

knowledgeable than, for example, Theology (6.15–16)—but she nevertheless controls the physical world of earthly matter. She is hailed by the
narrator as

> Quae tuis mundam moderans habenis,
> cuncta concordi stabilita nodo
> nectis, et pacis glutino maritas
> caelica terris.
> (7.9–12)

> You who, guiding the world with your reins, impose stability on all things
> in binding agreement, and unite heavenly to earthly in the closeness of
> peace.

Nature and the natural world hold out the promise of unity and peace
that human social and political structures strive to emulate, even if it is
impossible for them to achieve this ideal. At the same time, nature offers
a metaphor for individual self-governance that seems to be equally as difficult to maintain for the subjects and audiences of medieval allegorical
poetry. In this way, Alan finds it both lamentable and inevitable when
the vices and weaknesses of the human transgress against the perfection
of the natural system—a sentiment that will find new traction several
centuries later in the political allegories and complaints of the end of the
fourteenth and the beginning of the fifteenth centuries. These poets take
Alan's mirror of nature and establish the natural world as an idealized
measure for interpreting the characters and actions of political figures
in both their public and private personae, and like Alan, they figure the
natural world as enjoying an intimate relationship with divine truth and
virtue. The difference between the political use of nature and Alan's mirroring nature rests in the ways in which allegorized nature speaks in these
later texts to the matter and form of social and political unrest: rather
than extrapolating the decline of humanity from a highly abstracted personification of cosmic order and natural law, these texts embed the natural world inside the political world by tethering it to individual (and
very real) historical characters. The two compose, to borrow Timothy
Morton's vocabulary, a mesh wherein the human/political world exists
interdependently with the natural/ecological world.[5] As a result, we see

---

5. See, for example, Timothy Morton's first chapter of *The Ecological Thought* (Cambridge, MA: Harvard University Press, 2010), 1–19.

a kind of productive tension emerge between the rhetorical and poetic effects of abstract, allegorical figuration on the one hand, and the physical and embodied truths of the political on the other.

The relation between the form of the allegory and the matter of the political commentary takes advantage of Alan's notion of nature as a legible system of order: for him, the natural world is the sign system through which human reason can read the truth of God.[6] Reading the presence of God through human intellect requires understanding of *logos* as the bridge that enables the human to understand the divine but also as the means by which divine presence is made visible in the world. Eriugena explains the interaction between *logos* and the human as one of manifestation and assimilation:

> From this you are to understand that the divine essence is incomprehensible in itself, but when it is joined to an intellectual creature it becomes after a wondrous fashion manifest; so that the divine essence is seen alone in the intellectual creature. For the ineffable excellence of the former surpasses every creature which participates in it, so that in all things nothing but itself is presented to those that have understanding, while, in itself, it is not manifest in any fashion.[7]

> Ac per hoc intellige divinam essentiam per se incomprehensibilem esse, adjunctam uero intellectuali creaturae mirabilis modo apperere ita ut ipsa, diuina, dico essentia, sola in ea, creatura intellectuali uidelicet, appareat. Ipsius enim ineffablis excellentia omnem naturam sui participem superat ut nil aliud in omnibus praeter ipsam intelligentibus occurrat, dum per se ipsam, ut diximus, nullo modo appereat.

In a similar way, the natural world links the human intellect to the divine, functioning much like the *logos* through which divine will operates. The eco-logy of the natural world becomes a true question of interpretation; it acts as a *logos* originating from God—a *logos* which can, in turn, be read

---

6. "At an individual and a collective level, in a positive and a negative way, various notions of nature and the natural served to define, motivate, and evaluate the good. Always presumed to derive from a divine Creator, often modeled and powered by the celestial spheres, regularly associated with hierarchical order and harmonious proportion, nature was, at the very least, a reservoir for the formation of values and norms" (Cadden, "Trouble in the Earthly Paradise," 217).

7. Johannes Scotus Erigena, *Periphyseon*, vol. 161, Corpus Christianorum, Continuatio Mediaevalis (Turnhout, BE: Brepols, 2007), 14–15. Translation taken from John J. O'Meara, *Eriugena* (Oxford: Clarendon Press, 1988), 82–83.

by man. In defining nature in this way, political allegory casts *Natura* as governing a hermeneutic structure that is available to human intelligence but that remains perennially unaffected by it. In the same way, words on a page remain unaffected by the act of being read even if the meaning of the text is constructed in the intellect of the reader. The book of nature, then, can be read, but it has already been written. As a result, when political complaints use the natural world as grounds for their commentary, they frequently cast nature as an objective lens that reveals the truth of historical and social events. This is the logic behind allegories that equate the representation of a character as an animal, plant, or other natural element with the critique of that character.

To take a simple example, the political complaint poem from the reign of Edward I entitled "A Song of the Times" presents the king, who appears as a lion, passing judgment on three of his subjects: a fox and a wolf (both members of the nobility) and an ass (a simple peasant). The noble animals are pardoned on account of their flattery of the king, while the innocent ass, trusting to the king's judgment rather than bribery, is summarily condemned.[8] The untrustworthy nature of the flatterer is transformed into the predatory presence of the wolf and the fox, leading readers to pity the helpless and honest ass and thereby generating a critique of the king's misjudgment. The straightforward poetic commentary on the dangers of flattery and the privilege of royal judgment use the ability of the natural to signify meaning as a cue for the political commentary.

And yet, the connection between the form of the allegory and the matter of the critique is nevertheless an artful one, established and managed by the skill of the poet. Nature has provided the characteristics of the wolf, fox, and ass, but they come together here only at the behest of the author and in the interpretive ability of the reader. The poet, in accessing the representational powers of nature, has cultivated his allegory to reflect the matter he wishes readers to glean. Conversely, the poetic language opens the necessity of interpretation for its readers—a skill that is husbanded as a marker of prudence and wisdom.[9] The ability

---

8. Thomas Wright, ed. *Political Songs of England: From the Reign of King John to That of Edward II* (1839; Cambridge: Cambridge University Press, 1996), 195–205.

9. A very Augustinian position inasmuch as the ability to interpret ambiguous signs leads to an exercise of reason and will that brings the human closer to the divine. As Marcia Colish notes, Augustinian sign theory "fused a classical conception of words, both literal and figurative, as the authentic, sensible signs of knowable realities, with the Christian belief that language, redeemed through the Incarnation, was both a necessary and an inadequate means to the knowledge of God": Marcia L. Colish, *The Mirror of Language: A*

to draw connections between form and matter, between representation and meaning, becomes an indication of the higher order of reason that is the province of man. Animals are granted the estimative power, but to be rational is to be human.

For all its eco-logical power, then, a text requires interpretation to move beyond the immaterial world of poetry and become realized, political action. The poet might harness the signifying power of Nature, but any effect the text can have as complaint must pass through this eco-logical judgment. The reader's rational judgment allows texts to circulate beyond their lives as abstract objects; they can, in fact, become part of the biopolitical mesh—as long as they are first consumed as texts. In this way, the reader becomes, through the act of interpretation, a biopolitical agent, a bridge between the textual world and the political world.

Reliance on biopolitical interpretation as the bridge between the text and the world opens, however, an unavoidable problem for political allegories. Just as good readers have the ability to harmonize the natural and the political, bad readers have the ability to disorder the systems nature makes visible. This is as true for readers of texts as it is for readers of politics. Political action marked by foolishness or unrest shifts the eco-logical relationship: no longer right readers of nature, no longer able to control nature as a resource for political signification, men suddenly becomes subsumed by the nonhuman, nonrational world. Consequently, they lose their humanity and appear to physically transform in response, thereby creating the allegory of the text. John of Salisbury, for example, opens the first book of the *Policraticus* by describing how men who have lost their power of reason through self-indulgence and vice become less like God and more like animals: "Sic rationalis creatura brutescit, sic imago creatoris quadam morum similitudine deformatur in bestiam, sic a conditionis suae dignitate degenerat homo" [In this way the rational creature is rendered brutish; the image of the Creator is distorted into something resembling the character of a beast; and man degenerates from his condition of dignity].[10] In the same vein, when

---

Study in the Medieval Theory of Knowledge (Lincoln: University of Nebraska Press, 1983), 54. These poems use right reading as guides to prudent governmentality by drawing on the exercise of will as necessary for interpretation.

10. Latin text taken from John of Salisbury, *Joannis Saresberiensis Episcopi Carnotensis Policratici Sive De Nugis Curialium et Vestigiis Philosophorum Librui VIII; Recognovit et Prologomenis,* ed. Clement Charles and Julian Webb (Frankfurt am Main: Minerva, 1965), I.10. Translations are taken from John of Salisbury, *Policraticus: Of the Frivolities of Courtiers and the Footprints of Philosophers,* ed. Cary J. Nederman (Cambridge: Cambridge University Press, 1990). Several lines later, John will continue by equating reason with a stable

the biblical Nebuchadnezzar is transformed into his bestial self, it is
due to his inability to recognize the truth of his dream of the spreading
tree, even after Daniel translates the natural images of the dream into
political and devotional advice.[11] The suffering king is only returned to
his right reason (*sensus*) by the grace of God, as a gift that signifies that
his lesson had been learned and that his pride had been destroyed.[12]
The human-to-beast conversion illustrates the effects of misreading the
natural world by failing to adequately see its signifying power: either
an unwillingness or an inability to relate to the ordered form of nature
transforms man from his position as a rational reader of nature to an
unwilling and bestial participant in it.

   The order of *Natura*, when appropriately read, therefore offers a system of representation that constructs interpretation for political contexts.
It also makes visible the instance when biopolitical readers cross the line
between political action and political dissent, when they become "unnatural." But nature's order is always outside of the human, always beyond
the borders of the political. It is available as a resource to be used in
the quest for political signification, but its productivity does not require
human control to be ordered in the first place; Nature has her own order,
separate from that of man.[13] Because of this, the eco-logy of political alle-

---

social/political identity: "Who is so brutish as one who, because of defective reason and
libidinal impulses, forsakes his own proper sphere?" (1.19–20) [Quid eo brutius, qui ex
defectus rationis et impulso libidinis, dimissis propriis?].

   11. See Daniel 4:15–24. In the *Confessio Amantis*, Gower famously focuses on not
this dream but on Nebuchadnezzar's earlier vision in Daniel 2:31–35 of "the monster of
time," which speaks to the apocalyptic downfall of the human condition, rather than the
divine punishment of a single, proud ruler. (Prol.595 et pass.) For a fuller treatment of
Gower's use of these passages, see Russell A. Peck, "John Gower and the Book of Daniel," in *John Gower: Recent Readings. Papers Presented at the Meetings of the John Gower
Society at the International Congress on Medieval Studies, Western Michigan University,
1983–88*, ed. R. F. Yeager (Kalamazoo: Western Michigan University, 1989), 159–87.

   12. This is the reading Gower gives of the story in his Latin headnote to the passage
in the Prologue of the *Confessio*. The fact that Nebuchadnezzar regains *sensus* rather than
ratio or *mens* is telling inasmuch as it equates the way in which he physically relates to
the surrounding world with his ability to read and understand the surrounding world.
The word choice beautifully illustrates the completeness of his transformation and that his
physical separation from the human is the direct cause of his mental separation. Moreover,
they are both restored simultaneously when he is returned to himself.

   13. In political allegory, nature is not an appropriate escape from human order. This
is a very different textual situation from political songs that cast the natural world as an
escape from the harmful and dehumanizing effects of a corrupt government. See, for
example, "The Outlaw's Song of Trailbaston" (c. 1305), in which the speaker urges his
readers to abandon the perils of the law for the safety of the forest: "You who are indited,
I advise you, come to me,—to the green wood of Beauregard, there where there is no

gory has little space for a notion of nature as wild and untamed. However much the irrational is associated with the bestial, wildness is simply not a form of nature that finds a great deal of traction in these texts. Nature's role, as Boethius writes and as Alan will concur, is that of ordering the world she creates, even if that order looks unfamiliar and occasionally threatening to human eyes:

> What are the reigns of powerful Nature,
> Guiding the universe? By what statutes
> Does her Providence hold the infinite sphere,
> Binding and keeping this world of things
> In unbreakable bonds?[14]

> Quantas rerum flectat habenas
> natura potens, quibus immensum
> legibus orbem prouida seruet
> stringatque ligans inresoluto
> singula nexu. . . .
> (II.met.2, 1–5)

Boethius goes on to describe tame lions and caged birds breaking free of their restraints at the first opportunity, trees lifting themselves continually to the heavens despite attempts to subdue them, and the inexorable progress of the sun from east to west. Such impulses are the rule of nature, regardless of human attempts to corral or alter them. This vision of nature is one of a powerful order that occasionally touches on a man-made order but that is distinctly (and sometimes violently) independent of it. Importantly, the undercurrent of chaos and danger appears only when viewing nature's order from the human perspective: the unruly lion breaking free of his master, like the ungrateful bird who flees the safety

---

plea,—except wild beast and beautiful shade;—for the common law is too much to be feared" (53–56) [Vus qy estes endité, je lou, venez à moy, / Al vert bois de Belregard, là n'y a nul ploy, / Forque beste savage e jolyf umbroy; / Car trop est dotouse la commune loy]. From MS. Harl. 2253, fol. 113v and printed in Wright, *Political Songs*, 231–37.

14. Latin text from Boethius, *Anicii Manlii Severini Boethii Philosophiae Consolatio*, in *Corpus Christianorum. Series Latina 94*, ed. Ludwig Bieler (Turnholt, BE: Brepols, 1984). Translation taken from Boethius, *Consolation of Philosophy*, trans. Joel C. Relihan (Indianapolis, IN: Hackett, 2001). George Economou discusses the personification of Nature in this meter as a mediation between grammatical or rhetorical personification and poetic personification (*The Goddess Natura in Medieval Literature* (Cambridge, MA.: Harvard University Press, 1972), 38–40).

of its cage in favor of the uncertainty of the forest, is interpreted through the lens of human order and rationality. The human world values a concept of order grounded in social contracts (in the case of the lion), security and comfort (in the case of the bird), and domination (in the case of the trees). As such, it is inhuman for the bird to prefer the forest to the cage, where food is provided for her, but it is not unnatural.

The problem Boethius outlines in this meter illustrates a conflict between what Foucault sees as the primary epistemological order of the Middle Ages, the logic of similitude, and the human attempt to control the world by regulating nature as a source of production, something he finds beginning primarily in the sixteenth century.[15] For Foucault, representation and similitude create for the Middle Ages an interpretive system that works by analogy. Analogy draws connections between objects that are not "the visible, substantial ones between the things themselves; they need only be the more subtle resemblances of relations. Disencumbered thus, it can extend, from a single given point, to an endless number of relationships."[16] Nature offers a wealth of such analogies, and Foucault's notion of similitude works as a more abstracted version of the mirroring power of representation Alan offered. At the same time, however, Foucault adds that political discourse also clearly sees the natural world as part of its domain, as a resource of the state, and under the jurisdiction of the king's rule. The production of the land, the labor of the peasants, the natural resources of the kingdom—these all exist under the control of the state and are regulated through the power of the crown. The combination of these two seemingly conflicting possibilities—nature as representation/similitude and nature as political resource—represents what might best be described as the way biopolitics functions in the medieval period. It is less a question of population control, as we see in Foucault's formulation, and more a question of harmonizing the natural and political orders through the intervention of the state.[17] In this way, the

---

15. In the first lecture in *The Birth of Biopolitics,* for example, Foucault links his stated topic to the development of the *raison d'état,* something he locates as a phenomena of the early modern period inasmuch as this is when we see the state embracing a "self-limiting" position rather than the desire for universal sovereignty he finds in the Middle Ages. This rather flat reading of the medieval period leads to significant problems in reading the role of biopolitics before the sixteenth, seventeenth, and eighteenth centuries. Michel Foucault, *The Birth of Biopolitics: Lectures at the Collège De France, 1978–79,* ed. Michel Senellart, trans. Graham Burchell (New York: Palgrave Macmillan, 2008).

16. Michel Foucault, *The Order of Things: An Archaeology of the Human Sciences* (New York: Vintage Books, 1994), 21.

17. See, for example, the final lecture (17 March 1976) in *Society Must Be Defended:*

Neoplatonic argument that human nature, when functioning at its most rational, moves closer to nature and therefore to God works alongside the presentation of nature as dominated by industrious human cultivation and consumption.

The moral and political lessons of Nebuchadnezzar's dream and punishment, for example, appear in the space between the understanding of nature as similitude and nature as resource, dominated by human interests of governmentality and control. Negotiation of the two is a question of judgment: it is up to the reader of nature to determine the correct *telos* of a bird, a lion, or a king within the contingent political world, informed by immutable natural law. Nebuchadnezzar misinterprets his place in the divine order and his place in nature, and therefore, he is reduced to being controlled by (his) nature. He fails to understand the appropriate function of his political identity; he therefore becomes a reflection of his inability to negotiate biopolitical control and the analogy of natural order that tends toward God.

The same is true in the allegories that pointedly attach themselves to political events of the later fourteenth century—texts like *Richard the Redeless* or Gower's *Visio Anglie*. Moreover, they use eco-logy to validate political action. Nature allegory, working from the Neoplatonism of Alan de Lille and Boethius, establishes a hermeneutic order that, when laid over political and historical events, allows a reader to locate disorder and unnatural behavior. The charge of correcting such behavior, then, becomes the province of those readers whose judgment has, in turn, been validated by their facility with reading the allegory. This circular system of authorizing reading defines communities of political actors who understand their position as antiradical and inherently orthodox, as "natural," even when they urge violent political change. They are not rebels; they are advocates and loyalists, authorized by *Natura's* order. As such, they become sometime akin to "radical conservationists," working to return and restore the similitude between the natural world and the political landscape. Through this conservationist mode of interpretation, nature allows for a stabilizing hermeneutic that renders political action as a deeply conservative and therefore recognizable species of justice.

We can see this clearly in the short poem that James M. Dean entitles "There is a Busch that is Forgrowe" and that elsewhere appears as "On King Richard's Ministers."[18] This short complaint focuses on the after-

---

Lectures at the Collège De France, 1975–76, ed. Mauro Bertani and Alessandro Fontana, trans. David Macey (New York: Picador, 2003), 239–64.

18. I follow the text, published for TEAMS in association with the University of

math of the Merciless Parliament of 1388 and describes the 1397 execu-
tions of Thomas Woodstock, Duke of Gloucester; and Richard Fitzalan,
the Earl of Arundel; along with the banishment that same year of Thomas
Beauchamp, Earl of Warwick. The poem is particularly concerned with
exposing the corruption of some of Richard's most despised ministers,
Sir John Bushy—who served as the speaker of the Commons in 1394 and
1397—Sir Henry Green, and Sir William Bagot, while it presents Henry
of Derby's (later Henry IV's) return to England as a renewal of royal
justice that will return prosperity to the exploited common people. The
poem takes all of these political actors and transforms them into allegorized figurations that rely either on name-based *paranomasia* (in the case
of the ministers) or on heraldic talismans (in the case of the Appellant
Lords). In this way, Bushy becomes a bush, Henry Green becomes green
grass, and William Bagot a bag or purse, while Gloucester is cast as the
swan, Arundel as the steed, and Warwick as the bear—the latter three
following the established sigils of their respective houses. Importantly,
the instinct to allegorize is not an attempt by the poet to shield himself
from the potential repercussions of royal anger; indeed, there can be little
doubt as to which historical personage he references.[19] The poet is rather
relying on the sense that his language will be recognized as an accurate
depiction of the characters in play. As Ann Astell has noted, "The alle-
gorist fails in his communicative purpose if his intended audience misses

---

Rochester, in James M. Dean, ed. *Medieval English Political Writings*, TEAMS Middle
English Texts (Kalamazoo, MI: Medieval Institute Publications, 1996). The poem was
first printed by William Hamper as "Sarcastic Verses, Written by an Adherent to the
House of Lancaster, in the Last Year of the Reign of Richard the Second, AD 1399," *Archaeologia* 21 (1827): 88–91. It was then reprinted in Thomas Wright in *Political Poems
and Songs Relating to English History Composed during the Period from the Accession of
Edw. III. to That of Ric. III*, 2 vols. (London: Longman, Green, Longman, and Roberts,
1859), I.363–66. Hamper transcribed his version from a manuscript labeled "Deritend
House" (c. 1400) in a letter for the Society for Antiquities, but the manuscript unfortu-
nately cannot now be located. It appears to have been lost in the sale of Hamper's library
by the London auctioneer Robert Harding Evans in 1831.

    19. Thomas Kinney, in "The Temper of Fourteenth-Century English Verse of Com-
plaint," *Annuale Medievalae* 7 (1966): 85, n. 26, reads the complaint here as "clumsily
disguised" on account of the allegory, but Eric Stockton finds the allegorical appellations
to be far more clear: "Such terms, which now strike us as pointless circumlocutions or
subterfuges, constituted a literary motif or convention at that time, used in several other
poems. The labels derive largely from heraldic badges or cognizances and must have
been as familiar to Englishmen at the end of the fourteenth century" (*The Major Latin
Works of John Gower: The Voice of One Crying, and the Tripartite Chronicle*, trans. Eric W.
Stockton [Seattle: University of Washington Press, 1962], 38).

the message he has hidden,"[20] and there can be little doubt that the allegory here was intended to be legible enough for readers to understand the critique with little difficulty.[21]

The effect of the allegory is not, then, to occlude the truth of the poem from inappropriate or disapproving readers but is rather (1) to translate political events into a legible system of order and disorder and (2) to validate imagined political action on the part of the reader. In casting the political as the textual, in short, the allegory makes the political readable. We see this from the first lines of the poem:

> Ther is a busch that is forgrowe;
> Crop hit welle, and hold hit lowe,
> Or elles hit wolle be wilde.
> The long gras that is so grene,
> Hit most be mowe, and raked clene—
> For-growen hit hath the fellde.
> (1–6)

The "busch" and the "gras" here clearly correspond to Bushy and Green, and the poem complains of their faults in language that maintains the allegory while simultaneously chastising them for their inappropriate political ambitions. The bush is growing too wild, and therefore it must be cut back. The grass is overtaking the fields that would otherwise be producing valuable crops, and it must be mowed down and raked away. The sense of urgency here is striking, particularly when paired with the violence contained in the allegorical language. The imperative verbs "crop" and "hold," along with the hortatory insistence that the grass must be mown and raked, work off the paranomasia of Bushy's and Green's names, maintaining the natural imagery the language provides. Violently physical, the allegory here embraces a language of upheaval and rebellion:

---

20. Ann W. Astell, *Political Allegory in Late Medieval England* (Ithaca, NY: Cornell University Press, 1999), 39.

21. David R. Carlson disagrees, arguing that the elite and specialized study of heraldry would prevent most readers from easily translating the allegory. He cites the apologies that Gower and the poet of *Richard the Redeless* give at the beginnings of their poems for writing in such difficult and obscure terms, and he states emphatically that "the manner of speaking at issue was emphatically not popular, at least in this sense: no one could be expected to understand it without help, or few" (*John Gower, Poetry and Propaganda in Fourteenth-Century England* (Cambridge: D. S. Brewer [for The John Gower Society], 2012), 130. I contend that it is indeed the "few" who would be the target audience of this form of allegory.

it calls for the removal and possible execution of royal favorites, arguing that readers cross (even if only in the imagination) the political and social space between themselves and the ministers in an attempt to restore justice to the country. However, it does so through language that makes such dramatic action just, a return to the appropriate order. Because the political action described depends on the right judgment of the reader, as evidenced by their interpretation of the poem, the text here supports seeing Richard's downfall and Henry's subsequent rise as natural.

In part, this is possible because of the abstract nature of the allegory. The poem does not present Bushy to readers for execution (his eventual fate at Henry's hands in 1399); it presents a piece of unruly vegetation that needs trimming. Similarly, it is not Henry Green but a field of grass that requires cutting. Political justice thus becomes a kind of aggressive (but necessary) landscaping. And even though the poem gestures toward "the heron's" eventual removal of Richard's supporters, the allegory presents the need for their elimination without casting Henry as the only person qualified to judge and sentence them. They have become unnatural and unruly, and therefore, it is the responsibility of anyone capable of cultivating virtue to return the land to its proper, productive state. Not addressed directly to Henry but rather to an unspecified reader, the verbs in the poem both render the faults of the ministers as abstract enough to fit into an allegorical system and universally visible and accessible enough to demand readerly engagement. In this way, the allegory uses the transformation of a political figure into a natural object, using the structures of similitude and analogy to produce a vision of governmentality available in the text that is consistent with the biopolitical management of nature. Even if this management never physically appears in the world, the biopolitical rhetoric of the poem allows it to be concrete and available in the text. Thus, readers' biopolitical agency is validated by their position as radical conservationists. And, of course, their position as radical conservationists is constructed by their eco-logical reading of the poem.

To bolster the eco-logic of the text even further, the use of figurative language separates the allegorization of the ministers from the allegorization of the Appellant Lords by differentiating each group through the manner of their figuration. While Bushy, Green, and Bagot are transformed into vegetation and an inanimate object by the sound of their names, the Appellant Lords enjoy the privilege of being known by the sigils of their noble houses. In addition to presenting Gloucester/the swan, Arundel/the steed, and Warwick/the bear, the poem also works

its way through Henry's (here appearing as the heron) arrival at Ravens-
pur in 1397 with Arundel's son, Thomas (the colt), and his reception
by the Percy family (seen as geese) and the Neville family (peacocks).[22]
The stratification between characters appearing through paranomasia
and those coming from heraldic allegory is significant: by emphasizing
the nobility of Richard's adversaries and, in turn, the relatively low status
of their opponents, the poem heightens the sense of political and social
upheaval that must be corrected.[23] Identification of the groups through
the differences in their social circumstances makes the actions of each
group legible and available for interpretation; more importantly, in creating formal differences, it defines when those differences are unnaturally
or unethically collapsed. For example, the poem describes Bushy's role
in Gloucester's execution ("Thorw the busch a swan was sclayn" [13]),
Bagot's betrayal of Warwick ("Thorwe the bag the bereward is taken"
[28]), and Arundel's execution, as well as the seizure of his lands by the
crown as a result of Green's political influence ("The grene gras that
was so long, / Hit hath sclayn a stede strong" [19–20]). Each description strains the bounds of the allegory inasmuch as narrative logic is
displaced in favor of the factual reality of the historical events described.
Gloucester and Arundel are executed in 1397, but bushes are not usually
responsible for the deaths of swans, and grass is not frequently seen as a
cause of equine mortality. On the contrary, narrative logic and the order
of nature hold that the grass should nourish and sustain the horse, not
result in its untimely demise. Meaning is therefore encoded in the spaces
where historical events overrun the ability of nature to contain and narrate them. Unjust and unnatural, the deaths of the Appellants are legible
because the limits of the allegorical form and the natural world generate
the poem's political critique.

Helen Barr notes a similar pattern in the contemporary *Richard the
Redeless*, describing the inversion of the natural order in the poem as
Richard's political identity begins to evade his control.[24] The second pas-

---

22. Hamper, in his original printing of the poem, glossed the geese as "the Commons" and the peacocks as "the Lords."

23. Emily Steiner discusses in much greater detail the differences between patronymic
names and allegorical figuration. She argues for seeing the signifying power of names,
particularly in relation to the 1381 Uprising, as a powerful display of the ties between
documentary culture, historiography, and literary language. Emily Steiner, "Naming and
Allegory in Late Medieval England," *Journal of English and Germanic Philology* 106, no.
2 (2007): 248–75.

24. Helen Barr, *Socioliterary Practice in Late Medieval England* (Oxford: Oxford University Press, 2001), 63–79.

sus of the poem tells of the king's unpopular use of livery as a marker of
political favor and power; Richard established a visual representation of
his sovereignty though the use of his personal sigil, the white hart. In the
poem, this becomes a lively source of complaint, as the poet describes the
liberties that the mark of the king encouraged:

> Thus lyverez overeloked youre liegis ichonne;
> For tho that had hertis on hie on her brestis,
> For the more partie, I may well avowe,
> They bare hem the bolder for her gay broches,
> And busshid with her brestis and bare adoun the pouere
> Lieges that loved you the lesse for her yvell dedis.
> (II.35–40)

The liveried supporters are thus the "hertis" who begin to rampage
through the country, exploiting the power they gain on account of the
king's favor and oppressing the common people. Richard has lost the
"hearts" of his people by privileging his chosen "harts." Barr reads the
criticism here as the political order reverting to natural chaos: "The deer
are rampant, not in the heraldic, but in the bestial ('hornyd of kynde')
sense, of running wild through the kingdom, and oppressing the king's
subjects."[25] She also describes the unnaturalness of the deer by referring to the ways in which they appear in medieval bestiaries, noting that
deer were thought to renew themselves by feeding on the venom of ser-
pents; instead of renewing political power by searching out and eliminating the corrupt and the serpent-like, Richard's followers viciously attack
the wrong animals: the horse, swan, and bear.[26] The commentary here
depends on representation and similitude. Using punning and wordplay
(the hearts/harts confluence) to extend the metaphor rather than to
establish it, the analogies (to return to Foucault's term) established rely
not on spatial proximity or visual appearance but on figuration governed
by Richard's livery and his political identity—two things that must be
linguistically connected by the author of the poem.

   The artfulness of the allegory displays the limits of similitude when
applied to the political. It is one thing for the poet to connect hearts and
harts through linguistic suggestion and for the poem to use a reading of the
natural figure of the deer to argue for the errors of Richard's supporters,

---

25. Ibid., 70
26. Barr, "Treatment of Natural Law," 57.

but the thrust of the critique about the use of livery attacks Richard's failed attempts to use analogy and similitude as a political strategy. In multiplying the visibility of his figurative identity, Richard attempts to unnaturally augment his political power—a method that backfires both because he oversteps his ability to exert his will over his people and because, in the end, he is unable to control the ways in which his identity is read. The unnatural and ineffective exertion of Richard's political will relies on the seemingly limitless extension of similitudes, between himself and his self-figuration, for example, but it also betrays the limited control he has over the sociopolitical ecology created by the proliferation of his identity. As Barr notes, "[Richard's] faith in his white hart badge as a sign of the loyalty of his subjects is shown to be misplaced, for despite the king's attempt to govern the political valency of image-making, the signs are 'duble': capable of having more than one meaning, and untrustworthy."[27] The problem with the livery becomes, therefore, a question of biopolitical judgment and interpretation. Richard, the author of his expanded political persona, is thwarted in his attempt to establish unnatural similitude thanks to the right reading of the poem's (and his people's) critique.

The control that the political exercises over the natural extends from the sovereign and affects the people of the kingdom as an expression of the state's authority and power. Furthermore, it does so as part of a system of governmentality that, in this period, is deeply committed to seeing the natural world as both a reflection of divine goodness and as a means of social and political control.[28] In this, appropriate application of political power over the natural world is read as harmonizing the natural and political orders—a prime example of how biopolitics works to maintain political stability through a discourse of the natural world. When the political system fails to maintain and further this harmony, usually due to the fault of corrupt officials and monarchs, the results become an index for reading the errors of those in power. In the closing stanzas of

---

27. Barr, *Socioliterary Practice*, 71.

28. This clearly follows François Debrix and Alexander D. Barder's definition of bio-politics: "Biopolitics is not about the power or control of the one and only centralized sovereign, even if it is a so-called benign sovereign that would have surrendered its power to put to death, to the benefit of a power to make life live and thrive. Rather, it is about a proliferation of practices, expert knowledges, technologies, dispositifs, and exercises of government (what Foucault at times refers to as 'governmentality') that invest, dissect, enhance, and reproduce individual and collective life across the social spectrum, and often beneath and beyond the level of the state itself," in *Beyond Biopolitics* (London: Routledge, 2012), 9.

"There is a Busch," we see that the poem is expressly concerned over the "hunger" the kingdom suffers under the dishonest direction of Richard's ministers. The poet complains of Bagot's miserly ways and his lack of generosity:

> The bag is ful of roton corne,
> So long ykep, hit is forlorne;
> Hit wille stonde no stalle.
> The pecokes and the ges all so,
> And odor fowles mony on mo,
> Schuld be fed withalle.
>
> (67–72)

Figuration, here again, falls neatly within the poem's nature allegory, maintaining the characterizations previously associated with aristocratic power struggles as the text shifts to focus on the plight of those suffering under the government's misrule. The greedy Bagot withholds sustenance from the people in keeping with his greedy "bag" persona. Likewise, the bush is bare and dry, no longer able to produce leaves, and the grass is rotten and unfit for consumption.

In using the image of "roton corne" and the threat of hunger as a marker of political failing on the part of Richard's ministers, the poem moves away from chronicling historical events and begins to generate critique through a discourse of human justice and political order. Just as the allegory crosses over from natural figuration to unnatural action, Richard's favorites move from representing a political order founded on counsel and justice to unjustly exploiting their positions for political and economic gain. This happens first and foremost on a figurative level, as the natural allegory renders critique primarily through representation. The intensity with which hunger appears in the poem speaks to the vehemence of the poet's criticisms, but the historical contexts of the poem simply do not match the hardships it relates. The Great Famine in the early part of the century is a documented source of economic distress that resulted in food shortages and increased illness across the country. Chroniclers note the strain related to hunger between 1315 and 1325, describing immense amounts of rainfall that inundated fields and destroyed harvests.[29] Widespread animal epidemics also depleted food

---

29. For example, the chronicle of Meaux Abbey reports, "In one year, because of a dry autumn and a great amount of rain, flooding, from Pentecost until Easter the grain failed in the earth" [Eo anno, propter serum autumnum et copiosam pluvias inundatio-

supplies. However, by the end of the century, population reduction from the Black Death sufficiently shifted the demand for food to the extent that, as Christopher Dyer notes, the peasantry of late-fourteenth-century England had far more plentiful and varied meals than did the preceding generations.[30] Peasant complaints were more likely to center on the abuse of noble privilege in terms of money: purveyance, tallage, and wage restrictions.[31] The lower classes had a great deal to protest about, as evidenced by the amount of complaint literature that survives, but hunger is not frequently cited as a popular grievance.[32] The use of it in "There is a Busch," therefore, is more figurative than historical, more representation than reportage. The hunger of the poem is purely fictional; it cannot be part of the biopolitical mesh of the political world.

Moreover, the problem of hunger and want that we see cited in "There is a Busch" is connected to the natural inasmuch as the materials for feeding the hungry have their origins in nature, but the poem is careful to cast the utilization of those materials as the province of the human world. Unlike in the early modern period, when nature becomes

nem, a Pentecoste usque ad Pascha blada terras defecerunt] resulting in dramatic food shortages. Thomas de Burton, *Chronica Monasterii De Melsa a Fundatione Usque Ad Annum 1396, Auctore Thoma De Burton, Abbate, a Monacho Quodam Ipsius Domus*, ed. Edward Augustus Bond (Burlington, ON: Tanner Ritchie, 2009), 332. See also Richard C. Hoffmann, "*Homo et Nature, Homo in Natura*: Ecological Perspectives on the European Middle Ages," in *Engaging with Nature: Essays on the Natural World in Medieval and Early Modern Europe*, ed. Barbara A. Hanawalt and Lisa J. Kiser (Notre Dame, IN: University of Notre Dame, 2008), 18; P. R. Schofield, "Medieval Diet and Demography," in *Food in Medieval England: Diet and Nutrition*, ed. C. M. Woolgar, D. Serjeantson, and T. Waldron (Oxford: Oxford University Press, 2006), 246–48.

30. See Christopher Dyer, "Did the Peasants Really Starve in Medieval England?" in *Food and Eating in Medieval Europe*, ed. Martha Carlin and Joel Thomas Rosenthal (London: Hambleton Press, 1998), 53–71. Also helpful in laying out the changing economic conditions in the wake of the Black Death is Dyer's chapter "The Black Death and Its Aftermath" in *Making a Living in the Middle Ages: The People of Britain, 850–1520* (New Haven, CT: Yale University Press, 2002), 271–97.

31. See Wendy Scase, *Literature and Complaint in England, 1272–1553* (Oxford: Oxford University Press, 2007), 5–41.

32. The most well-documented complaints come from the uprisings beginning in 1381: the *Anonimalle Chronicle* records the rebels demanding to be ruled only by Westminster law, that lordship was the sole province of the king, and that the practice of binding peasants to the land be abolished, in *The Anonimalle Chronicle, 1333 to 1381*, ed. V. H. Galbraith (Manchester, UK: Manchester University Press, 1927), 147. See also Scase, *Literature and Complaint*, 83–87. On the other hand, conspicuous excess was often used as a political and social display. See the discussion in C. M. Woolgar, "Fast and Feast: Conspicuous Consumption and the Diet of the Nobility in the Fifteenth Century," in *Revolution and Consumption in Late Medieval England*, ed. M. A. Hicks (Suffolk, UK: Boydell & Brewer, 2001), 7–26.

a symbol of fecundity and nourishment, the medieval idea of generative nature focuses primarily on procreation and birth rather than sustenance and livelihood.[33] Sustaining life by providing adequate food falls not on *Natura,* or even on God, but on the efforts of diligent, physical labor— an inherently human prerogative. In his commentary on Matthew in the *Catena Aurea,* Aquinas cites Pseudo-Chrysostom's privileging of work as the source of food: "Bread may not be gained by carefulness of spirit, but by toil of body; and to them that will labour it abounds, God bestowing it as a reward of their industry; and is lacking to the idle, God withdrawing it as punishment of their sloth."[34] In these terms, if the people of England are lacking the necessary means to sustain themselves, the fault should lie with their reluctance to engage in virtuous labor, preferring instead the solace of sloth. "There is a Busch," conversely, presents the fiction of starvation as a failing of those in power to allow the fruits of virtuous labor to reach the deserving and the needy. Aristocratic greed is the culprit, not a lack of industry. Hunger becomes then an effect of political injustice on the part of the nobility rather than either a question of agricultural sloth on the part of the peasantry or the failure of the natural world to produce expected fruits.

In the same way that the narrative limits of the animal allegory reveal the failings of Richard's ministers, the threat of hunger in the text pointedly argues for the political shortcomings of Bushy, Green, and Bagot. The well-known use of agriculture and cultivation as metaphors in devotional and political texts attests to their clear availability as figures of

---

33. "The power of Renaissance Nature was in and of the body, as represented by her nakedness, her breasts, and her inexhaustible lactation, rather than in intention, voice, or will. She provided value in the most basic sense of raw materials and resources, rather than values in the sense of ethical and religious norms. No longer the active shaper of natural phenomena and spokespersons for the moral order of God's creation, she became increasingly identified with its corporeal stuff: the inventory of God's creatures and the physical fabric of the natural world" (Katharine Park, "Nature in Person: Medieval and Renaissance Allegories and Emblems," in *The Moral Authority of Nature,* ed. Lorraine Daston and Fernando Vidal [Chicago: University of Chicago Press, 2004], 68).

34. "Non enim sollicitationibus spiritualibus, sed laboribus corporalibus acquirendus est panis, qui laborantibus pro praemio diligentiae, Deo praestante, abundat, et negligentibus pro poena, Deo faciente, subducitur" (*Catena Aurea,* Caput 6, Lectio 17). Aquinas gives this as the gloss to Matt. 6:25, "I say to you, then, do not fret over your life, how to support it with food or drink." Latin text from Thomas Aquinas and Enrique Alarcón, *Catena Aurea,* Corpus Thomisticum, http://www.corpusthomisticum.org/. Translation taken from Thomas Aquinas, *Catena Aurea: Commentary on the Four Gospels; Collected out of the Works of the Fathers,* trans. John Henry Newman (Oxford: J. H. Parker, 1841), I.251.

industry and virtue in the period;[35] their presence in "There is a Busch" builds on this by blending cultivation with the animal allegory as a commentary on the limits of political and social order. This argument is made all the more effective by the concrete and physical resonance hunger threatens—the appropriate response to which is equally located in physical action. The failure of the ministers to produce appropriate sustenance (leafy bushes and lush grass) for England's animals allows for violent human intervention; readers are advised that the only solution for a barren bush is to prune it down "crop and rote" in order to take it into town for sale. The same is recommended for the grass: it should be raked clean so new, healthful crops might grow in its place, while the bag must be hung up to dry before any attempt to mend it. The ministers have become unnatural and disorderly in the natural environment as well as in the political environment, and as a result, their crimes justify their permanent removal from both arenas. If necessary, God will be the one to restore justice in the closing lines of the poem:

> Now God that mykelle is of myght,
> Grant us grace to se that syght,
> Yif hit be thy wille.
> Our lene bestes to have reste
> In place that hem lyketh beste,
> That were in point to spylle.
> (85–90)

The suffering people are being punished as sinful through the ravages of hunger, but without cause; governmental corruption has therefore

---

35. The most commonly discussed of these is, of course, the Hunger episode in *Piers Plowman*. See, for example, Kathleen M. Hewett-Smith's argument about Hunger as a concrete sensation rather than an allegorical abstraction in "Allegory on the Half-Acre: The Demands of History," *Yearbook of Langland Studies* 10 (1996): 1–22. For Hunger as the longing for a system of justice in terms of spiritual satisfaction, see R. E. Kaske, "The Character Hunger in *Piers Plowman*," in *Medieval English Studies Presented to George Kane,* ed. Donald Kennedy, Ronald Waldron, and Joseph S. Wittig (Cambridge, UK: D. S. Brewer, 1972), 187–97. For readings that find Hunger to be an economic control over peasant labor, see Steven Justice, *Writing and Rebellion: England in 1381* (Berkeley: University of California Press, 1994), and Margaret Kim, "Hunger, Need, and the Politics of Poverty in Piers Plowman," *Yearbook of Langland Studies* 16 (2002): 131–68. An excellent overview of much of the scholarship on the topic can be found in David C. Benson and Elizabeth S. Passmore, "The Discourses of Hunger in *Piers Plowman*," in *Satura: Studies in Medieval Literature in Honour of Robert R. Raymo,* ed. Nancy M. Reale and Ruth E. Sternglantz (Donington, UK: Shaun Tyas, 2001), 150–63.

disrupted the rewards of virtue and the punishments of sin. Like the unnatural fates of Gloucester, Arundel, and Warwick, the presence of the injustice of hunger here validates any action taken against the corrupt ministers, action that is made permissible by the right reading and sanctioned judgment of the poem's audience. The readers, in their role as biopolitical agents, are wise enough to see the allegorical and political problem with hunger, and they are thus legitimated in acting to correct the problem.[36]

The poetic intervention in "There is a Busch" lies in the text's ability to draw on the representational power of nature, harnessing the hermeneutic of the natural world in service of the political critique. The subsequent exercise of judgment on the part of the reader then legitimates the political action the poem sanctions (or suggests) on account of the wisdom and prudence needed to negotiate between the form of the allegory and the matter of the complaint. For this short text, the paired mechanisms of (1) analogous similitude and (2) human command over nature form the biopolitical eco-logy of the poem. Moreover, when this same eco-logical system appears in longer, more complicated texts, the effect is much the same. In Gower's *Cronica Tripertita*, for example, the same events are discussed through the same system of allegorical figuration. In Gower's poem, however, the effect of authorial intervention is far more prominent and the artfulness of the allegorical language far more pronounced. Whereas the allegorical connections in "There is a Busch" are available to the poet who, in turn, makes them available to the reader, Gower's version of events encodes the connections he draws among historical characters, their allegorical figurations, and the political and ethical interpretation of both in the form of poetic language in addition to the form of the allegory. In the case of a text like the *Cronica*, the biopolitical reader simply has more work to do, but the mechanics of biopolitical agency remain the same. The text cannot enter into the biopolitical mesh without the eco-logical judgment of the reader.

As a construct for medieval political allegory, then, the natural world is both constantly present and constantly absent. Successful harmonizing of the political and the natural can only ever be a future proposition, or occasionally, a memory of the distant past, casting the appearance of

---

36. This is particularly pertinent, as David Carlson points out, as this poem was likely composed after Henry's execution of Bushy, Green, and Bagot in 1399 (*John Gower, Poetry and Propaganda in Fourteenth-Century England,* 122–33, 138–39. In justifying past action, it shores up Henry's claim that the vengeance he takes against his political enemies be read as righteous political action.

nature in poetic complaints as a device that allows for an abstract under-standing of concrete, historical events. Their appeal for prudent political action follows from this negotiation of the abstract and concrete, the authorial and the readerly. These texts thus work both as goads to action and as justification of loyalties that resist other political discourses of suc-cession and royal prerogative. The role of nature in political complaint poetry of this period permits this dual effect and promotes a biopolitical rhetoric of interpretation, activism, and critique for authors as well as for their audiences. To read politically in these texts is, in the end, to read naturally.

9

Lost Geographies, Remembrance, and *The Awntyrs off Arthure*[1]

Kathleen Coyne Kelly

Even Memorie her selfe, though striving, would come short
But the material things Muse helpe me to report
—*Michael Drayton,* Poly-Obion *(1612–13)*

And so there grew great tracts of wilderness,
Wherein the beast was ever more and more,
But man was less and less, till Arthur came.
—*Alfred, Lord Tennyson, "The Coming of Arthur" (1859)*

IN "AFTER THE GOLD RUSH," Neil Young combines a streak of medievalism with a cry of distress:

Well, I dreamed I saw the knights
In armor coming,

1. This chapter attests to the efforts of professional and amateur historians to remember and preserve local lore. I would like to acknowledge the help of Mrs. Margaret Edwards of the Cumberland and Westmorland Antiquarian and Archaeological Society, who was kind enough to pass on my questions about Inglewood Forest and the Tarn to Mrs. Sheila Fletcher and Dr. June C. F. Barnes, both of whom generously shared their memories, their expertise, and their materials. Dr. Barnes took photographs of the site for me in 2010, which were helpful in establishing the *mise-en-scène*. Mrs. Fletcher has been quite patient in responding to my many emails. I thank Susan Wall for her gentle and insightful intervention and Barry Hoberman for his microreading.

Saying something about a queen.
There were peasants singing and
Drummers drumming
And the archer split the tree.

. . . .

Look at Mother Nature on the run
In the nineteen seventies.[2]

Young has a dream vision of a pop-medieval past, then slips, in stream-of-consciousness fashion, into a bleak present, in which a personified, feminized nature recedes. The repetition of the line, coupled with Young's trebly tenor—the embodiment, quite literally, of anguish, heartbreak, and misery—underscores the nature of the song as a lament, a song of grief and mourning. (In concert, Young has revised the lyrics, first from "the nineteen seventies" to "the twentieth century," and, because he still rocks, to "the twenty-first century.") Mourning the environment has a refresh button.

In both high and low culture, we can trace out a deep and wide river of mourning, together with what Freud calls melancholy—grief dammed—in Western cultural representations of human engagements with nature. Examples of depictions of grieving beings range from the forlorn Adam and Eve shut out of the garden in the fifteenth-century *Tres Riches Heures* to the lonely robot sorting mountains of garbage in the 2008 film *Wall-E*. Then there is the real world: the trauma experienced by victims of the Tōhoku earthquake and tsunami in 2011 and of Hurricane Sandy in 2012 (traumas with which many of us sympathized); the sadness that many of us experience upon learning that one more species is now extinct; the rage mixed with sorrow when one flies over a mountaintop blasted open for coal; the surge of regret at seeing a silver maple cut down next door.[3] One might argue that mourning is a—if not *the*—major trope of nature writing and of environmentalism itself. The stories that construct our understanding of what David Abram calls the "more-than-human world"[4] are, often, stories of what-used-to-be, of change, separation, and

---

2. Neil Young, "After the Gold Rush," *After the Gold Rush* (1970), Reprise B000002KD9, 1990, compact disc.

3. See Ashlee Cunsolo Willox, "Climate Change as the Work of Mourning," *Ethics and the Environment* 17, no. 2 (2012): 138–64, for a discussion of mourning as a response to real-world ecological degradation. Her arguments parallel mine.

4. David Abram, *The Spell of the Sensuous: Perception and Language in a More-Than-Human World* (New York: Vintage, 1997).

loss. The book of nature, as medievals liked to imagine it and into which they peered to learn about themselves, contains erasures, palimpsests, and missing pages.

In what follows (and as a prolegomenon to a larger project titled "Lost and Invented Ecologies"), I begin with Freud's question: "In what, now, does the work which mourning performs consist" with respect to our relation with the natural world and to its past?[5] In Freud's view, mourning is "reality-testing": the acceptance of the loss of the beloved object, "bit by bit, at great expense of time and cathectic energy. . . . Each one of the memories and expectations in which the libido is bound to the object is brought up and hypercathected."[6] In Western culture, "bringing up," as in remembering, is often a narrative process, sometimes private, sometimes shared: we pore over photographs, we tell stories about the lost beloved, we write an elegy. We also peruse archives, drill down into ice caps, and count tree rings. We tell tales of the fireflies that used to swarm in our backyard on a summer night, or of the fox that we saw trotting down our snowy suburban street, never to be seen again. Whether such narrative processes perform the work of mourning or leave us stuck in melancholy, unable to reconcile a loss and, as Freud puts it, unable to "see clearly what it is that has been lost," is another question.[7]

In the twenty-first century, studying ecological change and degradation entails a different consciousness altogether, one that twentieth-century conservationist thinking and even the more progressive forms of environmentalist thinking no longer serve—hence, as I shall discuss shortly, the importance of new materialist thinking in the humanities and sciences. Moreover, new approaches to studying ecological change have entailed developing a new vocabulary that is predicated on loss. For example, the word *Anthropocene* is now firmly established in public discourses on the environment (as a Google Ngram exercise dramatically attests). We are now in a new geological era, defined by the degree to which humans dominate and shape the world's ecosystems. Landscape ecologist Eric Sanderson, using a Geographical Information Systems program (GIS), created what he calls a "Human Footprint Map," which demonstrates that "83 percent of the earth's land surface is influenced by people (and that without considering climate change or air pollution,

---

5. Sigmund Freud, "Mourning and Melancholia" (1914), *The Standard Edition of the Complete Psychological Works of Sigmund Freud, Volume XIV (1914–1916)*, trans. and ed. James Strachey (London: Heath, 1957), 244.

6. Ibid.

7. Ibid., 245.

which nevertheless touches the remaining 17 percent)."[8] More recently, geographer Erle C. Ellis and his coresearchers have coined a new word to replace the term *biome*—the name for a geographic area that shares similar communities of flora and fauna. They argue that the term *anthrome* more accurately captures the astonishing degree to which the planet is now comprised of anthropogenic landscapes. Ellis and his colleagues created a series of global maps at century intervals (1700, 1800, 1900, and 2000) based on population density and land use. Their conclusion:

> In 1700, nearly half of the terrestrial biosphere was wild, without human settlements or substantial land use. Most of the remainder was in a semi-natural state (45%) having only minor use for agriculture and settlements. By 2000, the opposite was true, with the majority of the biosphere in agricultural and settled anthromes, less than 20% seminatural and only a quarter left wild.[9]

Ellis argues that because of anthropogenic change, "global patterns of the form and function of terrestrial ecosystems are no longer accurately depicted by the now classic approach to mapping and modelling biomes as a function of climatic and physiographic variables."[10] *Homo sapiens* is now the biggest variable to be accounted for.

The maps that Ellis produced make me sad. But Ellis has no patience for such a response. He says,

> It's time for a "postnatural" environmentalism. Postnaturalism is not about recycling your garbage, it is about making something good out of grandpa's garbage and leaving the very best garbage for your grand-children. Postnaturalism means loving and embracing our human nature, the nature we have created to feed ourselves, the nature we live in. What good is environmentalism if it makes you depressed about the future?[11]

I would argue, however, that a "good" environmentalism going into the future actually requires an essay into the past—not only studying and

---

8. Eric A. Sanderson, *Mannahatta: A Natural History of New York City* (New York: Abrams, 2009), 31.

9. Erle C. Ellis et al. "Anthropogenic Transformation of the Biomes, 1700 to 2000," *Global Ecology and Biogeography* 19 (2010): 589.

10. Ibid., 590.

11. Erle Ellis, "Stop Trying to Save the Planet," *Wired*, 6 May 2009, http://www.wired.com/wiredscience/2009/05/ftf-ellis-1/. It's worth scrolling through the comments to read poet and environmental activist Gary Snyder's response.

discovering and explaining the past of the natural world, but also studying and discovering and writing the stories that we tell ourselves about
the natural world and examining what is often a structuring grief in such
narratives. But grief, or depression, as Ellis formulates it, does not always
lead to sorrow and despair, to *wanhope,* "un-hope," as Middle English
expresses it, which is a state of un- or no action. There are other, real
questions to ask instead of Ellis's antagonistic, rhetorical one: to begin
with, what might we discover about the natural world if we explored that
river of mourning and melancholy that runs through our stories? Does,
or how might, mourning get in the way of engaging with the material
world, and inhibit, both locally and globally, the development of a fully-
realized environmentalist agenda? Or facilitate it? Melancholy is not just
a (Freudian) modern problem; it has a history: Giorgio Agamben notes
that melancholy had a "double polarity" in antiquity and the Middle
Ages—the melancholic may have been "sad, envious, malevolent," but
she or he was also associated with "genius."[12] This twofold nature of
melancholy is perhaps worth recuperating. Hugh of St. Victor, Agamben
points out, speaks of a *tristitia utilis* (useful sorrow); melancholy (given
bodily materiality as black bile) enables the exercise of the intellect and
the imagination.[13] To ask a question by way of Hugh of St. Victor: what
kinds of cultural and ideological work might melancholy and mourning
accomplish as we attempt to understand our enmeshment (to use Timothy Morton's term) with "nature"?—a word that now lives in perpetual
scare quotes. Does, or could, mourning advance an activist agenda, one
in which "biopower" is not the imposition of the state on all beings,
their bodies, and their relations (as Foucault first conceived of it) but
instead is the willing and ethical embracing of *all* beings, bodies, and
relations? I should note here that current attempts to rehabilitate Foucault's and Agamben's biopolitics as an "affirmative" biopolitics is limited
in its anthropocentricity, in that Foucault's and Agamben's conception
of the "animal" is folded into the human, and thus in effect serves as a
metaphor in a binarized system. Instead of thinking in terms of animal/
human, what would happen if we were to think of ourselves as *biocul-
tural* beings—no slash between bio and cultural? As Stacy Alaimo says,
"New materialist theories should not divide human corporeality from
a wider material world, but should instead submerse the human within

---

12. Giorgio Agamben, *Stanzas: Word and Phantasm in Western Culture,* trans. Ronald
R. Martinez (Minneapolis: University of Minnesota Press, 1992), 12.

13. As qtd. in Agamben, "Melancholia," 13: "render[ing] men now somnolent, now
vigilant, that is, now grave with anguish, now vigilant and intent on celestial desires."

the material flows, exchanges, and interactions of substances, habitats, places, and environments."[14] A true affirmative biopolitics, in its commitment to achieving the full realization of the individual in society, would extend the same urgent agenda to the more-than-human world. Thus my question: how might one *productively* explore grief and loss (*tristitia utilis*) not only of animals but of "habitats, places, and environments"?[15]

These are questions that we could answer (though perhaps never completely) in a variety of ways. In what follows, I trace one path that might help us arrive at some answers—a path that is familiar to medievalists, for it takes us through the past. I offer a thick description— a lament, in a way—of two lost places that medievalists know mainly through a few identifying footnotes; namely, Inglewood Forest and the Tarn Wathelan (or Wadling) in *The Awntyrs off Arthure* (c. 1400–1430) and its companion poems (*The Avowing of King Arthur, Sir Gawan, Sir Kaye, and Sir Bawdewyn* [c. 1400–1425], *The Weddynge of Sir Gawen and Dame Ragnell* [c. 1450], and *The Marriage of Sir Gawaine* [1500s?]). At the same time, the story that I want to tell about several square miles in the north of England is also an act of recovery—while acknowledging the problems, ideological and otherwise, inherent in the very idea of recovery. "Mourning," says Kathleen Biddick, "does not find the lost object; it acknowledges its loss, thus suffering the lost object to be lost while maintaining a narrative connection to it."[16] I argue that maintaining narrative connections—storytelling—is a process that can contribute to attaining an ethical and enmeshed understanding of the natural world and our place within it. In my larger project, "Lost and Invented Ecologies," I have multiple goals, to which I can only gesture here: to recover and/or reexamine the histories of lost and invented ecologies as they are

---

14. Stacy Alaimo, "New Materialisms, Old Humanisms, or, Following the Submersible," *NORA: Nordic Journal of Feminist and Gender Research* 19, no. 4 (2011): 281.

15. See Svetlana Boym's distinction between two kinds of nostalgia: "Restorative nostalgia protects the absolute truth, while reflective nostalgia calls it into doubt. . . . Reflective nostalgia does not follow a single plot but explores ways of inhabiting many places at once and imagining different time zones; it loves details, not symbols . . . reflective nostalgia can present an ethical and creative challenge, not merely a pretext for midnight melancholias," *The Future of Nostalgia* (New York: Basic Books, 2001), xviii. Also C. Nadia Seremetakis: "In English the word nostalgia (in Greek *nostalghía*) implies trivializing romantic sentimentality. . . . This reductiveness of the term confines the past and removes it from any transactional and material relation to the present," (*The Senses Still: Perception and Memory as Material Culture in Modernity* [Chicago: University of Chicago Press, 1996], 4).

16. Kathleen Biddick, *The Shock of Medievalism* (Durham, NC: Duke University Press, 1998), 10.

attested to in medieval literary texts and other documents; to emphasize the value of looking anew at any place and any historical period with a more ecologically informed perspective; and to contribute to ongoing conversations in environmental studies by considering not only human interventions into the natural world throughout the historical past but also taking into account the attitudes and beliefs formed about such interventions that continue to influence our beliefs and practices today. As environmental historian William Cronon says: "Nature alone cannot explain [a] landscape. You need history too."[17] And, I would add, *poetry*.

Writing of "traditionally oral peoples," David Abram says: "*The land . . . is the primary mnemonic, or memory-trigger, for recalling the ancestral stories*" (emphasis in the original).[18] The reverse is also true: stories, narratives, historical or imaginative accounts, written or oral or visual, are the primary means by which we remember and memorialize a given place in the natural world. I hope to complement the conception of a historicized, synchronic biopolitics that investigates how medievals made various interventions into their ecosystems—and the social and cultural consequences thereof (the subject of this volume)—with a diachronic biopolitics (in which the human is decentered) that traces anthropogenic changes from the Middle Ages to the present. In other words, exploring how medievals constructed and imagined the natural world is one kind of crucial and generative narrative; attending to how that natural world changed over the centuries is another.

My premise is that there is no single ecological thought (versus Timothy Morton) but many thoughts worth weighing together. At this moment, the master narrative of humanism is under attack not only for its hierarchizing of certain humans over others (through sexism, racism, imperialism, and other exploitative discourses and practices) but also because of its privileging of the human over all else in the natural world. Still, as we engage in debates about what many call posthumanism—that is, as we rethink agency, consciousness, embodiment, identity, and subjectivity, not only of human beings but of everything else with which we are enmeshed—I want to argue that there are some humanist ideas worth retaining. Poststructuralism taught us that all binaries are limiting in their absolutism: to binarize humanism/posthumanism potentially constrains the ways in which we might act across time—*just in time*—with respect

---

17. William Cronon, The Riddle of the Apostle Islands," *Orion*, May/June 2003, http://www.williamcronon.net/writing/Cronon_Riddle_Apostle_Islands.htm.

18. David Abram, "Between the Body and the Breathing Earth: A Reply to Ted Toadvine," *Environmental Ethics* 27 (2005): 177.

to the environment. Here is one thought, then, and mainly an anthro-pocentric one: Cronon states that "environmental histories transform ecosystems into the scenes of human narratives."[19] However, Cronon is not offering up a modern equivalent to the medieval book of nature in which human concerns trump all else. Yes, histories *for* humans: Cronon, like Sanderson and Ellis, argues that we have transformed much more of the planet's ecosystem than we realize and—also important—remember. But Cronon also rejects binary formulations like nature/culture, wild/tame, and pristine/spoiled-by-human-hands. For Cronon, studying the past, which includes our telling "stories about stories about nature," is a supremely humanist endeavor:

> The special task of environmental history is to assert that stories about the past are better . . . if they increase our attention to nature and the place of people within it . . . if environmental history is successful in its project, the stories of how different peoples have lived in and used the natural world will become one of the most basic and fundamental narratives in all of history, without which no understanding about the past could be complete.[20]

It is *this* version of a humanist environmentalism that I wish to keep, while at the same time exploring another, perhaps even oppositional thought: the importance of attending to the physicality of the past; that is, to the material reality and history of the natural world, not on our terms but as it "is." Histories *for* and *of* every thing. I find Karen Barad's rethinking of representationalism particularly useful in this respect: she argues for what she calls "agential realism," an understanding of the world in which both the material and the discursive "intra-act," and out of those intra-actions, "objects" emerge.[21] Imagine expanding "the archive"—at the moment, an anthropocentric metaobject in which other objects are nested—in a way that decenters the human and recognizes various material-discursive processes, including earth, water, stone, and tree. We humans do not have a monopoly on this archive—however often we tell stories about the more-than-human world. (Nature, Cronon wryly notes,

---

19. Willam Cronon, "A Place for Stories: Nature, History, and Narrative," *The Journal of American History* (1992): 1372.

20. Ibid., 1375.

21. Karen Barad, *Meeting the Universe Halfway: Quantum Physics and the Entanglement of Matter* (Durham, NC: Duke University Press, 2007), see chapter 1.

cannot read, but he argues that it still is a part of our audience.[22]) Our
default position—what we cannot escape—is that we always already produce anthropocentric texts: there is no other kind. To interpret such texts
inevitably *re*produces such anthropocentrism. Mindful of this problem, in
what follows, I explore how the nonhuman and the human have, and are,
"entangled" in a very particular part of the world from the eleventh century through the twenty-first. We grant other humans their histories and
"life-stories" as crucial to their identities and our relations with them;
why not do the same for the more-than-human world?

∞

It is worth pausing for a moment to offer a brief history of how some of
us in literary studies have arrived at attempting to apprehend the more-
than-human world, given our prior and ongoing investments in representationalism. I invoke feminist criticism as an analogy—and in doing
so, I by no means intend to suggest that the history of feminism or of
the humanist study of the natural world represents any sort of harmonious progression. Feminist critics (most notably, Elaine Showalter) have
identified three phases of Anglo-American feminist literary theory (which
nevertheless overlap), beginning with, first, critiquing representations of
women in canonical literature while simultaneously expanding that canon
to include "lost" or marginalized women writers; second, encouraging
and publishing new women writers; and third, developing a wide range
of approaches to literature that takes into account differences such as
race, class, and sexual identity. Ecocriticism, I would argue, also has had
its parallel first and second phases, captured by Cheryl Glotfelty's much-
quoted definition: ecocriticism is "the study of the relationship between
literature and the physical environment." Glotfelty adds:

> Ecocritics and theorists ask questions like the following: How is nature
> represented in this sonnet? What role does the physical setting play in the
> plot of this novel? Are the values expressed in this play consistent with
> ecological wisdom? How do our metaphors of the land influence the way
> we treat it?[23]

---

22. William Cronon, "The Uses of Environmental History," *Environmental History
Review* 17, no. 3 (1993): 7.

23. Cheryl Glotfelty, "Introduction: Literary Studies in an Age of Environmental Crisis," *The Ecocriticism Reader: Landmarks in Critical Ecology,* ed. Cheryl Glotfelty and
Harold Fromm (Athens: University of Georgia Press, 1996), xviii–ix.

An ecocritical reading is thus a potentially activist endeavor as well as a literary one. In addition to a set of questions, early ecocritical studies established a methodology and a vocabulary, a canon and an institutional presence. Humanist ecocritical studies, sometimes gathered under the umbrella term "environmental studies," continue to flourish and increase in Anglo-American universities. However, these days, one can identify a shift in emphasis within first- and second-phase ecocriticism. I borrow from Lynne Bruckner and Dan Brayton's discussion of the history of ecocriticism: "As nature itself has become a more problematic and less conceptually transparent ground of meaning," ecocriticism has broadened from an aesthetic appreciation of nature writing to encompass "anthropocentrism, ecocentrism, living systems, environmental degradation, ecological and scientific literacy, and an investment in expunging the notion that humans exist apart from other life forms."[24] I particularly like the agency attributed to "nature itself" in the assemblage we call *theory*.

An emerging third phase—and a phase that many experience as a break with earlier ecocritical theory and practice—falls under the rubric of *ecomaterialism*. (Or materialist ecocriticism, as it is also called. I must note that, in the interests of brevity, I am bracketing—while assuming—the influence of object-oriented ontologies and speculative realisms on ecocriticism and ecomaterialism.) Ecomaterialism, when considered from the vantage point of literary studies, generates a different set of questions and different ways of reading from those of ecocriticism, not about the meaning of a poem *for us humans* but about the liveliness and agency—the "thinginess" of the things—being represented in the text. And, in the context of the intra-action between nature and theory, let us not forget the liveliness and agency of the "poem itself," which, albeit in the context of the only-human world, Cicero partially glimpsed and expressed through his affective triad, *docere, delectare, movere*.

Medievalists who are interested in pursuing an ecocritical reading are immediately faced with a problem. In general, landscape and geography in medieval literature are not described but simply noted, participating in what might be called a "trope of location" rather than in concrete or imaginative representations. In *The Awntyrs,* for instance, this knight meets that one in a wood or by a lake, with no attention given to detail. This is geography "by adventure," as medieval authors of romance so often describe chance meetings (and as the author or scribe of the Thorn-

24. Lynne Bruckner and Dan Brayton, eds., "Introduction: Warbling Invaders," *Ecocritical Shakespeare* (London and New York: Ashgate, 2011), 3.

ton ms. who gave *The Awntyrs* [1440] its title emphasizes). Apart from
the most basic observations about the relationship between setting and
plot in *The Awntyrs* (or the interpretations that we make to connect setting and plot), the poem simply doesn't allow us to come up with satis-
factory, interesting answers to Glotfelty's questions. "Nature" in this text
is merely indexical. To repurpose Myra Jehlen on feminist criticism, *one
has to read for nature;* "Unless it figures explicitly in story or poem, it will
seldom read for itself."[25]

On the other hand, medieval literary texts, maps, and other documents, as well as painting and the plastic arts, are full of *things* for us to
discover. Bill Brown states that "we look *through* objects (to see what
they disclose about history, society, nature, or culture—above all, what
they disclose about us)" but hardly see the things themselves.[26] When we
take a *literary* thing *literally,* however, we may discover meanings that lie
outside the realm, perhaps even the tyranny, of the figural. Overdetermined as an allegory and as an example of the *locus classicus* of romance,
a place such as Inglewood Forest in *The Awntyrs* is therefore underde-
termined as a thing. Telling its story may help us recover some of its
real, material, historical thinginess. Granted, the very act of *reading (for)*
nature, materiality, and the agency of the more-than-human world is,
inevitably, to figure nature and to use nature to figure *us.* In this essay so
far, I have described mourning as a *river,* and ecocritical studies as *flour-
ishing* and *increasing.* Ecomaterialism doesn't get us out of the bind of
representationalism; rather, it makes us constantly mindful of it. Donna
Haraway argues:

> We must find another relationship to nature besides reification, posses-
> sion, appropriation, and nostalgia. No longer able to sustain the fictions
> of being either subjects or objects, all the partners in the potent conversa-
> tions that constitute nature must find a new ground for making meaning
> together.[27]

I agree with Haraway wholeheartedly. Still, Haraway's statement drama-
tizes the difficulties of language and praxis. Also, note that she repro-

---

25. Myra Jehlen, "Gender," in *Critical Terms for Literary Study,* 2nd ed., ed. Frank
Lentriccia and Thomas McLaughlin (Chicago: University of Chicago Press, 1995), 273.

26. Bill Brown, "Thing Theory," *Critical Inquiry* 28, no. 1 (2001): 4.

27. Donna Haraway, "Otherworldly Conversations: Terran Topics; Local Terms," in
Feminist Materialisms, ed. Stacy Alaimo and Susan Hekman (Bloomington and Indianapolis: Indiana University Press, 2008), 158.

duces the binary between nature and "us" by using the word *relationship*. We are *partners* (in dance? in business? in a monogamous relationship?), which is a human construct of affiliation. Nature is a *conversation,* a human-centered metaphor; we must find new *ground,* a metaphor appropriated from nature. Perhaps *ground* is also a pun, which, as far as I know, only humans are capable of recognizing. I don't mean to nitpick. No one is more acutely aware of the ethical and environmental consequences resulting from the split that we have made (or, at least, come to believe in) between nature and culture than Haraway. However, while such exhortations as Haraway's are welcome and inspiring, we need more praxis. (For one example of a practice that destabilizes anthropocentricism, consider how some people, albeit self-consciously, are attempting to replace the word *pet* with *companion species,* signaling a shift in thinking about what it means to live with animals—a shift in which Haraway plays a major role.) The difficulties in "finding a way" seem almost insurmountable given that we cannot (as Haraway's passage demonstrates) escape the prison house of language. Put another way: at this point in our intellectual and philosophical history, we seem unable to avoid acting as if being locked up in language were *true.*

୦୨

In what follows, I treat *The Awntyrs off Arthure* and its companion poems as witnesses to a lost geography and as an occasion for a foray into environmental history. By chronicling an occluded or ignored narrative; namely, the story of Inglewood Forest and the Tarn that together furnish the setting for these poems, I hope to recuperate their lively thinginess as well as to enrich our experience of them. Let us begin with the story of *The Awntyrs,* here summarized:

> From Carlisle Arthur goes out hunting at Tarn Wadling. Accompanied by Gawain, Guenevere [Gaynour] rides out and rests under a laurel. During a sudden, violent storm, a terrifying apparition ["Al biclagged in clay" and "Serkeled with serpentes" ll. 106, 120[28]] rises from the lake—the

---

28. Ll. 1–8. 1400–30; preserved in four MSS of the fifteenth century: Bodleian 21898 MS Douce 324 [D] (which I quote), by a scribe from NE Derbyshire (late-fifteenth century); Lincoln Cathedral MS 91 [T] by Robert Thornton of Ryedale, N. Yorkshire (c. 1440); Ireland Blackbourne MS [Ir], at Hale in North Lancashire/Cumbria (c. 1450–75); Lambeth Palace MS 491 [L] (copied in London in the dialect of Rayleigh, Essex, by a professional scribe before 1450). Only Thornton furnishes a title: "Here By gynnes The

ghost of Guenevere's mother. The spirit warns Guenevere of the consequences of sin, prophecies the downfall of Arthur and the Round Table,
and appeals to her daughter for thirty masses to be said for the salvation
of her soul. The ghost vanishes, the storm subsides, and when the king
and his company return, they are informed of the wondrous occurrence.
When Arthur and his court are at table, a gorgeously attired lady leads
into the hall Sir Galaron of Galloway, a Scottish knight who challenges
one of Arthur's knights to combat to correct an injustice. Gawain under-
takes the duel, but just as he is about to win the victory, the distraught
lady appeals to the queen to ask Arthur to intervene. Galaron yields his
claim, acknowledging Gawain's superiority. By Arthur's decree, Galaron
becomes a knight of the Round Table and marries the Lady.[29]

The pertinent part of the text:

> In the tyme of Arthur an aunter bytydde,   [adventure occurred]
> By the Turne Wathelan, as the boke telles,
> Whan he to Carlele was comen, that conquerour kydde,   [famous]
> With dukes and dussiperes that with the dere dwelles.   [beloved king]
> To hunte at the herdes that longe had ben hydde,   [i.e., in the wild]
> On a day thei hem dight to the depe delles,
> To fall of the femailes in forest were frydde,   [enclosed]
> Fayre by the fermesones in frithes and felles.
> (1–8)

In *The Avowing of King Arthur*, Arthur, Gawain, Kay, and Bawdwyn hunt
a fierce boar ("well grim gryse") in "Ingulwode," and tarry by "Tarne
Wathelan." Kay meets a mysterious knight "Atte the anturis hoke" (mar-
velous oak).[30] In *The Weddynge of Sir Gawen and Dame Ragnell*, Arthur
again goes hunting "in Ingleswod, / With alle his bold knyghtes good."[31]
Finally, in *The Marriage of Sir Gawaine*, Arthur encounters a woman

---

Awntyrs off Arthure At the Terne Wathelyn." All poems edited by Thomas Hahn in *Sir
Gawain: Eleven Romances and Tales* (Kalamazoo: Medieval Institute Publications, 1995),
169–226. Online at http://www.lib.rochester.edu/camelot/teams.htm. Hahn's glosses.

29. Helaine Newstead, summary, in *A Manual of the Writings in Middle English,
1050–1500*, vol. 1, J. Burke, Severs, gen. ed. (New Haven: Connecticut Academy of Arts
and Sciences, 1967), 61.

30. Ll. 32; 500, composed c. 1400–25; Ireland Blackburn [c. 1450–75], in the same
hand as *Awntyrs*.

31. Ll. 17–18, composed c. 1450 in the East Midlands; single MS of the sixteenth
century: Bodleian 11951 (Rawlinson C.86).

"betwixt an oke and a greene hollen" in the forest, as well as a bellicose baron at "Tearne Wadling" who demands an answer to the question, "What do women desire most?"[32]

Until the forest and the tarn got tangled up in Arthurian romance, they had their own reputations. We find a passing reference to the Tarn in the *Pleas of the Forest of Inglewood* (1285). One Adam Felton, Prior of St. Mary's, Carlisle, put forth a claim to the tithe caught for fish "in lacu de Terwathelan qui dicitur Laykebrayt."[33] Gervase of Tilbury, in *Otia Imperialia* (1210–14), explains this name:

> In the greater British isle, there is a forest with copious species of different game, which overlooks the city of Carlisle. Almost in the middle of this is a valley, surrounded by mountains, near the public road [now the A6]. In this valley, I say, daily at daybreak a sweet trumpet of bells is heard. Hence, the natives in this desolate place call it, in the French language, "Laikibrait" ["the lake that cries"].[34]

Andrew of Wyntoun, in the *Orygynale Chronykil of Scotland* (c. 1420) connects another legendary figure besides Arthur to Inglewood:

> Lytil Jhon and Robyne Hude
> Wayth-men ware commendyd gude
> In Yngil-wode and Barnysdale
> Thai oysyd all this tyme thare trawale.   [for the year 1283][35]

These poems and tales are set in what is now Cumbria (formerly Cumberland), on the border between Scotland and England. The *Awntyrs* poems share a number of northern dialect features and are distinctly regional in their point of view. Patricia Ingham and Randy Schiff, among others, have argued for ways to understand these poems in their historical and geopolitical context. *The Awntyrs* and *The Avowing*, in particular, offer pointed lessons in border politics, ethnic identities, and the strife caused

---

32. Composed c. fifteenth century? Single copy in the Percy Folio MS 55.

33. *Pleas of the Forest of Inglewood* for 1285, as qtd. in R. C. Cox, "Tarn Wadling and Gervase of Tilbury's 'Laikibrait,'" *Folklore* 85, no. 2 (1974): 128–31. Translation adapted by me.

34. *Gervaise of Tilbury: Otia Imperialia; Recreation for an Emperor,* ed. S. E. Banks and J. W. Binns (London and Oxford: Oxford UP, 2002), LXIX.

35. Andrew of Wyntoun, *Orygynale Cronykil of Scotland,* ed. David Laing, 1872. X.263. Also in *Robin Hood and Other Outlaw Tales,* ed. Stephen Knight and Thomas H. Ohlgren (Kalamazoo: Medieval Institute Publications, 1997), 24.

by what Schiff calls "the ceaseless cycle of possession and dispossession that haunted the Anglo-Scottish borderlands."[36] Ingham argues that *The Avowing* "links military brotherhoods to the materiality of region and land"; she says of *The Awntyrs* that it "depicts the pleasures of a centralized sovereign who rests his practices of land distribution and regional control on the violence of knightly rivalry."[37] More recently, Joseph Taylor, following Agamben, reads *The Avowing* as "illustrat[ing] the bare life of the borderlands as it confronts the biopolitical sovereign signified . . . by Arthur."[38]

In these poems, natural places are experienced—that is, made intelligible—as *possessions*. Consider how, in *The Awntyrs*, Guenevere's mother's ghost laments all that she has lost:

> al gamen or gle that on grounde growes
> . . . of garson and golde,
> Of palaies, of parkes, of pondes, of plowes,   [enclosures, estates]
> Of townes, of toures, of tresour untolde,
> Of castelles, of contreyes, of cragges, of clowes.   [lands]
> (147–51)

The land itself is catalogued as treasure; specifically, domesticated and settled spaces and sites of managed natural resources. Such transformations of the land turn natural places into commodities. Moreover, Guenevere's mother says, "Quene was I somwile" (145) not only of *castelles* and *parkes,* but of natural geographical features—*cragges* and *clowes*.[39] Her sovereignty extends even into "waste" places; that is, as Gillian Rudd says, not "barren" places but "uncultivated land whose potential fertility and use (from a human point of view) was thus being wasted."[40] If such places can be owned, they can be gifted, inherited, sold—or, as here, "lost." These places endure even as they change; they change as they endure. It is Guenevere's mother who does not endure. She grieves her losses even after death.

---

36. Randy Schiff, "Borderland Subversions: Anti-Imperial Energies in *The Awntyrs off Arthure* and *Galagros and Gawane*," *Speculum* 84, no. 3 (2009): 632.

37. Patricia Clare Ingham, *Sovereign Fantasies: Arthurian Romance and the Making of Britain* (Philadelphia: University of Pennsylvania Press, 2001), 179–80.

38. Joseph Taylor, "Sovereignty, Oath, and the Profane Life in the *Avowing of Arthur*," *Exemplaria* 25, no. 1 (2013): 37.

39. *Pondes* could be either natural places or managed resources.

40. Gillian Rudd, *Greenery: Ecological Readings of Late Medieval Literature* (Manchester, UK: Manchester University Press, 2007), 41.

Later in *The Awyntrs,* the Scottish knight Sir Galaron of Galwey hopes to recover the lands that once belonged to him. He accuses Arthur: "Thou has wonen hem in werre with a wrange wile" (unjust trick) and then given the lands to Gawain (421). Sir Galaron challenges Arthur to furnish a champion for single combat so that he may win his lands back. Galaron ritually recites a catalogue: "Of Connok, of Conyngham, and also Kyle, / Of Lomond, of Losex, of Loyan hills" (424–25). Gawain takes up the challenge. However, at the point at which Gawain is about to defeat Galaron, Guenevere begs Arthur to intervene. Gawain restores the lands to Galaron, and, in another ritual recital that can be read as a confirmation of land rights, repeats Galaron's list—with embellishments:

> Al the londes and the lithes fro Lauer to Layre,
> Connoke and Carlele, Conyngham and Kile;
> Yet, if he of chevalry chalange ham for aire,
> The Lother, the Lemmok, the Loynak, the Lile,
> With frethis and forestes and fosses so faire.   [woods, moats]
> (686–90)

Galaron's recital is a narrative of injustice, a plaint about the loss of ancestral lands. His and Gawain's repetition of place names—a small pleasure as well as a political act for a local poet and local listeners—functions as a hedge against forgetting and chaos. Just as boundary stones or walls mark ownership, the remembrance of and the recitation or reading of place names ratifies local claims. These poems share an interest in local geography and geopolitics in their naming and listing of topographical features. It might be asked if the land that Galaron "lost" and Arthur "gifted" to Gawain, who in turn "restored" it to Galaron, is *the same thing each time.*[41] To itself, perhaps, it is, if the land ignores us. And to speculate so is to indulge in the trope of the timeless indifference of the land to the small humans who walk it—a trope that has its uses in the only-human world, both as a lesson in humility and as a license to do what we want to earth, air, stone, and tree.

In addition to the above natural places in *The Awynters,* we find other places, mainly of manors, estates, castles, and shires—places such as Rondoles Halle and Plumton Land (340, 480). A good deal of scholarly effort has gone into identifying the poems' place names with their

---

41. I thank Joseph Taylor for his comments on this section, which led me to the conundrum concerning lively mutability.

actual sites, a task made especially difficult with respect to *The Awntyrs* because of variants among the four manuscripts. As Susan Kelly puts it, "It is unfortunate that because of scribal error, emendation and dele- tion, no more than a tentative identification can be assigned to some of the localities."[42] Correctly identifying place names may seem a pedant's task—and a melancholy one. The whole field of toponymy, the scien- tific study of place names, is rather Borgesian in that toponymists con- stantly deal with corrupt texts, damaged images or maps, and faulty oral accounts that misremember, lead astray, or dead end. Toponymists, along with historians, literary historians, environmental historians and others, perform important work as they attempt to sort out what it is, pre- cisely, a given culture has chosen to memorialize and commemorate— and to preserve. Or not, as is the case with respect to Inglewood Forest and Tarn Wadling. Both are lost, vanished, the forest cut down and the Tarn drained. These natural places live on only in the archive and on the brown road signs of English Heritage. (And again, even accounting for ecological change, are we preserving or marking *the same thing*?)

Inglewood Forest once covered an area sixteen miles long and ten miles wide between the rivers Petteril and Eden, and reached from Carl- isle to Penrith; it was the second largest royal hunting forest in England.[43] Inglewood was not a "wild" forest in the Middle Ages but a managed forest cleared for hunting and for other uses, such as timber-harvesting, charcoal burning, cow grazing, and pig foraging.[44] In the past fifty years, developments in the field of environmental history have revealed a very different picture of the medieval landscape. The English medieval forest was not the forest primeval that supposedly stretched from coast to coast; that forest had been largely cleared by 8000 BCE. The enchanted wild forest and the wood pasture of medieval romance were really a patchy series of succession forests.

Earlier, I made a point of quoting when a tree was named in the *Awntyrs* poems—oaks and hollies, two trees often reduced to their sym- bolic import for humans. Such textual pointing is the extent of any

---

42. Susan Kelly, "Place-Names in the *Awntyrs off Arthure*," *Literary Onomastics Stud- ies* 6 (1979): 162–99, 189 [28]. Also see Rosamund Allen, "Place-Names in *The Awntyrs off Arthure*: Corruption, Conjecture, Coincidence," in *Arthurian Studies in Honour of P. J. C. Field*, ed. and foreword, Bonnie Wheeler (Cambridge, UK: D. S. Brewer, 2004), 181–98.

43. Mrs. Sheila Fletcher, personal email, 20 November 2010.

44. F. H. M. Parker, "Inglewood Forest IV" [Revenues], *Transactions of the Cum- berland & Westmoreland Antiquarian & Archeological Society* n.s. 9 (1909): 24–37, and "Inglewood Forest VII" [on the Huttons, Hereditary Foresters], *Transactions of the Cum- berland & Westmoreland Antiquarian & Archeological Society* n.s. 11 (1910): 1ff.

description that we find in these texts, which perhaps explains a reader's resorting to the figural. But let us try to see past what the tree *stands for* to the tree itself, *standing,* to the root and trunk and branch and leaf, as we consider the story of the demise of the "last" oak of Inglewood Forest. In 1840, the antiquarian Samuel Jefferson wrote:

> On Wragmire moss [low-lying soggy ground], until the year 1823, there was a well-known oak, known as *the last tree of Inglewood Forest,* which had survived the blast of 700 or 800 winters. This "time-honored" oak was remarkable, not only for the beauty of the wood . . . but as being a boundary-mark . . . and was noticed as such for upwards of 600 years. This "gnarled and knotted oak," which had weathered so many hundred stormy winters, became considerably decayed in its trunk. It fell not, however, by the tempest or the axe, but from sheer old age; this happened on the 13th of June, 1823. . . . Xerxes, who cared not for the sacrifice of human life, would not suffer his army to destroy trees, and halted his mighty host for three days that he might repose beneath the Phrygian plane; and yet, probably, that tree had not numbered half the years of this relic of Inglewood, under whose spreading branches may have reposed the victorious Edward I, who is said to have killed 200 bucks in this ancient forest.[45]

It is a rich little anecdote. It is an example of a history *for* humans (the tree serves as a boundary marker; the forest is a source of game for humans), and into which another history, or legend, of a tree has been embedded. By alluding to Xerxes, Jefferson inserts another, perhaps "real" tree into the endless loop of the signifier and the signified, canceling out the material liveliness of both the Inglewood oak and the plane tree. The story of Xerxes and the plane tree (perhaps a sycamore) is found in Herodotus and Aelian. Aelian condemns Xerxes for *hybris,* immoderation; in this case, tree-love: apparently Xerxes commanded that the tree be decorated with golden baubles, "as if it were a woman he loved." But, Aelian says: "The beauty of a tree consists of fine branches, abundant leaves, a sturdy trunk, deep roots, movement in the wind, shadow spreading all around, change in accordance with the passing of seasons."[46] Surely this is an anthropocentric aesthetic judgment intended

---

45. Samuel Jefferson, *The History and Antiquities of Leath Ward, in the County of Cumberland: With Biographical Notices and Memoirs,* (Carlisle, 1840), 206–7. Online.

46. Aelian, *Variae Historiae,* trans. Nigel G. Wilson (Cambridge, MA: Loeb Classical Library, 1997), II.14, 85–86.

to rebuke Xerxes (nature has its ideological uses in the only-human world), but it also strikes me as a moment in a text in which the ontological autonomy of a tree might be apprehended. Perhaps earth, water, stone, and tree are not always trapped in the prison house of language. Perhaps they have ways to announce themselves.

Besides turning things into symbols, another way that humans draw nature into culture is by documenting its extremes: the biggest, deepest, widest, tallest, smallest, oldest. In Cambridge, Massachusetts, where I live, the city keeps a record of the oldest trees (mainly oaks); some residents advocate designating them as landmark trees that cannot be cut down, even on private land. Cumbria, centuries too late, has such a plan in effect under the title of "Tree Preservation Orders." The Białowieża Puszcza is a forest in Poland with oaks dating to the Middle Ages: because of its age and uniqueness, parts of the Białowieża Puszcza have been designated as UNESCO World Heritage Sites. But there are far older trees in the world: in England, the Ankerwycke Yew, Berkshire, is estimated to be two thousand years old; King John, the legend goes, signed the Magna Carta in 1215 under its branches. In Jefferson's account of Inglewood, we meet another extreme: the *last* tree of its kind, and one about to become a family heirloom, for Johnson adds (in third person): "The writer has a cabinet, the panels of which are made of the heart of this fine oak tree."[47] The last tree of Inglewood Forest, like slivers of the True Cross, like John Garrick's chest made of a mulberry tree allegedly planted by Shakespeare himself, lives on, a piece of nature enshrined in culture.

Inglewood Forest is pressed into ideological service once again in folklorist John S. Stuart Glennie's guide, "Arthurian Localities; their Historical Origin, Chief Country, and Fingalian Relations; with a Map of Arthurian Scotland," published in 1869. Glennie wished to lay claim to Arthur for "Fenians," and Inglewood is made into one more proof of his argument for a historical and Gaelic Arthur. Glennie describes climbing Blaze Fell and surveying what he called the "undulating, mountain bounded, and still finely wooded ancient forest of Inglewood."[48] He may have seen fine woods, but they were not ancient. One might understand Glennie's desire to map the medieval past onto the landscape that he actu-

---

47. Jefferson, *History and Antiquities, 207.*
48. J. S. Stuart Glennie, *Arthurian Localities: Their Historical Origin, Chief Country, and Fingalian Relations, with a Map of Arthurian Scotland* (1869) (Somerset, UK: Llanerch Publishers, facsimile reprinted, 1994), 74.

ally beheld: such an impulse is an example of nineteenth-century romanticized medievalism as much as it is motivated by his political interest.
Glennie also visited the Tarn:

> [In High Hesket] at the "White Ox" I had the good fortune to encoun-
> ter an intelligent old man, who, taking me to the back of the farmyard,
> pointed out, down in the hollow, what I was in search of, the famous
> Tarn Wahethelyne of Ballad and Romance. But Tarn Wadling, as it has
> been called in later times, has been for the last ten years a wide meadow,
> grazed by hundreds of sheep.[49]

Tarn is the Scottish word for "mountain lake"; it is what geologists call
a *cirque,* or *corrie,* a Scottish-Gaelic word meaning "kettle"; both words
denote a body of water formed by glacial runoff in an amphitheater-like
valley at the foot of a mountain. In 1839, when Frederic Madden edited
The Awntyrs off Arthure, he was able to turn to local guidebooks and histories for information on the lake and the forest. But in 1858, the Tarn
was drained by a local noble, Lord Lonsdale, in order to create, according
to local lore, a horse-racing track.[50] However, the evidence contradicts
this rather romantic idea; rather, Lonsdale, it seems, was much taken with
engineering schemes and was eager to create more arable and grazing
land under the Acts of Enclosure. Lonsdale took a government loan sub-
sidized by the Enclosure Commissioners and took on the Tarn, which,
as in the Middle Ages, was "a hundred acres of water and flow mosses
valued only for carp, snipe, and wildfowl" (medieval fishers, happy to
catch such, would have taken exception to the "only").[51] The lake was
drained through a cut made to the River Eden, and conduits were built
to keep it drained. However, nineteenth-century drainage practices were

---

49. Ibid., 72. Glennie continues: "Of the draining of it [the Tarn] the old man . . .
here for the last fifty years, had a great deal to say" (73). Swine, apparently, used to fish
for carp themselves.

50. The Woodland Trust's *Tarn Wadling Management Plan 2014–2019* states: "Records make mention of the lake at Tarn Wadling up to the late 1850s when it is suggested
that Lord Lonsdale drained the area to create a training ground for racehorses." 2008.
http://www.woodlandtrust.org.uk/woodfile/785/management-plan.pdf?cb=06445ee21
f28419cb908d490e2fca45d. Intriguing as this idea is—the local lord intent upon wresting
the land to his own elitist purposes—the passive is not reassuring. Local historian Mrs.
Sheila Fletcher doesn't believe it, but does say that some land nearby was used for racing
in the past (email, 20 November 2010).

51. A. B. Humphries, "Agrarian Change in East Cumberland 1750–c. 1900," M. Phil.
thesis, University of Lancaster, 1993, 257. See 257–60 for details on the drainage and cost.

FIGURE 9.1. High Hesket, Armathwaite, Cumbria.
Tarn Wadling is the oval depression to the left of Tarn Wood (Google Maps).

not quite up to the task of keeping the Tarn a meadow. Laikibrait—"the lake that cries"—weeped and seeped through the ground and refused to stay dry enough for grazing, and it had to be drained sporadically through the years. (It was located, after all, in a glacial valley, and continued to absorb runoff from the mountains surrounding it.)

The Tarn was drained more successfully in the 1940s, this time by German prisoners of war.[52] After this last effort, the Tarn shrunk into an oval-shaped hollow that still occasionally flooded and froze; locals used to ice-skate there (see figure 9.1). Today, if one takes the pilgrimage to High Hesket, one can see the remains of an old stone building thought to have been a boathouse: if so, it now sits incongruously in a field at the edge of a stand of trees, surrounded by oblivious grazing cows.[53]

52. Mrs. Sheila Fletcher, email, 20 November 2010. German prisoners of war were housed in a few different places in Cumbria from 1939–44, including Penrith, which is close to Inglewood and the Tarn.

53. As in a 2110 annotated photograph of the building, courtesy of Dr. June C. F. Barnes.

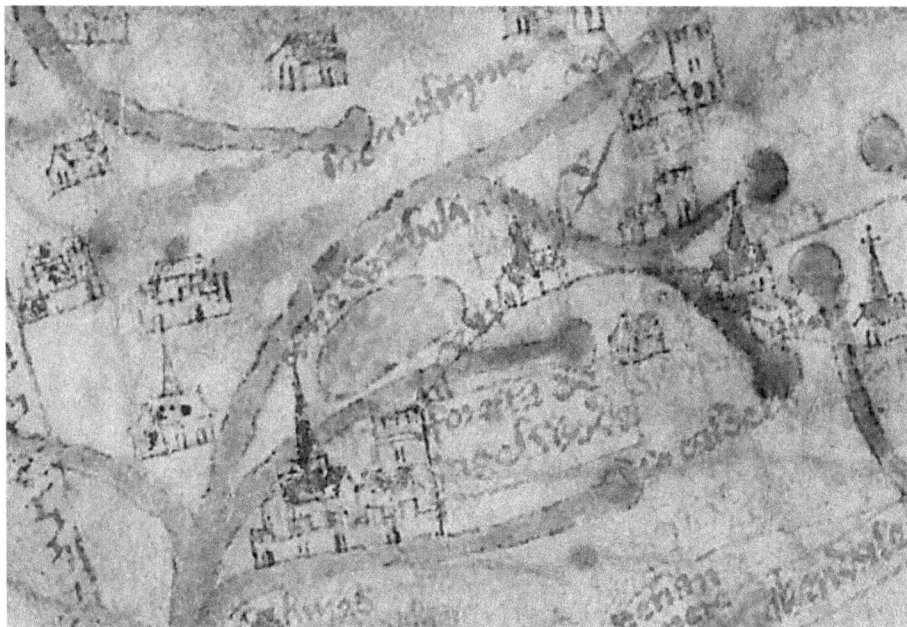

FIGURE 9.2. Detail, The Gough Map, *Tern [or the] Wathelan* and *Foresta de Inglewode* (1350–60). The tarn is the large oval above and to the right of the building with a steeple; the forest is spelled out just south of the tarn.
(Courtesy of Bodleian Libraries, University of Oxford.)

In addition to traces on Google Maps and in extant written accounts, Inglewood Forest and Tarn Wadling persist on maps of parchment and paper, as on the mid-fourteenth-century Gough Map (see figure 9.2).[54] In *The King's Two Maps,* Daniel Birkholz examines the Gough Map in the context of Edward I's interest in empire, especially insofar as Wales and Scotland are concerned. Birkholz argues that the map is "essentially a document of colonial administration and propaganda."[55] As evidence, Birkholz cites a tiny inscription on the map that claims that Brutus, together with his fellow Trojans, landed on the south Devon coast (*hic Brutus applicuit cum Troiani:* "here Brutus landed with his Trojans"). He also points to a few marked sites connected to the legend of Arthur: Glastonbury, Pendragon Castle, and Tintagel. Since Edward claimed descent from Brutus, and thus Arthur, the Gough Map, argues Birkholz, was intended to confirm Edward's ancestral rights to Scotland.

54. *Gough Map.* Online. http://www.goughmap.org/digital-map/.
55. Daniel Birkholz, *The King's Two Maps: Cartography and Culture in Thirteenth-Century England* (London: Routledge, 2004), 113.

In his dispute with Boniface VII over Scotland, Edward reminded the Pope that Brutus, the progenitor of "a race of kings," as Geoffrey of Monmouth puts it, was his ancestor, as was King Arthur, who, says Edward, "subjected to himself a rebellious Scotland."[56] Tarn Wadling (once c. 1/6 of a sq. mi., or 100 acres) is drawn on the Gough Map as if it were bigger than Windermere (almost 6 sq. mi., or 3,840 acres), which is actually thirty-eight times the area of the Tarn. This incongruity suggests the degree to which the Tarn mattered in the context of local history and legend, not only as a mysterious, haunted lake in and of itself but also as connected to Arthur.[57] Two centuries later, William Hole, who furnished the engravings for Michael Drayton's *Poly-Olbion* (I, 1612–13; II, 1622), depicts Inglewood Forest as a proud huntress or nymph (see figure 9.3). Drayton personifies Inglewood as a "Wood-Nymph" and turns the forest into a pastoral scene in which chaste nymphs and "shaggéd Satyrs" romp.[58] Drayton's topographical poem in praise of England and Wales is subtitled *A Chorographicall Description of All the Tracts, Rivers, Mountains, Forests, and other Parts of this Renowned Isle of Great Britain, with intermixture of the most Remarkeable Stories, Antiquities, Wonders, Rarities, Pleasures, and Commodities of the same.* Its alexandrine couplets are crowded with representatives of the more-than-human world who lament the changes wrought in the natural landscape by time and human depredations, particularly in the forests. Through their romanticized, celebratory descriptions of themselves—a mash-up of a very material but figural "Britain" that stretches from St. Michael's Mount (Book 1) to Mount Skiddaw (Book 30)—these personified topographical features in their representation of the past suggest that the land of Arthur and Merlin and other notable Britons is much more superior to "these yron times"; that is, contemporary "England" under James I (27.396).

---

56. Qtd. in Nick Millea, *The Gough Map: The Earliest Road Map of Great Britain?* (Oxford: Bodleian Library, 2007), 41, 43. On the Gough Map, Scotland is drawn as rather formless, even wrongly shaped. Perhaps the very vagueness of its contours is intended to reinforce the privileged position of England. See Ingham, *Sovereign Fantasies,* 188, on the use of Scotland as a metaphor.

57. Ralph Hanna III, in his edition of *The Awntyrs,* notes "Tarn Wadling almost certainly should be understood as a place with spectral or magical connotations, possibly as a place where transfer from the Other World (whether purgatory or Faery) is possible" (*The Awntyrs off Arthure at the Terne Wathelyn: An Edition Based on Bodleian Library MS. Douce 324* [Manchester, UK: Manchester University Press, 1974] 34).

58. Map, Michael Drayton, *Poly-Olbion.* Online. http://www.geog.port.ac.uk/webmap/thelakes/html/lgaz/dry508.htm.

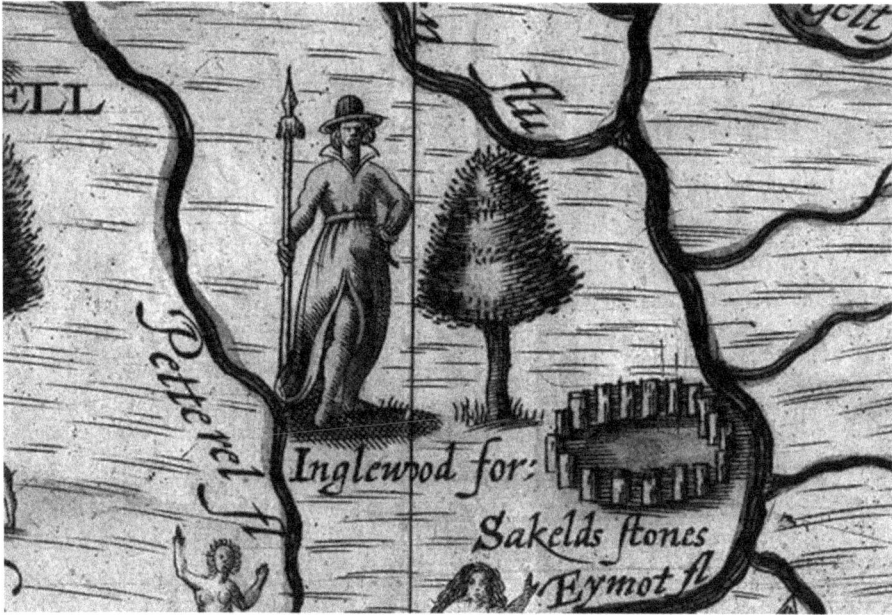

FIGURE 9.3. Detail, *Poly-Olbion* (1622).
(Courtesy of Jean and Martin Norgate,
Geography Department, Portsmouth University, U.K.)

FIGURE 9.4. Detail, *Westmorland and Cumberland* (1695).
(Courtesy of Jean and Martin Norgate,
Geography Department, Portsmouth University, U.K.)

Some decades later, bookseller and mapmaker Robert Morden, in his map of Westmorland and Cumberland (1695), marks the forest as an array of tiny trees (see figure 9.4).[59] Parts of the forest apparently still existed in the seventeenth century, but here, Inglewood is memorialized because it used to be a royal forest; at this time and to this day, Inglewood is merely the name for a district. Note that there is no Tarn Wadling on these two maps. The Gough Map is the only medieval map to depict it, and it is not marked on any of the other extant early modern or eighteenth-century maps that I have been able to examine. (However, the Tarn is often remarked upon in eighteenth- and nineteenth-century guidebooks as a fine fishing spot—as the thirteenth-century Prior of St. Mary's with a taste for fish mentioned in *The Pleas of the Forest* apparently knew.) Perhaps that the Tarn fell out of history—was "lost"— is simply due to the vagaries of mapmaking. Perhaps as cartographers began to use triangulation to record topographical features more accurately, the Tarn, compared to those large glaciated lakes south of it, no longer mattered. As the Lake District *became* the Lake District—that is, beginning in the late eighteenth century, became that quintessential English tourist site—mapmakers had no reason to remember a small regional lake. To adapt Philip Schwyzer, perhaps the Tarn was lost to an impulse to suppress the local in order to "rationaliz[e] . . . national space."[60] Remember that the Tarn was first drained because it suited a local lord. Whether national or local, the Tarn simply stopped mattering.

But see the Ordnance Survey of 1898 (see figure 9.5). The Tarn is back: it matters once more. British Ordnance maps have their origins in the Jacobite Rising of 1745. In the aftermath of that struggle, in which the exiled Charles Stuart tried to regain the British throne, one Lieutenant Colonel David Watson proposed a strategic and accurate mapping of the Scottish Highlands. George II then commissioned a military survey. In the early nineteenth century, the fear of an invasion by Napoleon spurred an urgent interest in mapping all of Great Britain.[61] Thus the Tarn enters back into recorded history at the point at which the military decides to map the nation. One would not want to bivouac by mistake in a lake or a bog while defending the country. And around forty years

---

59. Map, Robert Morden, *Westmorland and Cumberland*. Online. http://www.geog. port.ac.uk/webmap/thelakes/html/morden/md12ny44.htm.

60. Philip Schwyzer, "The Beauties of the Land: Bale's Books, Aske's Abbeys, and the Aesthetics of Nationhood," *Renaissance Quarterly* 57, no. 1 (2004): 106.

61. See Rachel Hewett, *Map of a Nation: A Biography of the Ordnance Survey* (London: Granta Books, 2010).

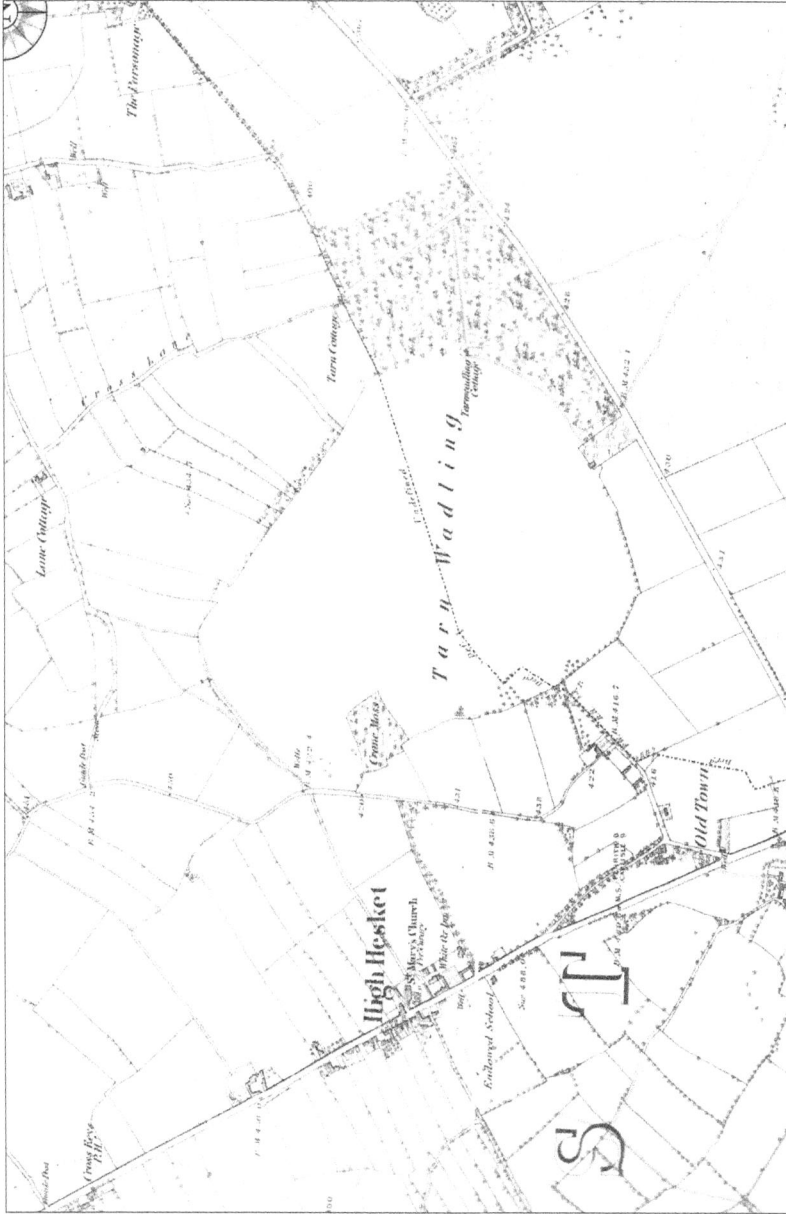

FIGURE 9.5. Detail, Ordnance Survey (1861; 1898–99). (Author's Collection)

later, as I have previously noted, the Tarn matters once again, as occupational therapy for German prisoners of war. And now, the Tarn really and finally matters no longer. Today, as we have seen, the Tarn is visible only as a trace in a high-resolution satellite image—and even then, to paraphrase Jehlen again, only if one knows how to look for it. Look again at figure 9.1, and note the small rectangle of woods in the lower right-hand corner. Compare this small stand to the more extensive stand of trees on the Ordnance Survey. What can be seen in the Google Maps image is a Woodland Trust property called "Tarn Wadling" (to the northeast is a much bigger Tarn Wood, as well). Glennie passed through this "fir wood" on his way to Blaze Fell.[62] Thus the Tarn is memorialized in a curious toponymic transfer, in which a succession forest of mainly 120-year-old Scots pine now goes by the name of an ancient lake.

There are other ways to remember Inglewood and the Tarn besides the poetic, the cartographic, and the historical: geologist D. Walker, for example, describes the Tarn in this way, which I quote in full:

> Tarn Wadling is a saucer-shaped depression, about 600 yd. (0.5 km) in diameter, immediately east of the village of High Hesket in Cumberland and 1 mile (2 km) west of the River Eden (Nat. Grid. Ref. 485445). The hollow lies in the sandy boulder clay of the Main Glaciation of the district (Hollingworth, 1931), the edge at about 410 ft. (124 m) O. D. To the south-east the drift-smeared slopes of Blaze Fell rise steeply to 792 ft. (240 m) O. D. but northward lies an undulating area of boulder clay at about 450 ft. (136 m) O. D. Eastward, beyond the narrow rim of boulder clay, the ground falls to the River Eden at about 180 ft. (55 m) O. D. Westwards, drumlins rising to between 450 ft. (136 m) and 500 ft. (152 m) O. D. lie between Tarn Wadling and the River Petteril. The site of Tarn Wadling was vacated by the ice by retreat stage H of Hollingworth (1931) but at that time melt-water probably poured through the region from west to east. With the opening of drainage down the Petteril valley (stage LM) the Tarn Wadling hollow would contain an isolated lake. During the Scottish Readvance Glaciation, the ice front of which lay 2 miles (3 km) north-west of the site, Tarn Wadling was probably temporarily incorporated in the large pro-glacial lake which stood at 440 ft. (133 m) O. D. Tarn Wadling has been identified with the Tern Wathelyne of Arthurian legend (Armstrong *et al.*, 1952) and was doubtless a lake over a very long period. It probably drained though the low

---

62. Glennie, *Arthurian Localities,* 73.

col and trough at its south-western end, a course still marked by the parish boundary. Many attempts have been made to drain the basin but, although some of these may have been temporarily successful, the present dry state was achieved as recently as the Second World War. The bed of the tarn is now overgrown by *Juncus effusus* and a well-marked break in slope around its edge probably indicates the former shore-line. The wooded Crane Moss at the north-western end is a small bog lying above the old shore-line. In a similar position at the southern end of the basin lies a narrow band of fen deposits. (232)[63]

Here, the Tarn matters differently; it is brought back as an object of scientific study, and we learn about its past before it entered into written history. Walker dates the Tarn to at least 3000 BCE. Walker also charts the history of the surrounding area from pollen layers, documenting changes from grasslands to open woodlands of birch supplemented by hazel thickets and then increasing numbers of oak and alder; he speculates that this stage of afforestation may have been either "natural or induced" (234)—perhaps, I would further speculate, at the time that Inglewood ("wood of the Angles") was declared a Royal Forest after the Norman Conquest. Finally, Walker documents the introduction of pine into the area in relatively modern times and notes that pollen layers indicate that the water in the Tarn rose after these pine forests were established. This corresponds, I would add, with the ending of the Little Ice Age (c. 1350–1850): as the weather warmed, more runoff from the mountains accumulated at their feet and flooded the Tarn.

The Tarn may not have moved the authors of *The Awntyrs* and its companion poems to lyric description, but I detect a thread of poetry running through this geological study, from the invocation of Arthurian romance to the quasi-Homeric epithets: "saucer-shaped depression" and the "drift-smeared slopes of Blaze Fell" (which, I like to think, also reads like an alliterative half-line). The word *drift* is a geological term, denoting the tumble of rocks and sediments deposited by retreating glaciers. It is a highly evocative adjective: *drift-smeared.* The phrase captures the sheer enormity of those glaciers leaving behind parts of themselves. Even scientific writing has its Lacanian unconscious. Writing in the early sixties, Walker documents the still-swampy nature of the Tarn through his identification of *Juncus effusus,* or rush, a wetland plant (see figure 9.6).

---

63. D. Walker, "Post Glacial Deposits at Tarn Wadling," *New Phytologist* 63, no. 2 (1964): 232.

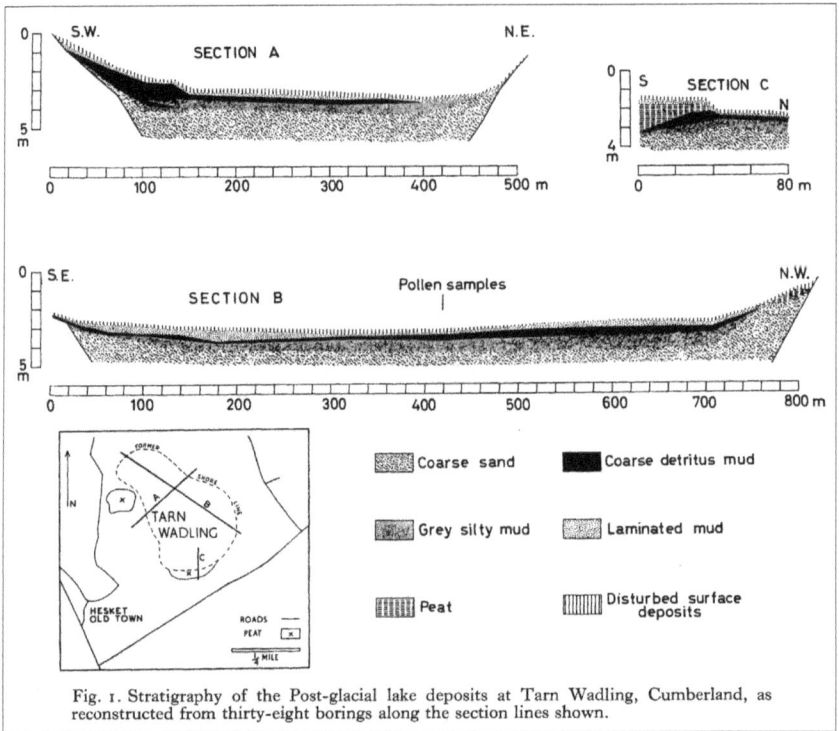

Fig. 1. Stratigraphy of the Post-glacial lake deposits at Tarn Wadling, Cumberland, as reconstructed from thirty-eight borings along the section lines shown.

FIGURE 9.6. Tarn Wadling, as seen by geologists.
Stratigraphy of the post-glacial lake deposits at Tarn Wadling, Cumberland,
as reconstructed from thirty-eight borings along the section lines shown.
(D. Walker, "Post-Glacial Deposits at Tarn Wadling")

Walker offers a narrative that is both synchronic and diachronic; that is, he writes of the condition of the Tarn at the moment he observed it as well as of its past. His observations are based on borings that, in their sediment layers—literally, as opposed to the use of the term in literary criticism—attest to a history of the natural world compressed into striations and ready to be deciphered by the poet, the farmer, and/or the geologist. And now, Walker's moment past, the Tarn has changed once more, from a partial swamp to a dry cultivated field, rich in organic material precisely because it was once swampy. The very fertility of the land is a witness to change over time. Nature may not read, but it writes.

I began by observing that mourning and melancholy underpin many of the stories we tell about the natural world of the past, wondering what we might gain if we explored the traces and effects of such grieving. I conclude by suggesting that we think about mourning and melancholy as legitimate, appropriate, and strategic affective responses to ecological change. Recognizing how sadness and grief over ecological change and degradation often structures narratives about the natural world helps us confront and manage our own grieving, which in turn influences how we might read or write tales in the future. Any work of fiction is inevitably mediating and representational, but a mindful reader who understands this is not prevented from extending her or his sympathies, as George Eliot describes the exercise of the imagination, into the more-than-human world through the pleasures and rigors of reading. Moreover, reading and writing about the natural world of the past, to continue to draw on Eliot's language, amplifies our experiences of—and extends our contact into—the world, because what once was abstract or Other or invisible begins to take on particular, solid being.[64] Once one looks with an ecological eye (to paraphrase Loren Eiseley) at the past and comprehends how that past informs and constitutes the present, then the world will never be experienced in the same way again.

In this essay, I turned an ecological eye on a story that is not about the natural world at all but is instead an invented tale about King Arthur in which the natural world is simply named and never described. I took the footnotes on Inglewood Forest and the Tarn found in standard editions as an invitation to write a hypertextual history, as it were. I began with the idea that the significance of Inglewood Forest and the Tarn Wadling in these Arthurian border poems is political and historical—but historical in the sense of the only-human world, within a very specific temporality and topicality (in both senses of the word). Let us dwell here a moment and attend to *The Awntyrs* once again. Within the temporality of the poem, it is as if time slows and stops as we listen to Gaynour's mother's ghost grieve her earthly losses: "Hit waried, hit wayment (wailed) like a woman" and "Hit marred (grieved), hit memered, hit mused for madde" (107, 110). As is Gaynour and Gawain, we are arrested by her horrific noncorporeal corporeality, as frozen to the spot as is the Wedding Guest by the Ancient Mariner. Gaynour's mother's ghost is a classic melancholic who is fixated, as we saw ear-

---

64. George Eliot, "The Natural History of German Life" (1856), in *The Essays of "George Eliot,"* 1883, ed. Nathan Sheppard (New York: Funk and Wagnalls, 2013), 81–101.

lier, on her possessions: "Of palaies, of parkes, of pondes, of plowes" (149). She further grieves the bodily pleasures that she has lost but that Gaynour currently enjoys, such as "fressh foroure" (fur garments) and "riche dayntés" (166, 183). The mother's advice to her daughter, to eschew the things of this world, is based on her own experience with fleshly and spiritual excess: what remains of her body is blackened, the color of melancholia, because of "luf paramour, listes and delites" (213). Biddick describes melancholy as "mimetic; it refuses loss and mimics the lost object."[65] We might see Gaynour's mother as practicing a kind of mimesis through her recitation of places and pleasures no longer hers. She attempts repossession through repetitive incantation. However, her warnings to her daughter and her prophecies to Gawain are not cathartic enough, it seems, to move her out of melancholy; instead, she is compelled to return to the Tarn "in wo for to welle" (boil in woe) (316), until Gaynour offers up enough masses and prayers to free her soul.

Along with Gaynour's mother, the knight Galaron also seems to be fixated on his earthly possessions. He, too, goes through a recitation exercise: "Of Connok, of Conyngham, and also Kyle," and more (424). But Galaron has an ally in the "lady lufsom of lote" (manner) (344). If I were offering a traditional ecocritical reading of *The Awntyrs,* at this point I would pay attention to the narrator's description of the lady's dress as "glorious and gay, of a gresse grene" (366). Perhaps the lady represents nature? And why does Gawain wear green? I would note that the lady and Gaynour intercede for Galaron and might venture that feminine "nature" has claims on culture, here represented by the masculine king. Moreover, I would want to note that if Galaron is to overcome his melancholic attachment to his lands, he needs culture too, and, thus in his public recognition of Gawain as the better knight, Galaron takes his proper place in the homosocial order. Yet I am more interested in how, in ceding his lands to Gawain, Galaron demonstrates that he has learned to do the work necessary to mourning. He lets go. He moves from melancholy to mourning, which, to quote Biddick again, is "performative. . . . It is only in action . . . that the suffering of the loss of the lost object can be constituted."[66] Still, the end of Galaron's tale extends beyond the ends of mourning. Amazingly, Galaron *gets the lost object back.* (Though perhaps not the "same"; moreover, it certainly was never "his.")

---

65. Biddick, *Shock of Medievalism,* 10.
66. Ibid., 11.

The narratives of both Guenevere's mother's ghost and Galaron are contained within the diegetic frame of the poem. Within this frame, the story of Galaron contravenes Freud's criterion of reality-testing for successful mourning; that is, that one must accept that one can never recover the lost object; one must move on to form a new attachment to a new object. *The Awntyrs* has an inside and an outside, as it were, in which two different logics prevail.

I have been reading the tales of Gaynour's mother and Galaron as stories of a very particular and passionate attachment to land, tales of a melancholic and a mourner, respectively. These tales are fixed in time—that is, in the imaginary time of the poet who invented them and in the fantasy time of the scholar who studies them. Inside the poem Galaron might gallop away on his "freson" a fully-cathected knight, but outside the poem, we remain behind in a melancholic fantasy. If we are committed to reality-testing, the poem functions only as a fairy tale, a fantasy, a wish-fulfillment dream, in which we, too, get the lost object back. This is a synchronic reading of the human characters in the poem, and one in which nature is simply background, and lands have value only as possessions that constitute one's identity and status—and even one's humanness. However, in this essay, I attempted to construct an extradiegetic and diachronic narrative for the trees and the water named in *The Awntyrs* and its companion poems. Melancholy, it might be said, stops time, while mourning keeps us moving through time—keeps us active, as Biddick suggests, as we continually renegotiate our relation to the past and to the natural world of the past.

In *Precarious Life: The Powers of Mourning and Violence,* Judith Butler, in contemplating marginalized sexed/gendered subjects and post-9/11 Muslim subjects, takes the absence or presence of grief as a sign of recognizing another's subjectivity. She asks why it is that "some lives are grievable and others are not," and "what counts as a livable life and a grievable death"?[67] Let us expand Butler's conception of the subject to include the lively beings of the more-than-human world. Let us be sad, but let us tell their stories—even if nature does not read them. Grief for all nonhuman beings is appropriate, and makes them present and *particular* to us. Like any lively being, Inglewood and the Tarn have histories (albeit entangled with human histories), and their stories constitute, in part, their agentive singularity. Aranye Fradenburg says: "What makes

---

67. Judith Butler, *Precarious Life: The Powers of Mourning and Violence* (New York: Verso, 2004), xvi.

grief agonizing is precisely that when someone or something particular has been lost, it cannot recur. . . . If the particular cannot be repeated, it remains forever lost; and this is why there can be no final closure to mourning."[68] Fradenburg's argument is crucial for two reasons: first, she reminds us that melancholy, the inability to move on, is not necessarily always negative (to return us to Hugh of St. Victor and *tristitia utilis*). Let me put it this way: we need both synchronic and diachronic narratives when it comes to ecological thinking. It is indisputably true that scientific insights into anthropogenic effects on ecosystems worldwide are needed if we are to devise an ethical, responsible, and effective response to such changes. However, we also need a more nuanced history of how humans have re-presented the natural world back to themselves in the arts. Instead of simply critiquing literature and other arts as mediating—even constituting—nature, instead of taking literature for granted as an object or thing, a phenomenon or an experience, let us think of it as a *movement* among the reader, the text, and the natural world along synchronic and diachronic axes. Literature *matters* to our understanding of the natural world and our experience of it.

Second, Fradenburg emphasizes the importance of the particular, and therefore what is grievable: not forests, but *this* forest that humans named Inglewood; not lakes, but *this* lake named Tarn Wadling. Erle Ellis and his coresearchers' mapping of global anthromes is too daunting a tale to narrate; the scale is too great—but we must not forget it for precisely that reason. Still, I prefer to follow Cronon, who, like Fradenburg, emphasizes the importance of "particularism" in storytelling.[69] (Act locally!) Inglewood. Tarn Wadling. As literal, material sites in and of themselves— *as was* and *as is*—and as literary and biopolitical historical sites that signify *for us humans,* they are case studies in the entanglements of both natural and social agencies. Inglewood and the Tarn have contributed to constituting us as much as we have constituted them—and now this essay and your reading of it are part of this material-discursive phenomenon.

What story do *you* have to tell?

David Hockney, for one, has a story to tell about a stand of beech trees in Yorkshire. A few years ago, Hockney planned to paint the trees in all seasons and had completed "Summer" and "Winter."[70] When he

---

68. L. O. Aranye Fradenburg, "Chaucer's Voice Memorial," *Exemplaria* 2, no. 1 (1990): 182–83.

69. Cronon, "Uses of Environmental History," 20.

70. Both paintings ("Bigger Trees Nearer Warter," Winter 2008, Summer 2008) can be seen here: http://artobserved.com/2009/11/go-see-new-york-david-hockneys-paintings-

returned in the spring to start the next in the series, the trees had been cut down, most likely for their timber. I give Hockney the last word:

> I admit this may matter only to me. Perhaps nobody else would feel like this. . . . There was something shocking about the scene. The landscape I remembered was gone completely. . . . To me even the approach to that little wood had a kind of grandeur, like the approach to some marvellous great temple, and the trees themselves were very large, very architectural, very majestic. . . . It was like coming into some little village or town and finding that overnight the people had obliterated a great church that had stood there for 900 years. . . . If they had pulled down a great church people would have seen and asked questions, but nobody asked about these trees. Nobody asks enough questions any more.[71]

2006-2009-at-pacewildenstein-through-december-24th-2009/. The Hockney saga is not over, and there is much more to explore—such as the tale of the tree stump that he was painting that was vandalized with pink spray paint. Moreover, there is now a "Hockney Trail" in the Yorkshire Wolds: http://www.yocc.co.uk.

71. As quoted in Maev Kennedy, "David Hockney, the Fallen Beech Trees and the Lost Canvas," *The Guardian*, 26 March 2009, http://www.guardian.co.uk/artanddesign/2009/mar/27/hockney-art-seasons-trees. Hockney (a confirmed smoker) wrote the following letter to *The Guardian*:

> I am aware of the constant change in nature and cutting down trees is a part of it. Indeed it has been a subject for me in the recent past and will be in the future. Nevertheless, a lot of trees were planted in East Yorkshire as windbreaks to protect the soil. I noticed on Friday morning how windy it was there, and now assume there will be a wind farm there. Admittedly it is a remote spot; all the hours I spent there, perhaps two cars an hour came by, if that. I admit no one else would be shocked. Most people think all trees look alike. I see them as all different (like people), especially in the winter when you see the branches pointing up searching for light. The first signs of spring are the tiny new branches at the tops that one hardly notices until they have buds. When a local wood was pruned, I found a good subject. Were the cut trees still alive? They seemed to be at first, but their horizontal position seemed to deny it. As some of the timber was left for two months, they lost their freshness, but gained grass and weeds growing through them (which naturally I drew). If they had been left longer they would have been covered over by nature's force. All living things die. Why, death awaits even if you do not smoke.

The Guardian, 31 March 2009, http://www.guardian.co.uk/artanddesign/2009/mar/31/hockney-cutting-down-trees.

Bibliography

Primary Works

Ælfric's Lives of the Saints. Vol. 1 and 2, edited by W. W. Skeat. London: EETS, 1881.

Aelian. *Variae Historiae*. Translated by Nigel G. Wilson. Cambridge, MA: Loeb Classical Library, 1997.

Alan de Lille. *Literary Works*. Dumbarton Oaks Medieval Library. Edited and translated by Winthrop Wetherbee. Cambridge, MA: Harvard University Press, 2013.

Albert the Great. *Man and the Beasts (De Animalibus, Books 22–26)*. Translated by James J. Scanlan. Binghamton, NY: Medieval & Renaissance Texts & Studies, 1987.

Andrew of Wyntoun. *Orygynale Cronykil of Scotland*. Edited by David Laing. Edinburgh: Edmonton and Douglas, 1872.

The Anglo-Saxon Chronicle. Edited and Translated by Michael James Swanton. New York: Routledge, 1998.

The Anglo-Saxon Chronicle. Edited and Translated by Dorothy Whitlock, David C. Douglas, and Susie Tucker. New Brunswick, NJ: Rutgers University Press, 1961.

The Anonimalle Chronicle, 1333 to 1381. Edited by V. H. Galbraith. Manchester, UK: Manchester University Press, 1927.

"Assize of Woodstock." *Early English Laws*. http://www.earlyenglishlaws.ac.uk/laws/texts/ass-wood/view/#edition,1/hv-im.

Bede's Ecclesiastical History of the English People. Edited by Bertram Colgrave and R. A. B. Mynors. Oxford: Clarendon Press, 1969.

Boethius. *Anicii Manlii Severini Boethii Philosophiae Consolatio*, Corpus Christianorum. Series Latina 94. Edited by Ludwig Bieler. Turnhout, BE: Brepols, 1984.

———. *Consolation of Philosophy*. Translated by Joel C. Relihan. Indianapolis: Hackett, 2001.

Calendar of the Close Rolls, Henry IV. London: H. M. Stationery Office, 1892–1963.

Calendar of the Letter-Books. Edited by Reginald Sharpe. London: Corporation of London, 1899–1912.

Calendar of the Patent Rolls, Edward III (1:34, membrane 18). London: H. M. Stationary Office, 1891–1916.

The Cambridge Illuminations: Ten Centuries of Book Production in the Medieval West. Edited by Paul Binski and Stella Panayotova. London: Harvey Miller, 2005.

Chaucer, Geoffrey. *The Riverside Chaucer.* Edited by Larry D. Benson, 3rd ed. Boston: Houghton Mifflin, 1987.

Chronicles of London. Edited by Charles L. Kingsford. London: Oxford University Press, 1905.

The Dream of the Rood. Edited by Michael Swanton. Exeter, UK: University of Exeter Press, 1996.

Dymmok, Roger. "Against the Twelve Errors and Heresies of the Lollards, part twelve." In *Richard Maidstone, Concordia (The Reconciliation of Richard II with London),* edited by David R. Carlson. Translated by A. G. Rigg, 109–28. Kalamazoo: Medieval Institute Publications, 2003.

———. *Liber contra XII errores et hereses lollardorum.* Edited by H. S. Cronin. London: Kegan Paul, Trench, Trübner, 1922.

Eadmeri Historia Novorum in Anglia, et Opuscula Duo De Vita Sancti Anselmi et Quibusdam Miraculis Ejus. Edited by Martin Rule. Wiesbaden, DE: n.p., 1965.

Edmer's History of Recent Events in England. Translated by Geoffrey Bosanquet. London: Cresset Press, 1964.

Edward of Norwich. *The Master of Game.* Edited by William A. and F. N. Baillie Grohman. Philadelphia: University of Pennsylvania Press, 2005.

English Law Suits from William I to Richard I. Vol. 2, edited by R. C. Van Caenegem. London: Selden Society, 1991.

"First Forest Assize." *Early English Laws.* http://www.earlyenglishlaws.ac.uk/laws/texts/ass-for/view/#edition,1/vu-image.

Fitzneale, Richard. *Dialogus de Scaccario.* Edited and Translated by Charles Johnson. 1950. rev. ed. edited by F. E. L. Carter, and Diana E. Greenway. Oxford: Oxford University Press, 1983.

Gervaise of Tilbury: Otia Imperialia; Recreation for an Emperor. Edited by S. E. Banks and J. W. Binns. London and Oxford: Oxford University Press, 2002.

Gower, John. *The Complete English Works.*Vol. 2. Edited by G. C. Macaulay. London: EETS, e.s., 82, 1901.

———. *John Gower: Poems on Contemporary Events; The Visio Anglie (1381) and Cronica Tripertita (1400).* Edited by David R. Carlson. Translated by A. G. Rigg. Toronto: Pontifical Institute of Mediaeval Studies, 2011.

———. *The Major Latin Works of John Gower: The Voice of One Crying, and the Tripartite Chronicle.* Translated by Eric W. Stockton. Seattle, WA: University of Washington Press, 1962.

Gregory, William. "Chronicle of London." In *The Historical Collections of a Citizen of London.* 1876, edited by James Gairdner, 55–239. New York: Johnson Reprints, 1965.

Guillaume de Lorris and Jean de Meun. *Le Roman de la Rose*. Edited and Translated by Armand Strubel. Paris: Librairie Générale Française, 1992.

Hardyng, John. *Chronicle*. Edited by Henry Ellis. London, 1812.

Henry, Archdeacon of Huntingdon. *Historia Anglorum*. Edited and Translated by Diana Greenway. Oxford: Clarendon Press, 1996.

Henry de Bracton. *On the Laws and Customs of England*. Edited by George E Woodbine. Translated by Samuel E. Thorne. 4 vols. Cambridge: Harvard University Press, 1968.

Horace. *Horace: Satires, Epistles, and Ars Poetica*. Translated by H. Rushton Fairclough. Cambridge, MA: Harvard University Press, 1929.

Hue de Rotelande. *Ipomedon*. Edited by A. J. Holden. Paris: Klincksieck, 1979.

Johannes Scotus Erigena. *Periphyseon*. Vol. 161, Corpus Christianorum Continuiatio Mediaevalist. Turnhout, BE: Brepols, 2007.

John of Salisbury. *Frivolities of Courtiers and Footprints of Philosophers, Being a Translation of the First, Second, and Third Books and Selections of the Seventh and Eighth Books of the Policraticus*. Translated by Joseph B. Pike. New York: Octagon Books, 1938.

———. *Joannis Saresberiensis Episcopi Carnotensis Policratici Sive De Nugis Curialium et Vestigiis Philosophorum Librui VIII; Recognovit et Prologomenis*. Edited by Clement Charles and Julian Webb. Frankfurt am Main: Minerva, 1965.

———. *Letters of John of Salisbury*. Vol. 2 (1163–80). Edited by W. J. Millor and C. N. L. Brooke. Oxford: Clarendon Press, 1979.

———. *Policraticus: Of the Frivolities of Courtiers and the Footprints of Philosophers*. Edited and translated by Cary J. Nederman. Cambridge: Cambridge University Press, 1990.

John of Worcester. *The Chronicle of John of Worcester*. Vol. 3, *The Annals from 1067–1140 with The Gloucester Interpolations and the Continuation to 1141*. Edited and Translated by P. McGurk. Oxford: Clarendon Press, 1998.

Livy. *History of Rome*. Vol. 2. Edited by T. E. Page et al. Translated by B. O. Foster. Loeb Classical Library. Cambridge, MA: Harvard University Press, 1922.

Mapes, Gualteri. *De Nugis Curialium, Distinctiones Quinque*. Edited by Thomas Wright. New York: AMS Press, 1968.

Orderic Vitalis. *The Ecclesiastical History of Orderic Vitalis*. Edited and Translated by Marjorie Chibnall. 5 vols. Oxford: Oxford University Press, 1981.

Ovid. *Metamorphoses*. Translated by Charles Martin. New York: W. W. Norton and Company, 2004.

The Parliament Rolls of Medieval England, 1275–1504 [PROME]. Edited by Christopher Given-Wilson. Leicester: Scholarly Digital Editions, 2005.

Pseudo-Cnut. *Über Pseudo-Cnuts Constitutiones de foresta*. Edited by Felix Liebermann, Halle: Max Niemeyer, 1894.

Roger of Howden. *Chronica*. Edited by William Stubbs. London: Longman, Green, Reader, & Dyer, 1868.

Sir Gawain and the Green Knight, rev. ed. Edited and Translated by William Vantuono. Notre Dame, IN: University of Notre Dame Press, 1999.

The Statutes at Large, from "Magna charta" to the End of the Last Parliament, 1761. Vol. 1. Edited and Translated by Owen Ruffhead. London: M. Basket, 1763.

The Statutes at Large from the Second Year of the Reign of King George the Third to the End of the Last Session of Parliament. Vol. 9. Edited by Owen Ruffhead. London: M. Basket, 1765.

The Statutes of the Realm. London: Dawsons, 1810–28.

Symeon of Durham. *A History of the Kings of England.* Translated by J. Stephenson. Dyfed, UK: Llanerch Enterprises, 1987.

Symeonis Monachi Opera Omnia. Vol. 2, *Historia Regum.* Edited by Thomas Arnold. London: Longmans & Co., 1885.

Thomas Aquinas. *Catena Aurea: Commentary on the Four Gospels: Collected out of the Works of the Fathers.* Translated by John Henry Newman. Oxford: J. H. Parker, 1841.

Thomas Aquinas and Enrique Alarcón. *Catena Aurea,* Corpus Thomisticum. http://www.corpusthomisticum.org/.

Thomas de Burton. *Chronica Monasterii De Melsa a Fundatione Usque Ad Annum 1396, Auctore Thoma De Burton, Abbate, a Monacho Quodam Ipsius Domus.* Edited by Edward Augustus Bond. Burlington, ON: Tanner Ritchie, 2009.

Thomas of Cantimpré. *Liber De Natura Rerum: Editio Princeps Secundum Codices Manuscriptos.* Edited by Helmut Boese. Berlin: W. De Gruyter, 1973.

Vincent of Beauvais. *Speculum Quadruplex Sive Speculum Maius: Naturale, Doctrinale, Morale, Historiale.* 1624. Reprint, Graz: Akademische Druck- u. Verlagsanstalt, 1964–65.

William of Malmesbury. *William of Malmesbury's Gesta Regum Anglorum.* Vol. 1. Edited and Translated by R. A. B. Mynors, R. M. Thomson, and Michael Winterbottom. Oxford: Clarendon Press, 1998.

William of Newburgh. *Historia Rerum Anglicarum.* Edited by Richard Howlett. London: Longman, 1884.

Secondary Works

Abram, David. "Between the Body and the Breathing Earth: A Reply to Ted Toadvine." *Environmental Ethics* 27 (2005): 171–90.

———. *The Spell of the Sensuous: Perception and Language in a More-Than-Human World.* New York: Vintage, 1997.

Adamson, Joni, Mei Mei Evans, and Rachel Stein, eds. *The Environmental Justice Reader.* Tucson: University of Arizona Press, 2002.

Agamben, Giorgio. *The Highest Poverty: Monastic Rules and Form-of-Life.* Translated by Adam Kotsko. Stanford, CA: Stanford University Press, 2013.

———. *Homo Sacer: Sovereign Power and Bare Life.* Translated by Daniel Heller-Roazen. Stanford: Stanford University Press, 1998.

———. *The Open: Man and Animal.* Translated by Kevin Attell. Stanford: Stanford University Press, 2004.

———. *Potentialities: Collected Essays in Philosophy.* Edited and Translated by Daniel Heller-Roazen. Stanford: Stanford University Press, 1999.

———. *Profanations*. Translated by Jeff Fort. New York: Zone Books, 2007.

———. *Stanzas: Word and Phantasm in Western Culture*. Translated by Ronald R. Martinez. University of Minnesota Press, 1992.

———. *State of Exception*. Translated by Kevin Attell. Chicago: University of Chicago Press, 2005.

Aguirre, Manuel. "The Riddle of Sovereignty." *Modern Language Review* 88 (1993): 273–82.

Alaimo, Stacy. *Bodily Natures: Science, Environment, and the Material Self*. Indianapolis: Indiana University Press, 2010.

———. "New Materialisms, Old Humanisms, or, Following the Submersible." *NORA: Nordic Journal of Feminist and Gender Research* 19, no. 4 (2011): 280–84.

———. *Undomesticated Ground: Recasting Nature as Feminist Space*. Ithaca, NY: Cornell University Press, 2000.

Allen, Rosamund. "Place-Names in *The Awntyrs off Arthure:* Corruption, Conjecture, Coincidence." In *Arthurian Studies in Honour of P. J. C. Field*, edited by Bonnie Wheeler, 181–98. Cambridge, UK: D. S. Brewer, 2004.

Almond, Richard. *Medieval Hunting*. Stroud, UK: History Press, 2003.

Anderson, Carolyn. "'Lady of the English' in the *History Novella*, the *Gesta Stephani*, and Wace's *Roman de Rou:* The Desire for Land and Other." *Clio* 29, no. 1 (1999): 47-67.

Anker, Suzanne, and Dorothy Nelkin. *The Molecular Gaze: Art in the Genetic Age*. New York: Cold Spring Harbor Laboratory Press, 2003.

Astell, Ann W. *Political Allegory in Late-Medieval England*. Ithaca, NY: Cornell University Press, 1999.

Bailey, Richard N. *England's Earliest Sculptors*. Toronto: Pontifical Institute for Medieval Studies, 1996.

Baker, Nancy V., Peter R. Gregware, and Margaret A. Cassidy. "Family Killing Fields: Honor Rationales in the Murder of Women." *Violence against Women* 5, no. 2 (1999): 164–84.

Barad, Karen. *Meeting the Universe Halfway: Quantum Physics and the Entanglement of Matter*. Durham, NC: Duke University Press, 2007.

Barlow, Frank. *William Rufus*. Berkeley: University of California Press, 1983.

Barr, Helen. *Socioliterary Practice in Late Medieval England*. Oxford: Oxford University Press, 2001.

———. "The Treatment of Natural Law in *Richard the Redeless* and *Mum and the Sothsegger*." *Leeds Studies in English* 23 (1992): 49–80.

Barron, Caroline. *London in the Later Middle Ages: Government and People, 1200–1500*. Oxford: Oxford University Press, 2004.

Bartlett, Robert. *England under the Norman and Angevin Kings, 1075–1225*. Oxford: Clarendon Press, 2000.

Batra, Nandita. "Dominion, Empathy, and Symbiosis: Gender and Anthropocentrism in Romanticism." *Interdisciplinary Studies in Literature and Environment* 3, no. 2 (1996): 101–20.

Beckett, Katharine Scarfe. *Anglo-Saxon Perceptions of the Islamic World*. Cambridge: Cambridge University Press, 2008.

Beirne, Piers. "A Note on the Facticity of Animal Trials in Early Modern Britain; or, The Curious Prosecution of Farmer Carter's Dog for Murder." *Crime, Law and Social Change* 55 (2011): 359–74.

Bellamy, J. G. *The Law of Treason in England in the Later Middle Ages*. Cambridge: Cambridge University Press, 2004.

Benjamin, Walter. "Theses on the Philosophy of History." In *Illuminations,* edited by Hannah Arendt. Translated by Harry Zohn, 253–64. New York: Schocken Books, 1969.

Bennett, Jane. "The Elements." *Postmedieval* 4, no. 1 (2013): 105–11.

———. "Systems and Things: A Response to Graham Harman and Timothy Morton." *New Literary History* 43 (2012): 225–33.

———. *Vibrant Matter: A Political Ecology of Things*. Durham, NC: Duke University Press, 2010.

Bennett, Michael. "Anti-Pastoralism, Frederick Douglass, and the Nature of Slavery." In *Beyond Nature Writing: Expanding the Boundaries of Ecocriticism,* edited by Karla Armbruster and Kathleen R. Wallace, 195–210. Charlottesville: University of Virginia Press, 2001.

Bennett, Michael, and D. W. Teague, eds. *The Nature of Cities: Ecocriticism and Urban Environments*. Tucson: University of Arizona Press, 1999.

Benson, David C., and Elizabeth S. Passmore. "The Discourses of Hunger in *Piers Plowman*." In *Satura: Studies in Medieval Literature in Honour of Robert R. Raymo,* edited by Nancy M. Reale and Ruth E. Sternglantz, 150–63. Donington, UK: Shaun Tyas, 2001.

Biddick, Kathleen. "Arthur's Two Bodies and the Bare Life of the Archive." In *Cultural Diversity in the British Middle Ages: Archipelago, Island, England,* edited by Jeffrey Jerome Cohen, 117–34. New York: Palgrave, 2008.

———. "Dead Neighbor Archives: Jews, Muslims, and the Enemy's Two Bodies." In *Political Theology and Early Modernity,* edited by Graham Hammill and Julia Reinhard Lupton, 124–42. Chicago: University of Chicago Press, 2012.

———. *The Shock of Medievalism*. Durham, NC: Duke University Press, 1998.

Binski, Paul. *Medieval Death: Ritual and Representation*. Ithaca, NY: Cornell University Press, 1996.

Birkholz, Daniel. *The King's Two Maps: Cartography and Culture in Thirteenth-Century England*. London: Routledge, 2004.

Birrell, Jean. "Deer and Deer Farming in Medieval England." *Agricultural History Review* 40, no. 2 (1992): 112–26.

———. "Procuring, Preparing, and Serving Venison." In *Food in Medieval England,* edited by C. M. Woolgar, D. Serjeantson, and T. Waldron, 176–88. New York: Oxford University Press, 2006.

———. "The Medieval English Forest." *Journal of Forest History* 24, no. 2 (1980): 78–85.

Blamires, Alcuin. *Chaucer, Ethics, and Gender*. Oxford: Oxford University Press, 2006.

Blark, James G., Frank T. Coulson, and Kathryn L. McKinley, eds. *Ovid in the Middle Ages.* New York: Cambridge University Press, 2011.

Bloch, R. Howard. *Medieval Misogyny and the Invention of Western Romantic Love.* Chicago: University of Chicago Press, 1991.

Bogost, Ian. *Alien Phenomenology, or, What It's Like to Be a Thing.* Minneapolis: University of Minnesota Press, 2012.

Borkowski, Andrew, and Paul du Plessis. *Textbook on Roman Law.* 3rd ed. Oxford: Oxford University Press, 2005.

Borsch, Stuart J. *The Black Death in Egypt and England: A Comparative Study.* Austin: University of Texas Press, 2009.

Bowers, John M. *The Politics of Pearl: Court Poetry in the Age of Richard II.* Cambridge: D. S. Brewer, 2001.

Boym, Svetlana. *The Future of Nostalgia.* New York: Basic Books, 2001.

Brockman, Sonya. "The Legacy of Jephthah's Daughter: Chastity, Sacrifice, and Feminine Complaint in Chaucer's *Franklin's* and *Physician's Tales.*" *Medieval Feminist Forum* 46, no. 2 (2010): 68–84.

Brown, Bill. "Thing Theory." *Critical Inquiry* 28, no. 1 (2001): 1–22.

Bruckner, Lynne, and Dan Brayton, eds. "Introduction: Warbling Invaders." In *Ecocritical Shakespeare,* edited by Lynne Bruckner and Dan Brayton, 1–12. London and New York: Ashgate, 2011.

Bruno, Michael J. S. "The Investiture Contest in Norman England: A Struggle between St. Anselm of Canterbury and the Norman Kings: Part II." *American Benedictine Review* 61, no. 3 (2010): 307–24.

Bryant, Levi R. *The Democracy of Objects.* Ann Arbor, MI: Open Humanities Press, 2011.

———. "The Interior of Objects." *Larval Subjects,* December 16, 2010, http://larvalsubjects.wordpress.com/2010/12/16/the-interior-of-objects/.

Buell, Lawrence. *The Environmental Imagination: Thoreau, Nature Writing, and the Formation of American Culture.* Cambridge, MA: Belknap Press, 1995.

———. *The Future of Environmental Criticism.* West Sussex, UK: Blackwell, 2005.

Bugge, John. "Fertility Myth and Female Sovereignty in *The Wedding of Sir Gawen and Dame Ragnelle.*" *Chaucer Review* 39, no. 2 (2004): 198–218

Bull, Malcolm. "Vectors of the Biopolitical." *New Left Review* 45 (2007): 7–25.

Burkholder, Kristen M. "Threads Bared: Dress and Textiles in Late Medieval English Wills." In *Medieval Clothing and Textiles,* edited by Robin Netherton and Gale R. Owen-Crocker, 133–53. Woodbridge, UK: Boydell & Brewer, 2005.

Butler, Judith. *Precarious Life: The Powers of Mourning and Violence.* New York: Verso, 2004.

Bynum, Caroline Walker. *Christian Materiality: An Essay on Religion in Late Medieval Europe.* Cambridge, MA: Zone Books, 2011.

———. *Fragmentation and Redemption: Essays on Gender and the Human Body in Medieval Religion.* New York: Zone, 1991.

Cadden, Joan. "Trouble in the Earthly Paradise: The Regime of Nature in Late Medieval

Christian Culture." In *The Moral Authority of Nature*, edited by Lorraine Daston and Fernando Vidal, 207–31. Chicago: University of Chicago Press, 2004.

Callon, Michel. "Some Elements of a Sociology of Translation: Domestication of the Scallops and the Fishermen of St. Brieuc Bay." In *Power, Action, and Belief: A New Sociology of Knowledge?* edited by John Law, 196–233. New York: Routledge, 1986.

Callon, Michael, and John Law. "After the Individual in Society: Lessons on Collectivity from Science, Technology and Society." *Canadian Journal of Sociology* 22, no. 2 (1997): 165–82.

Campbell, Emma. "Political Animals: Human/Animal Life in *Bisclavret* and *Yonec*." *Exemplaria* 25 (2013): 95–109.

Campbell, Timothy C. "*Bíos*, Immunity, Life: The Thought of Roberto Esposito." *Diacritics* 36, no. 2 (2006): 2–22.

Carlson, David R. *John Gower, Poetry and Propaganda in Fourteenth-Century England.* Cambridge: D. S. Brewer, 2012.

Carragáin, Éamonn Ó. *Ritual and the Rood: Liturgical Images and the Old English Poems of the Dream of the Rood Tradition.* Toronto: University of Toronto Press, 2005.

Carter, Susan. "Coupling the Beastly Bride and the Hunter Hunted: What Lies behind Chaucer's *Wife of Bath's Tale*." *Chaucer Review* 37, no. 4 (2003): 329–45.

Cartmill, Matt. *A View to a Death in the Morning: Hunting and Nature though History.* Cambridge, MA: Harvard University Press, 1993.

Cassidy, Brendan. "The Later Life of the Ruthwell Cross: From the Seventeenth Century to the Present." In *The Ruthwell Cross*, edited by Brendan Cassidy, 3–34. Princeton, NJ: Princeton University Press, 1992.

Chaganti, Seeta. "Vestigial Signs: Inscription, Performance and *The Dream of The Rood*." *PMLA* 125 (2010): 48–72.

Chenard, Marianne Malo. "King Oswald's Holy Hands: Metonymy and the Making of a Saint in Bede's Ecclesiastical History." *Exemplaria* 17, no. 1 (Spring 2005): 33–56.

Chibnall, Marjorie. *The World of Orderic Vitalis: Norman Monks and Norman Kings.* Woodbridge, UK: Boydell & Brewer, 1984.

Chow, Rey. *Entanglements, or Transmedial Thinking about Capture.* Durham, NC: Duke University Press, 2012.

Christie, Edward. "Self-Mastery and Submission: Holiness and Masculinity in the Lives of the Anglo-Saxon Martyr-Kings." In *Holiness and Masculinity in the Middle Ages*, edited by P. H. Cullum and Katherine J. Lewis, 143–57. Toronto: University of Toronto Press, 2004.

Clanchy, Michael. *From Memory to Written Record: England 1066–1307.* 2nd ed. Oxford: Blackwell, 1993.

Clemoes, Peter. "The Cult of Saint Oswald on the Continent." In *Bede and His World: The Jarrow Lectures 1958–1993*, edited by Michael Lapidge, 587–610. London: Aldershot, 1994.

Cohen, Jeffrey Jerome. "An Abecedarium for the Elements." *Postmedieval* 2, no. 3 (2011): 291–303.

———. "Ecology's Rainbow." In *Prismatic Ecology: Ecotheory beyond Green*, edited by Jeffrey Jerome Cohen, xv–xxxvi. Minneapolis: University of Minnesota Press, 2013.

———. "Introduction: All Things." In *Animal, Vegetable, Mineral: Ethics and Objects,* edited by Jeffrey Jerome Cohen, 1–8. Washington, DC: Oliphaunt Books, 2011.

———. *Medieval Identity Machines.* Minneapolis: University of Minnesota Press, 2003.

Cohen, Jeffrey Jerome, and Lowell Duckert. "Howl." *Postmedieval* 4, no. 1 (2013): 1–5.

Cole, Luke, and Sheila Fisher. *From the Ground Up: Environmental Racism and the Rise of the Environmental Justice Movement.* New York: New York University Press, 2000.

Colish, Marcia L. *The Mirror of Language: A Study in the Medieval Theory of Knowledge.* Lincoln: University of Nebraska Press, 1983.

Collingwood, W. G. *Northumbrian Crosses of the Pre-Norman Age.* London: Faber, 1927.

Cooper, Melinda. *Life as Surplus: Biotechnology in the Neoliberal Era.* Seattle: University of Washington Press, 2008.

Corsa, Helen Storm, ed. *A Variorum Edition of the Works of Geoffrey Chaucer, vol. 2, The Canterbury Tales, Part 17, The Physician's Tale.* Norman, OK: University of Oklahoma Press, 1987.

Cox, J. Charles. *The Royal Forests of England.* London: Methuen & Co., 1905.

Cox, R. C. "Tarn Wadling and Gervase of Tilbury's 'Laikibrait.'" *Folklore* 85, no. 2 (1974): 128–31.

Crane, Susan. *Animal Encounters: Contacts and Concepts in Medieval Britain.* Philadelphia: University of Pennsylvania Press, 2012.

———. "For the Birds." *Studies in the Age of Chaucer* 29 (2007): 23–41.

Cronon, William. "A Place for Stories: Nature, History, and Narrative." *Journal of American History* (1992): 1347–76.

———. "The Trouble with Wilderness, or, Getting Back to the Wrong Nature." In *Uncommon Ground: Rethinking the Human Place in Nature,* edited by William Cronon, 69–90. New York: W. W. Norton & Co., 1995.

———. "The Uses of Environmental History." *Environmental History Review* 17, no. 3 (1993): 1–22.

Crowe, Christopher. "Early Medieval Parish Formation in Dumfries and Galloway." In *The Cross Goes North: Processes of Conversion in Northern Europe ce 300–1300,* edited by Martin Carver, 195–206. Rochester, NY: Boydell & Brewer, 2003.

Cubitt, Catherine. "Sites and Sanctity: Revisiting the Cult of Murdered and Martyred Anglo-Saxon Royal Saints." *Early Medieval Europe* 9, no. 1 (2000): 53–83.

Cullum, Patricia H. "Leperhouses and Borough Status in the Thirteenth Century." In *Thirteenth-Century England III: Proceedings of the Newcastle Upon Tyne Conference, 1989,* edited by Peter R. Coss and Simon D. Lloyd, 37–46. Woodbridge, UK: Boydell & Brewer, 1991.

Cummins, John. *The Art of Medieval Hunting: The Hound and the Hawk.* Edison, NJ: Castle, 1988.

———. "Veneurs s'sen vont en paradis: Medieval Hunting and the 'Natural' Landscape." In *Inventing Medieval Landscapes: Senses of Place in Western Europe,* edited by John Howe and Michael Wolfe, 33–56. Gainesville: University Press of Florida, 2002.

Damon, John E. "*Desecto Capito Perfido:* Reciprocal Violence in Anglo-Saxon England." *Exemplaria* 13, no. 2 (2001): 399–432.

——. *Soldier Saints and Holy Warriors: Warfare and Sanctity in the Literature of Early England*. Burlington, VT: Ashgate, 2003.

Darwin, Erasmus. *Botanic Garden*. London: J. Nichols, 1790.

Davis, Kathleen. *Periodization and Sovereignty: How Ideas of Feudalism and Secularization Govern the Politics of Time*. Philadelphia: University of Pennsylvania Press, 2008.

Davis, Rebecca A. "More Evidence for Intertextuality and Humorous Intent in *The Weddynge of Syr Gawen and Dame Ragnelle*." *Chaucer Review* 35, no. 4 (2001): 430–39.

De Hamel, Christopher F. R. *Glossed Books of the Bible and the Origins of the Paris Booktrade*. Woodbridge, UK: Boydell & Brewer, 1984.

DeLoughrey, Elizabeth, and George B. Handley, eds. *Postcolonial Ecologies: Literatures of the Environment*. Oxford: Oxford University Press, 2011.

Dean, James M., ed. *Medieval English Political Writings*. TEAMS Middle English Texts. Kalamazoo, MI: Medieval Institute Publications, 1996.

Dean, Tim. "The Biopolitics of Pleasure." *South Atlantic Quarterly* 111 (2012): 477–95.

Debrix, François, and Alexander D. Barder. *Beyond Biopolitics*. London: Routledge, 2012.

Deleuze, Gilles. *Pure Immanence: Essays on a Life*. Translated by Anne Boyman. New York: Zone Books, 2001.

Deleuze, Gilles, and Félix Guattari. *A Thousand Plateaus: Capitalism and Schizophrenia*. Translated by Brian Massumi. Minneapolis: University of Minnesota Press, 1987.

——. *What Is Philosophy?* Translated by Graham Burchell and Hugh Tomlinson. New York: Verso, 1994.

Derrida, Jacques. *The Animal That Therefore I Am*. Edited by Marie-Louise Mallet. Translated by David Wills. New York: Fordham University Press, 2008.

——. *The Beast and the Sovereign*. Vol. 1. Edited by Michel Lisse, Marie-Louise Mallet, and Ginette Michaud. Translated by Geoffrey Bennington. Chicago: University of Chicago Press, 2009.

——. "Before the Law." In *Acts of Literature,* edited by Derek Attridge. Translated by Avital Ronell and Christine Roulston, 181–220. New York: Routledge, 1992.

——. "Force of Law: The 'Mystical Foundation of Authority.'" In *Acts of Religion,* edited by Gil Anidjar. Translated by Mary Quaintance, 230–98. New York: Routledge, 2002.

Didi-Huberman, Georges. *Confronting Images: Questioning the Ends of a Certain History of Art*. Translated by John Goodman. University Park: Pennsylvania State University Press, 2005.

Dinshaw, Carolyn. *Chaucer's Sexual Poetics*. Madison: University of Wisconsin Press, 1989.

Donoghue, Daniel. *Old English Literature: A Short Introduction*. Oxford: Wiley-Blackwell, 2004.

Donnelly, Colleen. "Aristocratic Veneer and the Substance of Verbal Bonds in 'The Weddynge of Sir Gawen and Dame Ragnell' and 'Gamelyn.'" *Studies in Philology* 94, no. 3 (1997): 321–43.

Douglas, David. C. *William the Conqueror: The Norman Impact upon England*. Berkeley: University of California Press, 1964.

Douglas, Mary. *Purity and Danger,* rev. ed. London: Routledge, 2002.

Duby, Georges. *A History of Private Life: Revelations of the Medieval World.* Translated by Arthur Goldhammer. Cambridge, MA: Belknap Press, 1988.

Duggan, Anne. "*De consultationibus:* The Role of Episcopal Consultation in the Shaping of Canon Law in the Twelfth Century." In *Bishops, Texts and the Use of Canon Law around 1100,* edited by Bruce C. Brasington and Kathleen G. Cushing, 191–214. Burlington, VT: Ashgate, 2008.

———. "Thomas Becket's Italian Network." In *Pope, Church and City: Essays in Honour of Brenda M. Bolton,* edited by Frances Andrews, Christoph Egger, and Constance M. Rousseau, 177–204. Leiden, NL: Brill, 2004.

Dyer, Christopher. "Did the Peasants Really Starve in Medieval England?" In *Food and Eating in Medieval Europe,* edited by Martha Carlin and Joel Thomas Rosenthal, 53–71. London: Hambledon Press, 1998.

———. *Making a Living in the Middle Ages: The People of Britain, 850–1520.* New Haven, CT: Yale University Press, 2002.

Economou, George D. *The Goddess Natura in Medieval Literature.* Cambridge, MA: Harvard University Press, 1972.

Elden, Stuart. *The Birth of Territory.* Chicago: University of Chicago Press, 2013.

Ellard, Donna Beth. "Going Interspecies, Going Interlingual, and Flying Away with the Phoenix." *Exemplaria* 23, no. 3 (2011): 268–92.

Ellis, Erle C., Kees Klein Goldwijk, Stefan Siebert, Deborah Lightman, and Naiv Ramankutty. "Anthropogenic Transformation of the Biomes, 1700 to 2000." *Global Ecology and Biogeography* 19 (2010): 589–606.

Eliot, George. "The Natural History of German Life." In *The Essays of "George Eliot."* 1883. Edited by Nathan Sheppard, 81–101. New York: Funk and Wagnalls, 2013.

Esposito, Roberto. *Bíos: Biopolitics and Philosophy.* Translated by Timothy C. Campbell. Minneapolis: University of Minnesota Press, 2008.

———. *Third Person: Politics of Life and Philosophy of the Impersonal.* Translated by Zakiya Hanafi. New York: Polity Press, 2012.

Ettinger, Bracha L. *The Matrixial Borderspace.* Minneapolis: University of Minnesota Press, 2006.

Evans, E. P. *The Criminal Prosecution and Capital Punishment of Animals.* London: William Heineman, 1906.

Evitt, Regula Meyer. "Eschatology, Millenarian Apocalypticism, and the Liturgical Anti-Judaism of the Medieval Prophet Plays." In *The Apocalyptic Year 1000: Religious Expectation and Social Change, 950–1050,* edited by Richard Landes, Andrew Gow, and David C. Van Meter, 205–29. Oxford: Oxford University Press, 2003.

Fabre-Vassas, Claudine. *The Singular Beast: Jews, Christians, and the Pig.* New York: Columbia University Press, 1997.

Fagan, Brian M. *The Little Ice Age: How Climate Made History, 1300–1850.* New York: Basic Books, 2000.

Farber, Lianna. "The Creation of Consent in the *Physician's Tale.*" *Chaucer Review* 39, no. 2 (2004): 151–64.

Farmer, Sharon. "Aristocratic Power and the 'Natural' Landscape: The Garden Park at Hesdin, ca. 1291–1302." *Speculum* 88, no. 3 (2013): 644–80.

Field and Stream Editors. "Deer: Hang Time." *Field and Stream*. January 11, 2006. http://www.fieldandstream.com/articles/other/recipes/2006/01/deer-hang-time

Finkelstein, J. J. "The Ox That Gored." *Transactions of the American Philosophical Society* 71 (1981): 64–81.

Flaubert, Gustave. *Dictionnaire des idées reçues,* edited by Étienne-Louis Ferrère. Paris: Louis Conard, 1913.

Fletcher, Angus. "The Sentencing of Virginia in the *Physician's Tale.*" *Chaucer Review* 34, no. 3 (2000) 300–308.

Forste-Grupp, Sheryl L. "A Woman Circumvents the Laws of Primogeniture in 'The Weddynge of Sir Gawen and Dame Ragnell.'" *Studies in Philology* 99, no. 2 (2002): 105–22.

Foucault, Michel. *The Birth of Biopolitics: Lectures at the Collège de France, 1978–79.* Edited by Michel Senellart. Translated by Graham Burchell. New York: Picador, 2010.

———. *The History of Sexuality: Volume 1, An Introduction.* Translated by Robert Hurley. New York: Vintage Books, 1980.

———. *The Order of Things: An Archaeology of the Human Sciences.* New York: Vintage Books, 1994.

———. *Security, Territory, Population: Lectures at the Collège de France, 1977–78.* Edited by Michel Senellart. Translated by Graham Burchell. New York: Picador, 2007.

———. *Society Must Be Defended: Lectures at the Collège de France, 1975–76.* Edited by Mauro Bertani and Alessandro Fontana. Translated by David Macey. New York: Picador, 2003.

Fradenburg, L. O. Aranye. "Beauty and Boredom in *The Legend of Good Women.*" *Exemplaria* 22, no. 1 (2010): 65–83.

———. "Chaucer's Voice Memorial." *Exemplaria* 2, no. 1 (1990): 169–202.

———. "'Fulfild of fairye': The Social Meaning of Fantasy in the *Wife of Bath's Prologue* and *Tale.*" In *The Wife of Bath: Complete, Authoritative Text with Biographical and Historical Contexts, Critical History, and Essays from Five Contemporary Critical Perspectives,* edited by Peter G. Biedler, 205–20. Boston: Bedford, 1996.

———. *Sacrifice Your Love: Psychoanalysis, Historicism, Chaucer.* Minneapolis: University of Minnesota Press, 2002.

Freud, Sigmund. "Mourning and Melancholia." In *The Standard Edition of the Complete Psychological Works of Sigmund Freud, Vol. 14 (1914–16).* Translated and edited by James Strachey 239–58. London: Heath, 1957.

Gaborit-Chopin, Danielle. "Walrus Ivory in Western Europe." In *From Viking to Crusader: The Scandinavians and Europe 800–1200,* edited by Else Roesdahl and David M. Wilson, 204–5. New York: Rizzoli, 1992.

Gamber, John. *Positive Pollutions and Cultural Toxins: Waste and Contamination in Contemporary U.S. Ethnic Literatures.* Lincoln: University of Nebraska Press, 2012.

Garrard, Greg. *Ecocriticism.* London: Routledge, 2004.

Geary, Patrick J. *Living with the Dead in the Middle* Ages. Ithaca, NY: Cornell University Press, 1994.

Giancarlo, Matthew. "Dressing up a 'Galaunt': Traditional Piety and Fashionable Politics in Peter Idley's 'Translacions' of Mannyng and Lydgate.'" In *After Arundel: Religious Writing in Fifteenth-Century England,* edited by Vincent Gillespie and Kantik Ghosh, 429–47. Turnhout, BE: Brepols, 2011.

Gilbert, John M. *Hunting and Hunting Reserves in Medieval Scotland.* Edinburgh: J. Donald, 1979.

Glennie, J. S. Stuart. *Arthurian Localities: Their Historical Origin, Chief Country, and Fingalian Relations, with a Map of Arthurian Scotland.* 1869. Somerset, UK: Llanerch Publishers, facsimile reprinted, 1994.

Glotfelty, Cheryl. "Introduction: Literary Studies in an Age of Environmental Crisis." In *The Ecocriticism Reader: Landmarks in Literary Ecology,* edited by Cheryl Glotfelty and Harold Fromm. xv–xxxvii. Athens: University of Georgia Press, 1996.

Godfrey, Mark. "Mirror of Displacements: On the Art of Zoe Leonard." *Artforum International* 46 (March 2008): 292–302.

Goldberg, Eric J. "Louis the Pious and the Hunt." *Speculum* 88 (2013): 613–43.

Green, Richard Firth. *A Crisis of Truth: Literature and Law in Ricardian England.* Philadelphia: University of Pennsylvania Press, 2002.

Griffin, Emma. *Blood Sport: Hunting in Britain since 1066.* New Haven, CT: Yale University Press, 2007.

Hahn, Thomas, ed. *Sir Gawain: Eleven Romances and Tales.* Middle English Texts. Kalamazoo, MI: Medieval Institute Publications, 1995.

Hallam, Elizabeth. "Royal Burial and the Cult of Kingship in France and England, 1060–1330." *Journal of Medieval History* 8, no. 4 (1982): 359–80.

Hamper, William. "Sarcastic Verses, Written by an Adherent to the House of Lancaster, in the Last Year of the Reign of Richard the Second, AD 1399." *Archaeologia* 21 (1827): 88–91.

Hanawalt, Barbara. *Of Good and Ill Repute: Gender and Social Control in Medieval England.* Oxford: Oxford University Press, 1998.

Hanna, III, Ralph. *The Awntyrs off Arthure at the Terne Wathelyn: An Edition Based on Bodleian Library MS. Douce 324.* Manchester, UK: Manchester University Press, 1974.

Hansen, Elaine Tuttle. *Chaucer and the Fictions of Gender.* Berkeley: University of California Press, 1992.

Hansen, Miriam Bratu. "Benjamin's Aura." *Critical Inquiry* 34 (2008): 336–75.

Hanson, Thomas B. "Chaucer's Physician as Storyteller and Moralizer." *Chaucer Review* 7, no. 2 (1972): 132–39.

Hanssen, Beatrice. *Walter Benjamin's Other History: Of Stones, Animals, Human Beings, and Angels.* Berkeley: University of California Press, 1998.

Haraway, Donna. "Otherworldly Conversations: Terran Topics; Local Terms." In *Feminist Materialisms,* edited by Stacy Alaimo and Susan Hekman, 157–86. Bloomington and Indianapolis: Indiana University Press, 2007.

———. *When Species Meet.* Minneapolis: University of Minnesota Press, 2007.

Harding, Vanessa. "Space, Property, and Propriety in Urban England." *Journal of Interdisciplinary History* 32, no. 4 (2002): 549–69.

Hardt, Michael, and Antonio Negri. *Multitude: War and Democracy in the Age of Empire.* London: Penguin, 2004.

Hare, Kent G. "Heroes, Saints and Martyrs: Holy Kingship from Bede to Ælfric." *The Heroic Age: A Journal of Early Medieval Northwestern Europe* 9 (Oct 2006): np.

Harman, Graham. *Prince of Networks: Bruno Latour and Metaphysics.* Melbourne: re.press, 2009.

———. *Towards Speculative Realism: Essays and Lectures.* Winchester, UK: Zero Books, 2010.

Harrison, Robert Pogue. *Forests: The Shadow of Civilization.* Chicago: University of Chicago Press, 1993.

Heise, Ursula K. "The Hitchiker's Guide to Ecocriticism." *PMLA* 121, no. 2 (2006): 503–16.

Helmholz, Richard H. *Roman Canon Law in Reformation England.* Cambridge: Cambridge University Press, 1990.

Henderson, Ernest F., trans. and ed. *Select Historical Documents of the Middle Ages.* London: George Bell & Sons, 1892.

Heng, Geraldine. *Empire of Magic: Medieval Romance and the Politics of Cultural Fantasy.* New York: Columbia University Press, 2003.

Herbert, William. *The History of the Twelve Great Livery Companies of London.* 1834–37. New York: Kelley, 1968.

Hewett, Rachel. *Map of a Nation: A Biography of the Ordnance Survey.* London: Granta Books, 2010.

Hewett-Smith, Kathleen M. "Allegory on the Half-Acre: The Demands of History." *Yearbook of Langland Studies* 10 (1996): 1–22.

Hieatt, Constance B., and Sharon Butler, eds. *Curye on Inglysch : English Culinary Manuscripts of the Fourteenth Century (including the Forme of Cury).* London: Oxford University Press, Early English Text Society, Special Series 8, 1985.

Hill, John M. "The Sacrificial Synecdoche of Hands, Heads, and Arms in the Anglo-Saxon Heroic Story." In *Naked before God: Uncovering the Body in Anglo-Saxon England,* edited by Jonathan Wilcox and Benjamin C. Withers, 116–137. Morgantown: West Virginia University Press, 2003.

Hilton, R. H., ed. *Peasants, Knights, and Heretics: Studies in Medieval English Social History.* Cambridge: Cambridge University Press, 1976.

Hoffman, Richard C. "*Homo et Nature, Homo in Natura:* Ecological Perspectives on the European Middle Ages." In *Engaging with Nature: Essays on the Natural World in Medieval and Early Modern Europe,* edited by Barbara A. Hanawalt and Lisa J. Kiser, 11–38. Notre Dame, IN: University of Notre Dame, 2008.

———. "Jephtha's Daughter and Chaucer's Virginia." *Chaucer Review* 2 (1967–68): 20–31.

Hollister, C. Warren. *Henry I.* Edited and completed by Amanda Clark Frost. New Haven, CT: Yale University Press, 2001.

———. "The Strange Death of William Rufus." *Speculum* 48, no. 4 (1973): 637–53.

Hollister, C. Warren, and John W. Baldwin. "The Rise of Administrative Kingship: Henry I and Philip Augustus." *American Historical Review* 83, no. 4 (1978): 267–305.

Holsinger, Bruce. "Of Pigs and Parchment: Medieval Studies and the Coming of the Animal." *PMLA* 124, no. 2 (2009): 616–23.

Hooke, Della. *Trees in Anglo-Saxon England: Literature, Lore, and Landscape.* Rochester, NY: Boydell & Brewer, 2010.

Hourihane, Colum. *Pontius Pilate, Anti-Semitism, and the Passion in Medieval Art.* Princeton, NJ: Princeton University Press, 2009.

Howe, John M. "The Conversion of the Physical World: The Creation of a Christian Landscape." In *Varieties of Religious Conversion in the Middle Ages,* edited by James Muldoon, 63–78. Gainesville: University Press of Florida, 1997.

Huggan, Graham, and Helen Tiffin. *Postcolonial Ecocriticism: Literature, Animals, Environment.* New York: Routledge, 2010.

Humphries, A. B. "Agrarian Change in East Cumberland 1750–c. 1900." M. Phil. thesis. University of Lancaster, 1993.

Huot, Sylvia. *Madness in Medieval French Literature: Identities Lost and Found.* Oxford: Oxford University Press, 2003.

Hurt, James Riggins. "Ælfric and the English Saints." PhD diss., Indiana University, 1965.

Ingham, Patricia Clare. *Sovereign Fantasies: Arthurian Romance and the Making of Britain.* Philadelphia: University of Pennsylvania Press, 2001.

Jandora, John W. "Archers of Islam: A Search for 'Lost' History." *Medieval History Journal* 13 (2010): 97–114.

Jasinski, Marek E., and Fredrik Søreide. "Norse Settlements in Greenland from a Maritime Perspective." In *Vinland Revisited: The Norse World at the Turn of the First Millennium,* edited by Shannon Lewis-Simpson, 123–32. St. John's, NL: Historic Sites Association of Newfoundland and Labrador, 2000.

Jefferson, Samuel. *The History and Antiquities of Leath Ward, in the County of Cumberland: With Biographical Notices and Memoirs.* Carlisle, 1840.

Jehlen, Myra. "Gender." In *Critical Terms for Literary Study,* 2nd ed., edited by Frank Lentriccia and Thomas McLaughlin, 263–73. Chicago: University of Chicago Press, 1995.

Joy, Eileen. "Like Two Autistic Moonbeams Entering the Window of My Asylum: Chaucer's Griselda and Lars von Trier's Bess McNeill." *Postmedieval* 2 (2011): 316–28.

Judkins, Ryan. "The Game of the Courtly Hunt: Chasing and Breaking Deer in Late Medieval English Literature." *Journal of English and Germanic Philology* 112 (2013): 70–92.

Jurasinski, Stefan. "The Rime of King William and Its Analogues." *Neophilologus* 88 (2004): 131–44.

Justice, Steven. *Writing and Rebellion: England in 1381.* Berkeley: University of California Press, 1994.

Kafka, Franz. *Franz Kafka: The Complete Stories.* Translated by Nahum Norbert Glatzer. New York: Schocken Books, 1995.

———. "Before the Law." In *Parables and Paradoxes: Bilingual Edition.* Edited by Nahum M. Glatzer, 60–65. New York: Schocken Books, 1975.

Kantorowicz, Ernst H. "The Archer in the Ruthwell Cross. *Art Bulletin* 42 (1960): 57–59.

————. *The King's Two Bodies: A Study in Mediaeval Political Theology*, rev. ed. Princeton, NJ: Princeton University Press, 1957; rev. 1997.

Kaske, R. E. "The Character Hunger in *Piers Plowman*." In *Medieval English Studies Presented to George Kane*, edited by Donald Kennedy, Ronald Waldron, and Joseph S. Wittig, 187–97. Cambridge: D. S. Brewer, 1972.

Keith, Alison, and Stephen Rupp. "After Ovid: Classical, Medieval, and Early Modern Reception of the Metamorphoses." In *Metamorphosis: Changing Face of Ovid in the Medieval, and Early Modern Periods*, edited by Alison Keith and Stephen Rupp, 15–32. Toronto: University of Toronto Press, 2007.

Kelly, Susan. "Place-Names in the *Awntyrs off Arthure*." *Literary Onomastics Studies* 6 (1979): 162–99.

Kempton, Daniel. "*The Physician's Tale:* The Doctor of Physic's Diplomatic 'Cure.'" *Chaucer Review* 19, no. 1 (1984): 24–38.

Kennedy, Maev. "David Hockney, the Fallen Beech Trees and the Lost Canvas." *The Guardian*, March 26, 2009. http://www.guardian.co.uk/artanddesign/2009/mar/27/hockney-art-seasons-trees.

Kerridge, Richard, and Neil Sammells. *Writing the Environment: Ecocriticism and Literature*. London: Zed Books, 1998.

Kim, Margaret. "Hunger, Need, and the Politics of Poverty in Piers Plowman." *Yearbook of Langland Studies* 16 (2002): 131–68.

Kingsford, Charles. *Prejudice and Promise in Fifteenth-Century England*. London: Oxford University Press, 1925.

Kinney, Thomas. "The Temper of Fourteenth-Century English Verse of Complaint." *Annuale Medievalae* 7 (1966): 74–89.

Kiser, Lisa, and Barbara Hanawalt, eds. *Engaging with Nature: Essays on the Natural World in Medieval and Early Modern Europe*. Notre Dame, IN: University of Notre Dame Press, 2008.

Kitzinger, Ernst. "Anglo-Saxon Vine Scroll Ornament." *Antiquity* 10, no. 37 (1936): 61–71.

Knight, Stephen. *Robin Hood: A Mythic Biography*. Ithaca, NY: Cornell University Press, 2003.

Knight, Stephen, and Thomas H. Ohlgren, eds. *Robin Hood and Other Outlaw Tales*. TEAMS Middle English Texts. Kalamazoo, MI: Medieval Institute Publications, 2000.

Koerner, Lisbet. *Linnaeus: Nature and Nation*. Cambridge, MA: Harvard University Press, 1999.

Kordecki, Leslie. *Ecofeminist Subjectivities: Chaucer's Talking Birds*. New York: Palgrave, 2011.

Lagueux, Robert C. "Sermons, Exegesis, and Performance: The Laon *Ordo Prophetarum* and the Meaning of Advent." *Comparative Drama* 43 (2009): 197–220.

Lancashire, Anne. "The Mayors and Sheriffs of London, 1190–1558." In *London in the Later Middle Ages: Government and People, 1200–1500*. Edited by Caroline M. Barron, 308–55. Oxford: Oxford University Press, 2004.

Langland, Sabrina. "Pilate Answered: What I Have Written I Have Written." *Metropolitan Museum of Art Bulletin* 26, no. 19 (1968): 410–29.

Larubia-Prado, Francisco. "Franco as Cyborg: 'The Body Re-formed by Politics: Part Flesh, Part Machine.'" *Journal of Spanish Cultural Studies* 1, no. 2 (2000): 135–52.

Latour, Bruno. "An Attempt at a 'Compositionist Manifesto.'" *New Literary History* 41 (2010): 471–90.

———. *Pandora's Hope: Essays on the Reality of Science Studies.* Translated by Catherine Porter. Cambridge, MA: Harvard University Press, 1999.

———. *The Pasteurization of France.* Translated by Alan Sheridan. Boston: Harvard University Press, 1993.

———. *Politics of Nature: How to Bring the Sciences into Democracy.* Translated by Catherine Porter. Cambridge, MA: Harvard University Press, 2004.

———. *Reassembling the Social: An Introduction to Actor-Network-Theory.* Oxford: Oxford University Press, 2005.

———. *Science in Action: How to Follow Science and Engineers through Society.* Cambridge, MA: Harvard University Press, 1987.

———. *We Have Never Been Modern.* Translated by Catherine Porter. Cambridge, MA: Harvard University Press, 1993.

Lambert, John James. *Records of the Skinners of London, Edward I to James I.* London: George Allen and Unwin, 1934.

Law, John. "Notes on the Theory of Actor-Network: Ordering, Strategy, and Heterogeneity." *Systems Practice* 5 (1992): 379–93.

Le Patourel, John. "The Norman Conquest of Yorkshire." *Northern History* 6 (1971): 1–21.

Lees, Clare, and Gillian Overing. "Anglo-Saxon Horizons: Places of the Mind in the Northumbrian Landscape." In *A Place to Believe In: Locating Medieval Landscapes,* edited by Clare Lees and Gillian Overing, 1–26. University Park: Pennsylvania State University Press, 2006.

LeGoff, Jacques. *The Medieval Imagination.* Chicago: University of Chicago Press, 1985.

Leicester Jr., H. Marshall. *The Disenchanted Self: Representing the Subject in the* Canterbury Tales. Berkeley: University of California Press, 1990.

Lemke, Thomas. *Biopolitics: An Advanced Introduction.* New York: New York University Press, 2011.

Liddell, George, and Robert Scott. *A Greek-English Lexicon,* revised by Sir Henry Stuart Jones, with the assistance of Roderick McKenzie. Oxford: Clarendon Press, 1940.

Loomis, R. S. "The Hag Transformed." *Celtic Myth and Arthurian Romance.* Chicago: Academy Chicago, 1997.

Lyon, Bruce. "Coup d'oeil sur l'infrastructure de la chasse au moyen âge." *Le Moyen Âge* 104 (1998): 224–25.

Maddicott, J. R. "Plague in Seventh-Century England." *Past and Present* 156 (1997): 7–54.

Maidstone, Richard. *Concordia (The Reconciliation of Richard II with London.* Edited by

David R. Carlson. Translated by A. G. Rigg. Kalamazoo MI: Medieval Institute Publications, 2003.

Mandel, Jerome. "Governance in the *Physician's Tale*." *Chaucer Review* 10, no. 4 (1976): 316–25.

Mann, Jill. *Chaucer and Medieval Estates Satire*. Cambridge: Cambridge University Press, 1973.

Margulis, Lynn, and Dorion Sagan. *What Is Life?* Berkeley: University of California Press, 2000.

Marsh, George Perkins. *Man and Nature: Or, Physical Geography as Modified by Human Action*. 1864. Edited by David Lowenthal. Cambridge, MA: Belknap Press, 1965.

Marvin, William Perry. *Hunting Law and Ritual in Medieval English Literature*. Woodbridge: Boydell and Brewer, 2006.

Mazzucco, Katia. "1941: English Arts and the Mediterranean: A Photographic Exhibition by the Warburg Institute in London." *Journal of Art Historiography* 5 (2011): 1–28.

Mbembe, Achille. "Necropolitics." *Public Culture* 15, no. 1 (2003): 11–40.

McLaren, Mary-Rose. *The London Chronicles of the Fifteenth Century: A Revolution in English Writing*. Woodbridge, UK: D. S. Brewer, 2002.

McNamer, Sarah. *Affective Meditation and the Invention of Medieval Compassion*. Philadelphia: University of Pennsylvania Press, 2010.

———. "Feeling." In *Middle English: Oxford Twenty-First Century Approaches to Literature*, edited by Paul Strohm. Oxford: Oxford University Press, 2007.

Medovoi, Leerom. "The Biopolitical Unconscious: Toward an Eco-Marxist Literary Theory." *Mediations* 24, no. 2 (2010): 123–38.

Meehan, Denis, ed. and trans. *Adamnan's De Locis Sanctis* (Scriptores Latini Hiberniae, 3). Dublin: Dublin Institute for Advanced Studies, 1958.

Merchant, Carolyn. *The Death of Nature: Women, Ecology, and the Scientific Revolution*. San Francisco: Harper and Row, 1990.

Michels, Karen. "Art History, German-Jewish Immigrants, and the Emigration of Iconology." In *Jewish Identity and the Emigration of Iconology*, edited by Catherine M. Soussloff, 167–79. Berkeley: University of California Press, 1999.

Middleton, Anne. "*The Physician's Tale* and Love's Martyrs: 'Ensamples Mo Than Ten' as a Method in *The Canterbury Tales*." *Chaucer Review* 8 (1973): 9–32.

Mileson, S. A. *Parks in Medieval England*. Oxford: Oxford University Press, 2009.

Millea, Nick. *The Gough Map: The Earliest Road Map of Great Britain?* Oxford: Bodleian Library, 2007.

Miller, Andrew G. "Knights, Bishops, and Deer Parks: Episcopal Identity, Emasculation, and Clerical Space in Medieval England." In *Negotiating Clerical Identities: Priests, Monks, and Masculinity in the Middle Ages*, edited by Jennifer D. Thibodeaux, 204–37. New York: Palgrave MacMillan, 2010.

Mills, Catherine. "Agamben's Messianic Politics: Biopolitics, Abandonment and Happy Life." *Contretemps* 5 (2004): 42–62.

Mills, Robert. "Judicial Violence, Biopolitics, and the Bare Life of Animals." *New Medieval Literatures* 12 (2010): 121–29.

Moisà, Maria. "The Rotten Gift: Caro Data Fuit Pauperibus." *Medieval Yorkshire* 26 (1997): 6–10.

Molnar, George. *Powers: A Study in Metaphysics.* Edited by Stephen Mumford. Oxford: Clarendon, 2006.

Mora, Necla. "Violence as a Communicative Action: Customary and Honor Killings." *International Journal of Human Sciences* 6, no. 2 (2009): 499–510.

Morden, Robert. *Westmorland and Cumberland* [map]. 1695. http://www.geog.port. ac.uk/webmap/thelakes/html/morden/md12ny44.htm.

Morrow, Karen Ann. "Disputation in Stone: Jews Imagined on the St. Stephen Portal of Paris Cathedral." In *Beyond the Yellow Badge: Anti-Judaism and Anti-Semitism in Medieval and Early Modern Visual Culture,* edited by Michael B. Merback, 64–86. Leiden, NL: Brill, 2007.

Morton, Timothy. *The Ecological Thought.* Cambridge, MA: Harvard University Press, 2010.

———. *Ecology without Nature: Rethinking Environmental Aesthetics.* Cambridge, MA: Harvard University Press, 2009.

Muir, John. *My First Summer in the Sierra.* Boston: Houghton Mifflin, 1911.

Murdoch, Jon. *Post-Structuralist Geography: A Guide to Relational Space.* London: Sage, 2006.

Nardizzi, Vin. "Medieval Ecocriticism." *Postmedieval* 4, no. 1 (2013): 112–23.

Nealon, Jeffrey T. *Foucault beyond Foucault: Power and Its Intensification Since 1984.* Stanford, CA: Stanford University Press, 2008.

Nederman, Cary, and Catherine Campbell. "Priests, Kings, and Tyrants: Spiritual and Temporal Power in John of Salisbury's *Policraticus.*" *Speculum* 66 (1991): 572–90.

Nicolle, David. *The Great Islamic Conquests, ad 637–750.* Oxford: Osprey Publishing, 2009.

Nilgen, Ursula. "Intellectuality and Splendour: Thomas Becket as a Patron of the Arts." In *Art and Patronage in the English Romanesque,* edited by Sarah Macready and F. H. Thompson, 145–58. London: Society of Antiquaries, 1986.

Nutt, Alfred. *Studies on the Legend of the Holy Grail.* London: Strand, 1888.

O'Loughlin, Thomas. *Adomnán and the Holy Places: The Perceptions of an Insular Monk on the Locations of the Biblical Drama.* London: T&T Clark, 2007.

O'Meara, John J. *Eriugena.* Oxford: Clarendon Press, 1988.

O'Reilly, Jennifer. "The Rough-Hewn Cross in Anglo-Saxon Art." In *Ireland and Insular Art ad 500–1200,* edited by Michael Ryan, 153–58. Dublin: Royal Irish Academy, 1987.

Ohberger, J., O. Langenen, N.C. Stenseth, and L. A. Volestad. "Community-Level Consequences of Cannibalism." *American Naturalist* 180, no. 6 (2012): 791–801.

Orchard, Andy. "The Dream of the Rood: Cross-References." In *New Readings in the Vercelli Book,* edited by Samantha Zacher and Andy Orchard, 225–53. Toronto: University of Toronto Press, 2009. Orlemanski, Julie. "Jargon and the Matter of Medicine in Middle English." *Journal of Medieval and Early Modern Studies* 42, no. 2 (2012): 395–420.

———. "How to Kiss a Leper." *Postmedieval* 3 (2012): 142–57.

Orton, Fred, and Ian Wood, with Clare A. Lees, eds. *Fragments of History: Rethinking the Ruthwell and Bewcastle Monuments.* Manchester, UK: Manchester University Press, 2007.

Otter, Monika. "1066: The Moment of Transition in Two Narratives of the Norman Conquest." *Speculum* 74, no. 3 (1999): 565–86.

Page, Sophie. *Astrology in Medieval Manuscripts.* London: British Library Press, 2002.

Park, Katharine. "Nature in Person: Medieval and Renaissance Allegories and Emblems." In *The Moral Authority of Nature,* edited by Lorraine Daston and Fernando Vidal, 50–73. Chicago: University of Chicago Press, 2004.

Parker, Elizabeth C. "Editing the *Cloisters Cross.*" *Gesta* 45 (2006): 147–60.

Parker, Elizabeth C., and Charles T. Little. *The Cloisters Cross: Its Art and Meaning.* New York: Metropolitan Museum of Art, 1994.

Parker, F. H. M. "I. Inglewood Forest, Part III: Some Stories of Deer-Stealers." *Transactions of the Cumberland & Westmorland Antiquarian & Archaeological Society,* n. s. 7 (1907): 1–30.

———. "Inglewood Forest IV [Revenues]." *Transactions of the Cumberland & Westmoreland Antiquarian & Archeological Society,* n. s. 9 (1909): 24–37.

———. "Inglewood Forest VII" [on the Huttons, Hereditary Foresters], *Transactions of the Cumberland & Westmoreland Antiquarian & Archeological Society,* n. s. 11 (1910): 1ff.

Parker, Patricia. *Literary Fat Ladies: Rhetoric, Gender, Property.* London: Methuen, 1987.

Pascua, Esther. "From Forest to Farm to Town: Domestic Animals from ca. 1000 to ca. 1450." In *A Cultural History of Animals in the Medieval Age,* edited by Brigitte Resl, 81–102. New York: Berg, 2007.

Patterson, Lee. *Chaucer and the Subject of History.* Madison: University of Wisconsin Press, 1991.

Pearsall, Derek, and Elizabeth Salter. "The Landscape of the Seasons." In *Landscapes and Seasons of the Medieval World,* 119–60. Toronto: University of Toronto Press, 1973.

Peck, Russell A. "John Gower and the Book of Daniel." In *John Gower: Recent Readings. Papers Presented at the Meetings of the John Gower Society at the International Congress on Medieval Studies, Western Michigan University, 1983–88,* edited by R. F. Yeager, 159–87. Kalamazoo, MI: Western Michigan University, 1989.

Pervukhin, Anna. "Deodands: A Study in the Creation of Common Law Rules." *The American Journal of Legal History* 47 (2005): 237–56.

Petit-Dutaillis, Charles, and George Lefebvre. *Studies and Notes Supplementary to Stubb's Constitutional History.* Manchester, UK: Manchester University Press, 1930.

Peyroux, Cathérine. "The Leper's Kiss." In *Monks and Nuns, Saints and Outcasts: Religion in Medieval Society,* edited by Sharon Farmer and Barbara H. Rosenwein, 172–88. Ithaca, NY: Cornell University Press, 2000.

Phillips, Kim M. "Masculinities and the Medieval English Sumptuary Laws." *Gender and History* 19 (2007): 22–42.

Platt, Colin. *The Black Death and Its Aftermath in Late-Medieval England.* Toronto: University of Toronto Press, 1997.

Plumwood, Val. *Feminism and the Mastery of Nature.* London: Routledge, 1993.

Pluskowski, Aleksander. "The Wolf." In *Extinctions and Invasions: A Social History of British Fauna,* edited by T. P O'Connor and Naomi Sykes, 71–73. Oxford: Windgather Press, 2010.

———. *Wolves and the Wilderness in the Middle Ages.* Woodbridge, UK: Boydell & Brewer, 2006.

Post, Gaines. *Studies in Medieval Legal Thought: Public Law and the State, 110–1322.* Princeton, NJ: Princeton University Press, 1964.

Prozorov, Sergei. *Foucault, Freedom and Sovereignty.* Aldershot, UK: Ashgate, 2007.

Ragan, Mark A. "Trees and Networks before and after Darwin." *Biology Direct* 4, no. 43 (2009). http://www.biology-direct.com/content/4/1/43.

Raines, Chris. "Aging of Venison—Sounds Like a Good Idea, Right?" *Meatblogger.* November 29, 2010. http://meatblogger.org/2010/11/29/aging-of-venison-sounds-like-a-good-idea-right/.

Raskolnikov, Masha. *Body against Soul: Gender and 'Sowlhele' in Middle English Allegory.* Columbus: The Ohio State University Press, 2009.

Rawcliffe, Carole. "Learning to Love the Leper: Aspects of Institutional Charity in Anglo-Norman England." In *Anglo-Norman Studies XXIII: Proceedings of the Battle Conference 2000,* edited by John Gillingham, 231–50. Woodbridge, UK: Boydell & Brewer, 2001.

———. *Leprosy in Medieval England.* Woodbridge, UK: Boydell & Brewer, 2006.

———. *Medicine and Society in Later Medieval England.* Phoenix Mill, UK: Sutton, 1995.

Reddy, William. *The Navigation of Feeling: A Framework for the History of Emotions.* Cambridge: Cambridge University Press, 2001.

Richard, Frances. "A Tree Dies in Brooklyn." *Art Forum* 36 (Feb. 1998): 19–20.

Riley, Henry Thomas. *Memorials of London and London Life.* London: Longmans, Green, and Co., 1868.

Rivard, Derek. "The Poachers of Pickering Forest, 1282–1338." *Medieval Prosopography* 17 (1996): 97–144.

Robertson, James Craigie, ed., *Materials for the History of Thomas Becket, Archbishop of Canterbury.* Vol. 1. London: Longman & Co., 1875.

Robertson, Kellie. "Medieval Things: Materiality, Historicism, and the Premodern Object." *Literature Compass* 5/6 (2008): 1060–80.

Roesdahl, Else. "Walrus Ivory." In *Ohthere's Voyages: A Late 9th-Century Account of Voyages along the Coasts of Norway and Denmark in its Cultural Context,* edited by Janet Bately and Anton Englert, 92–93. Roskilde, DK: Viking Ship Museum, 2007.

Rooney, Anne. *Hunting in Middle English Literature.* Woodbridge, UK: Boydell & Brewer, 1993.

Rosenwein, Barbara, ed. *Anger's Past: The Social Uses of an Emotion in the Middle Ages.* Ithaca, NY: Cornell University Press, 1999.

———. *Emotional Communities in the Early Middle Ages*. Ithaca, NY: Cornell University Press, 2006.

———. "Worrying about Emotions in History." *American Historical Review* 107 (2002): 821–45.

Roskell, J. S., Linda Clark, and Carole Rawcliffe, eds. *The History of Parliament: The House of Commons, 1386–1421*. Stroud: Alan Sutton Publishing, 1992.

Rossello, Diego. "Hobbes and the Wolf-Man: Melancholy and Animality in Modern Sovereignty." *New Literary History* 43 (2012): 255–79.

Rotherham, I. D. "The Ecology and Economics of Medieval Deer Parks." *Landscape Archaeology and Ecology* 6 (2007): 86–102.

Rowland, Beryl. "The Physician's 'Historial Thyng Notable' and the Man of Law.'" *English Literary History* 40 (1973): 165–78.

Rudd, Gillian. *Greenery: Ecological Readings of Late Medieval Literature*. Manchester, UK: Manchester University Press, 2007.

Sack, Robert. *Human Territoriality: Its Theory and History*. Cambridge: Cambridge University Press, 1986.

Salisbury, Joyce. *The Beast Within: Animals in the Middle Ages*. London: Routledge, 2010.

Sanderson, Eric A. *Mannahatta: A Natural History of New York City*. New York: Abrams, 2009.

Sanok, Catherine. "The Geography of Genre in the *Physician's Tale* and *Pearl*." *New Medieval Literatures* 5 (2002): 177–201.

Santner, Eric L. *On Creaturely Life: Rilke, Benjamin, Sebald*. Chicago: University of Chicago Press, 2006.

———. *The Royal Remains: The People's Two Bodies and the Endgames of Sovereignty*. Chicago: University of Chicago Press, 2011.

Sassen, Saskia. *Territory, Authority, Rights: From Medieval to Global Assemblages*, rev. ed. Princeton, NJ: Princeton University Press, 2008.

Saunders, Corinne J. *The Forest of Medieval Romance: Avernus, Broceliande, Arden*. Cambridge: Brewer, 1993.

Saxl, Fritz. "The Ruthwell Cross." *Journal of the Warburg and Courtauld Institutes* 6 (1943): 1–19.

Scarry, Elaine. *The Body in Pain: The Making and Unmaking of the World*. New York: Oxford University Press, 1985.

Scase, Wendy. *Literature and Complaint in England, 1272–1553*. Oxford: Oxford University Press, 2007.

Schama, Simon. *Landscape and Memory*. New York: A. A. Knopf, 1995.

Schapiro, Meyer. "The Religious Meaning of the Ruthwell Cross." *Art Bulletin* 26 (1944): 232–45.

Schiff, Randy P. "Borderland Subversions: Anti-Imperial Energies in *The Awntyrs off Arthure* and *Galagros and Gawane*." *Speculum* 84, no. 3 (2009): 613–32.

———. "The Loneness of the Stalker: Poaching and Subjectivity in *The Parlement of the Thre Ages*." *Texas Studies in Language and Literature*. 51, no. 3 (2009): 263–93.

———. *Revivalist Fantasy: Alliterative Verse and Nationalist Literary History*. Columbus: The Ohio State University Press, 2011.

———. "Sovereign Exception: Pre-National Consolidation in *The Taill of Rauf Coilyear*." In *The Anglo-Scottish Border and the Shaping of Identity, 1300–1600*, edited by Mark Bruce and Katherine Terrell, 33–50. New York: Palgrave Macmillan, 2012.

Schmitt, Carl. *Political Theology: Four Chapters on the Concept of Sovereignty*, rev. ed. 1934. Translated by George Schwab. Chicago: University of Chicago Press, 2005.

Schofield, P. R. "Medieval Diet and Demography." In *Food in Medieval England: Diet and Nutrition*, edited by C. M. Woolgar, D. Serjeantson, and T. Waldron, 239–53. Oxford: Oxford University Press, 2006.

Scholem, Gershom, and Theodor W. Adorno, eds. *The Correspondence of Walter Benjamin, 1910–1940*. Translated by Manfred R. Jacobson and Evelyn M. Jacobson. Chicago: University of Chicago Press, 1994.

Schwyzer, Philip. "The Beauties of the Land: Bale's Books, Aske's Abbeys, and the Aesthetics of Nationhood." *Renaissance Quarterly* 57, no.1 (2004): 99–125.

Seremetakis, C. Nadia. *The Senses Still: Perception and Memory as Material Culture in Modernity*. Chicago: University of Chicago Press, 1996.

Serres, Michael. *Le parasite*. Paris: Grasset, 1980.

Seshadri, Kalpana Rahita. *HumAnimal: Race, Law, Language*. Minneapolis: University of Minnesota Press, 2012.

Shaviro, Steve. "Panpsychism And/Or Eliminativism." *The Pinocchio Theory* (blog). October 24. 2011, http://www.shaviro.com/Blog/?p=1012.

Shaw, Brent D. "Raising and Killing Children: Two Roman Myths." *Mnemosyne* 54, no. 1 (2001): 31–77.

Siewers, Alfred K. "Ecopoetics and the Origins of English Literature." In *Environmental Criticism for the Twenty-First Century*, edited by Stephanie LeMenager, Teresa Shewry, and Ken Hiltner, 105–20. New York: Routledge, 2011.

———, ed. *Re-Imagining Nature: Environmental Humanities and Eco-Semiotics*. Lewisburg, PA: Bucknell University Press, 2013.

———. *Strange Beauty: Ecocritical Approaches to Early Medieval Landscape*. New York: Palgrave Macmillan, 2010.

Sinclair, John. *Statistical Account of Scotland*. Vol. 10. Edinburgh, UK: William Creech, 1794.

Siraisi, Nancy G. *Medieval and Early Renaissance Medicine: An Introduction to Knowledge and Practice*. Chicago: University of Chicago Press, 1990.

Smithson, Robert. *Robert Smithson: The Collected Writings*. Edited by Jack Flam. Berkeley: University of California Press, 1996.

Somerset, Fiona. *Clerical Discourse and Lay Audience in Late Medieval England*. Cambridge: Cambridge University Press, 1998.

Sponsler, Claire. *Drama and Resistance: Bodies, Goods, and Theatricality in Late Medieval England*. Minneapolis: University of Minnesota Press, 1997.

Stagg, Donald G. *A Calendar of New Forest Documents, 1244–1344*. Winchester, UK: Hampshire County Council, 1979.

Stanbury, Sarah. "Ecochaucer: Green Ethics and Medieval Nature." *Chaucer Review* 39, no. 1 (2004): 1–16.

Staniland, Kay. "Extravagance or Regal Necessity? The Clothing of Richard II." In *The Regal Image of Richard II and the Wilton Diptych,* edited by Dillian Gordon, Lisa Monnas, and Caroline Elam, 85–93. London: Harvey Miller, 1997.

Steel, Karl. *How to Make a Human: Animals and Violence in the Middle Ages.* Columbus: The Ohio State University Press, 2011.

———. "How to Make a Human." *Exemplaria* 20 (2008): 3–27.

———. "With the World, or Bound to Face the Sky: The Postures of the Wolf-Child of Hesse." In *Animal, Vegetable, Mineral: Ethics and Objects,* edited by Jeffrey Jerome Cohen, 9–34. Washington, DC: Oliphaunt Books, 2011.

Stein, Robert. *Reality Fictions: Romance, History, and Governmental Authority, 1025–1180.* Notre Dame, IN: University of Notre Dame Press, 2006.

Steiner, Emily. "Naming and Allegory in Late Medieval England." *Journal of English and Germanic Philology* 106, no. 2 (2007): 248–75.

Steiner, Emily, and Candace Barrington, eds. *The Letter of the Law: Legal Practice and Literary Production in Medieval England.* Ithaca, NY: Cornell University Press, 2002.

Stirnemann, Patricia. "The Study of French Twelfth-Century Manuscripts." In *Romanesque Art and Thought in the Twelfth Century,* edited by Colum Hourihane, 82–94. Princeton, NJ: Princeton University Press, 2008.

Stock, Brian. *The Implications of Literacy: Written Language and Models of Interpretation in the Eleventh and Twelfth Centuries.* Princeton, NJ: Princeton University Press, 1983.

———. *Listening for the Text: On the Uses of the Past.* Baltimore, MD: Johns Hopkins University Press, 1990.

Strasser, Bruno J. "Who Cares about the Double Helix?" *Nature* 422 (24 April 2003): 803–4.

Strohm, Paul. *England's Empty Throne: Usurpation and the Language of Legitimation, 1399–1422.* New Haven, CT: Yale University Press, 1998.

Stubbs, William, ed. *Selected Charters and Other Illustrations of English Constitutional History from Early Times to the Reign of Edward the First.* Oxford: Clarendon Press, 1870.

Sturgeon, Noel. *Ecofeminist Natures: Race, Gender, Feminist Theory and Political Action.* New York: Routledge, 1997.

Sutton, Teresa. "The Deodand and Responsibility for Death." *Journal of Legal History* 18 (1997): 44–55.

Swiffin, Amy. *Law, Ethics, and the Biopolitical.* London: Routledge, 2011.

Sykes, Naomi. "Deer, Land, Knives and Halls: Social Change in Early Medieval England." *Antiquaries Journal* 90 (2010): 175–93.

———. "Taking Sides: The Social Life of Venison in Medieval England." In *Breaking and Shaping Beastly Bodies: Animals as Material Culture in the Middle Ages,* edited by Aleksander Pluskowski, 149–60. Oxford: Oxbow, 2007.

Sykes, Naomi, Ruth Carden, and Kerry Harris. "Changes in the Size and Shape of Fallow

Deer—Evidence for the Movement and Management of a Species." *International Journal of Osteoarchaeology* 23, no. 1 (2011): 55–68.

Taylor, Joseph. "Sovereignty, Oath, and the Profane Life in the *Avowing of Arthur.*" *Exemplaria* 25, no. 1 (2013): 36–57.

Thomas, Kate. "Post Sex: On Being Too Slow, Too Stupid, Too Soon." In *After Sex?: On Writing Since Queer Theory,* edited by Janet Halley and Andrew Parker, 66–75. Raleigh, NC: Duke University Press, 2011.

Thomson, R. M. *William of Malmesbury,* rev. ed. Woodbridge, UK: Boydell & Brewer, 2003.

Thrupp, Sylvia. *The Merchant Class of Medieval London.* Ann Arbor: University of Michigan Press, 1948.

Tolan, John V. *Saracens: Islam in the Medieval European Imagination.* New York: Columbia University Press, 2002.

Toller, T. Northcote, and Joseph Bosworth. *An Anglo-Saxon Dictionary Based on the Manuscript Collections of the Late Joseph Bosworth.* Oxford: Clarendon Press, 1898.

Traherne, Elaine, ed. *Old and Middle English: An Anthology c. 890–c. 1400.* Oxford: Blackwell, 2004.

Tubbs, Coin R. *The New Forest: An Ecological History.* Newton Abbot, UK: David and Charles, 1968.

Turner, G. J., ed. *Select Pleas of the Forest.* London: Selden Society, 1901.

Turner, Ralph. *Judges, Administrators, and the Common Law in Angevin England.* London: Hambledon Press, 1994.

Twiti, William. *The Middle English Text of The Art of Hunting.* Edited by David Scott-Macnab. Heidelberg, DE: Heidelberg University Press, Winter 2009.

Van Dyke, Carolynn, ed. *Rethinking Chaucerian Beasts.* New York: Palgrave, 2012.

Vanacker, Janis. "'Why Do You Break Me'? Talking to a Human Tree in Dante's *Inferno.*" *Neophilologus* 95 (2011): 431–45.

Veale, Elspeth. *The English Fur Trade in the Later Middle Ages.* Oxford: Clarendon Press, 1966.

Wadmore, James. *Some Account of the Worshipful Company of Skinners of London, Being the Guild or Fraternity of Corpus Christi.* 1876. London: Blades, East, and Blades, 1902.

Walker, D. "Post Glacial Deposits at Tarn Wadling." *New Phytologist* 63, no. 2 (1964): 232–35.

Warren, Michelle R. "Lydgate, Lovelich, and London Letters." In *Lydgate Matters: Poetry and Material Culture in the Fifteenth Century,* edited by Lisa Cooper and Andrea Denny-Brown, 113–38. New York: Palgrave Macmillan, 2008.

Warren, Wilfred Lewis. *King John.* Berkeley: University of California Press, 1978.

Watson, Charlie. *Seahenge: An Archaeological Conundrum.* Swindon, UK: English Heritage, 2005.

Westerhof, Daniel. *Death and the Noble Body in Medieval England.* Woodbridge, UK: Boydell & Brewer, 2008.

White, Hugh. *Nature, Sex, and Goodness in a Medieval Literary Tradition.* Oxford: Oxford University Press, 2000.

Whittock, Trevor. *A Reading of the Canterbury Tales*. Cambridge: Cambridge University Press, 1968.

Wickens, Jim. "Sick as a Pig." *The Ecologist*. March 26, 2009. http://www.theecologist. org/investigations/food_and_farming/268866/sick_as_a_pig.html.

Wilks, Michael. *The Problem of Sovereignty in the Late Middle Ages: The Papal Monarchy with Augustinus Triumphus and the Publicists*. Cambridge: Cambridge University Press, 1963.

Willox, Ashlee Cunsolo. "Climate Change as the Work of Mourning." *Ethics and the Environment* 17, no. 2 (2012): 138–64.

Wolfe, Cary. *Before the Law: Humans and Other Animals in a Biopolitical Frame*. Chicago: University of Chicago Press, 2013.

———. "Human, All Too Human." *PMLA* 124, no. 2 (2009): 564–75.

Wood, Christopher S. "Art History's Normative Renaissance." In *The Italian Renaissance in the Twentieth Century*, ed. Allen Gerieco, Michael Rocke, and Superbi Fiorella, 65–92. Florence, IT: Olschki, 2002.

Woodland Trust. *Tarn Wadling: Management Plan 2014-2019*. 2008. http://www.woodlandtrust.org.uk/woodfile/785/management-plan.pdf?cb=06445ee21f28419cb908d490e2fca45d

Woolgar, C. M. "Fast and Feast: Conspicuous Consumption and the Diet of the Nobility in the Fifteenth Century." In *Revolution and Consumption in Late Medieval England*, edited by M. A. Hicks, 7–26. Woodbridge, UK: Boydell & Brewer, 2001.

Wright, Edith Armstrong. *Dissemination of the Liturgical Drama in France*. Geneva, CH: Slatkine Reprints, 1980.

Wright, Laura. *Wilderness into Civilized Shapes: Reading the Postcolonial Environment*. Athens: University of Georgia Press, 2010.

Wright, Thomas. *Political Poems and Songs Relating to English History Composed during the Period from the Accession of Edw. III. to That of Ric. III*. 2 vols. London: Longman, Green, Longman, and Roberts, 1859.

———, ed. *Political Songs of England: From the Reign of King John to That of Edward II*. 1839. Cambridge: Cambridge University Press, 1996.

Wrottesley, George, ed., "An Account of the Family of Okeover of Okeover, County Stafford, with Transcripts of the Ancient Deeds of Oceover." In *Collections for a History of Staffordshire*. Vol. 7, n. s., 4–187. London: Harrison and Sons, 1904.

———, ed. and trans. "The Pleas of the Forest, *Temp*. Henry III and Edward I." In *Collections for a History of Staffordshire*. Vol. 5. London: Harrison and Sons, 1884.

Young, Charles R. "English Royal Forests under the Angevin Kings." *Journal of British Studies* 12 (1972): 1–14.

———. *The Royal Forests of Medieval England*. Philadelphia: University of Pennsylvania Press, 1979.

Young, Neil. "After the Gold Rush." *After the Gold Rush* (1970). Reprise B000002KD9, 1990. Compact Disc.

Contributors

Stephanie L. Batkie is Assistant Professor of English and Director of Medieval Studies at the University of Montevallo. Her essays have appeared recently in *Chaucer Review* and in the collection *John Gower, Trilingual Poet: Language, Translation, and Tradition* (D. S. Brewer, 2010). Her work examines how medieval authors engage hermeneutic labor to produce social, devotional, and political change, and she is currently completing a monograph, *Forms of Authority: Gower's Political Ethic in the Cronica Tripertita.*

Kathleen Biddick is Professor of History at Temple University. Her books include *The Typological Imaginary: Circumcision, Technology, and History* (University of Pennsylvania Press, 2003), *The Shock of Medievalism* (Duke UP, 1998), and *The Other Economy: Pastoral Husbandry on a Medieval Estate* (University of California Press, 1989). Her numerous articles have appeared in *Rethinking History, GLQ, JMEMS, Speculum,* and the *Journal of British Studies.*

Mary Kate Hurley is Assistant Professor of English at Ohio University. Her articles have appeared in *JEGP, The Heroic Age,* and *The Old English Newsletter,* and she is a contributor to the medievalist blog "In the Middle."

Kathleen Coyne Kelly is Professor of English at Northeastern University. She is the author of *Performing Virginity and Testing Chastity in the Middle Ages* (Routledge, 2000) and coeditor of the collections *Queer Movie Medievalisms* (w. Tison Pugh; Ashgate, 2009) and *Menacing Virgins: Representing Virginity in the Middle Ages and Renaissance* (w. Marina Leslie; Delaware, 1999). Her essays have appeared in *Allegorica, Arthuriana, Exemplaria, Paerergon,* and *Studies in Philology.*

Jeanne Provost is Assistant Professor of English at Furman University. Having presented conference papers on *William of Palerne,* Gower, and Chrétien de Troyes, Provost is currently completing a bookproject, *Illicit Country: The Loathly Lady and the Imaginary Foundations of Medieval English Land Law.*

Randy P. Schiff is Associate Professor of English at SUNY Buffalo. He is the author of *Revivalist Fantasy: Alliterative Verse and Nationalist Literary History* (Ohio State UP, 2011), as well as articles appearing in journals such as *Exemplaria, Speculum,* and *Texas Studies in Language and Literature.*

Karl Steel is Assistant Professor of English at Brooklyn College. He is the author of *How to Make a Human: Animals and Violence in the Middle Ages* (Ohio State UP, 2011). He also coedited the "animal turn" issue of the journal *Postmedieval,* and his articles have appeared in *Exemplaria* and various essay collections. He contributes regularly to the medievalist blog "In the Middle."

Joseph Taylor is Assistant Professor of English at the University of Alabama in Huntsville. He is currently working on a book examining the North-South divide in medieval England. His articles have appeared in *JEGP, Modern Philology,* and *Exemplaria.*

Michelle R. Warren is Professor of Comparative Literature at Dartmouth College, and the author of both *Creole Medievalism: Colonial France and Joseph Bédier's Middle Ages* (University of Minnesota Press, 2011) and *History on the Edge: Excalibur and the Borders of Britain* (University of Minnesota Press, 2000). She has also coedited two books—*Postcolonial Moves: Medieval through Modern* with Patricia Clare Ingham (Palgrave-Macmillan, 2003) and *Arts of Calculation: Quantifying Thought in Early Modern Europe* with David Glimp (Palgrave-Macmillan, 2004)—and is the author of numerous articles and essays.

Index

INTERVENTIONS: NEW STUDIES IN MEDIEVAL CULTURE
Ethan Knapp, Series Editor

Interventions: New Studies in Medieval Culture publishes theoretically informed work in medieval literary and cultural studies. We are interested both in studies of medieval culture and in work on the continuing importance of medieval tropes and topics in contemporary intellectual life.

www.ingramcontent.com/pod-product-compliance
Lightning Source LLC
Chambersburg PA
CBHW030641270326
41929CB00007B/161